AFTER SUBURBIA

Urbanization in the Twenty-First Century
Edited by Roger Keil and Fulong Wu

After Suburbia presents a cross-section of state-of-the-art scholarship in critical global suburban research and provides an in-depth study of the planet's urban peripheries to grasp the forms of urbanization in the twenty-first century.

Based on cutting-edge conceptual thought and steeped in richly detailed empirical work conducted over the past decade, *After Suburbia* draws on research from Asia, Africa, Australia, Europe, and the Americas to showcase comprehensive global scholarship on the urban periphery. Contributors explicitly reject the traditional centre-periphery dichotomy and the prioritization of epistemologies that favour the Global North, especially North American cases, over other experiences. In doing so, the book strongly advances the notion of a post-suburban reality in which traditional dynamics of urban extension outward from the centre are replaced by a set of complex contradictory developments. *After Suburbia* examines multiple centralities and diverse peripheries which mesh to produce a surprisingly contradictory and diverse metropolitan landscape.

(Global Suburbanisms)

ROGER KEIL is a professor in the Faculty of Environmental and Urban Change at York University.

FULONG WU is Bartlett Professor of Planning at University College London.

GLOBAL SUBURBANISMS

Series Editor: Roger Keil, York University

Urbanization is at the core of the global economy today. Yet, crucially, suburbanization now dominates twenty-first-century urban development. This book series is the first to systematically take stock of worldwide developments in suburbanization and suburbanisms today. Drawing on methodological and analytical approaches from political economy, urban political ecology, and social and cultural geography, the series seeks to situate the complex processes of suburbanization as they pose challenges to policymakers, planners, and academics alike.

For a list of the books published in this series see p. 459.

After Suburbia

Urbanization in the Twenty-First Century

EDITED BY ROGER KEIL AND FULONG WU

UNIVERSITY OF TORONTO PRESS
Toronto Buffalo London

ISBN 978-1-4875-0487-8 (cloth)   ISBN 978-1-4875-3107-2 (EPUB)
ISBN 978-1-4875-2353-4 (paper)   ISBN 978-1-4875-3106-5 (PDF)

Global Suburbanisms

Library and Archives Canada Cataloguing in Publication
Title: After suburbia : urbanization in the twenty-first century / edited by
    Roger Keil and Fulong Wu.
Names: Keil, Roger, 1957– editor. | Wu, Fulong, editor.
Series: Global suburbanisms.
Description: Series statement: Global suburbanisms | Includes bibliographical
    references and index.
Identifiers: Canadiana (print) 20220241422 | Canadiana (ebook) 20220241503 |
    ISBN 9781487504878 (cloth) | ISBN 9781487523534 (paper) |
    ISBN 9781487531072 (EPUB) | ISBN 9781487531065 (PDF)
Subjects: LCSH: Urbanization – Case studies. | LCSH: Suburbs – Case
    studies. | LCSH: City planning – Case studies. | LCGFT: Case studies.
Classification: LCC HT361.A38 2022 | DDC 307.76 – dc23

We wish to acknowledge the land on which the University of Toronto Press
operates. This land is the traditional territory of the Wendat, the Anishnaabeg,
the Haudenosaunee, the Métis, and the Mississaugas of the Credit First
Nation.

University of Toronto Press acknowledges the financial support of the
Government of Canada, the Canada Council for the Arts, and the Ontario Arts
Council, an agency of the Government of Ontario, for its publishing activities.

Canada Council   Conseil des Arts
for the Arts   du Canada

ONTARIO ARTS COUNCIL
CONSEIL DES ARTS DE L'ONTARIO
an Ontario government agency
un organisme du gouvernement de l'Ontario

Funded by the   Financé par le
Government   gouvernement
of Canada   du Canada

Contents

Openings

Land | Infrastructure | Governance

After Suburbia | The Path Ahead

Conclusion

Figures and Tables

Figures

Tables

Images

Openings

Image 1. Public sculpture in Lingang, China

1 Beyond Suburban Stereotypes: Urban Peripheries in the Twenty-First Century

FULONG WU AND ROGER KEIL

"If (Biden) ever got to run this country and they ran it the way he wants to run it, our suburbs would be gone, our suburbs would be gone," Trump said at the first presidential debate. "And you would see problems like you have never seen."

– LeBlanc, "Trump Uses Outdated Thinking"

It was not just Black and brown cities in swing states that brought Democrats their victories; it was also their suburbs. Across the nation, suburbs moved further blue than in 2016, particularly in key battleground states.

– Lung-Amam, "Why Trump's Suburban Strategy Failed"

Introduction

The title of this volume, *After Suburbia,* plays on the notion that extended urbanization as it exists today is a far cry from what the literature on suburbanization (and the former American president) treated as suburban in the past. But *After Suburbia* is also recognition of an unprecedented, and now concluded, collective effort at suburban scholarship of which it is a record. This volume, then, has been generated from concluding work done under the project "Major Collaborative Research Initiative (MCRI) Global Suburbanisms: Governance, Land and Infrastructure in the 21st Century" (see www.yorku.ca/suburbs for an archive of the work that was produced; also see Keil & Lynch, 2019, for an overview). Drawn chiefly from presentations at the final conference of the initiative held in York University in 2017, the volume also contains work by associated research teams in Europe, Asia, Africa, Australia, and the Americas to offer a global showcase of scholarship on the urban periphery around the world. These studies advance the understanding of the

reality of urban peripheries where traditional dynamics of urban extension moving outward from the centre are replaced by a set of complex
processes in which multiple centralities and diverse peripheries produce new diversified and mixed metropolitan landscapes. The MCRI
aimed for a global perspective of suburbanization, by which we mean
"the combination of non-central population and economic growth with
urban spatial expansion" (Ekers et al., 2015, p. 22). Recent publications
demonstrate our approach to understand the peripheral areas after suburbia, examining peripheral urban changes without anchoring the vantage point to the conventional notion of the suburb (e.g., Güney et al.,
2019; Harris & Vorms, 2017; Hamel & Keil, 2015; Keil, 2018a, 2018b;
Wu & Keil, 2020).

This book examines the planet's urban peripheries to grasp the
processes and forms of urbanization in the twenty-first-century. The
authors in this book and the scholarship they represent defy the common tendency in urban studies to treat the suburban as somewhat epi-
phenomenal and mostly unmentioned in a review of theory on "global
urbanism" (McFarlane & Lancione, 2021), a phase (Taylor, 2021), a transitional moment of urban taking of agricultural land for industrialization and new town development (Ren, 2020), a "hinterland" (Brenner &
Katsikis, 2020), an undefined filler in a megaurban world (Sorensen &
Labbé, 2021). Instead, *After Suburbia* invites "the suburban" into the
centre of urban scholarship and urban theory. Not a special, secondary,
and subordinated domain, suburban studies have asserted themselves
as a font of conceptual and methodological insight and innovation. The
vantage point of this "suburban revolution" is not confined within the
periphery, no matter how we define or name the "suburb" (Harris &
Vorms, 2017). Instead, the book places "suburbia" in multiple contexts
in the world – before and after "classical" suburbanization and inside
and outside suburban areas. We stress historical and geographical contingency. This world of urban peripheries is no longer attached to an
Anglo middle-class suburban way of life (Walks, 2013; Nijman, 2020a).
Rather, we see greater diversity, higher density, and significant complexity (Charmes & Keil, 2015; for studies beyond stereotypical suburbs, see
Hern in this volume, and Maginn and Steinmetz in this volume), massive development and large-scale housing projects in many contexts
around the globe (Güney et al., 2019), the assemblage of villages, irregular settlements, new towns, and gated estates in "a world of villages"
between the urban and rural areas (Gururani, 2020; Kuson, 2020; Kusno
in this volume; Li et al., 2020; Shen & Wu, 2020). Understanding changes
in today's urban peripheries requires a vantage point beyond the suburb. More attention needs to be paid to local histories (see Indigenous

communities, Shields in this volume) and geographies as well as interconnected and related development across scales (for example, the role of diasporas in gated communities; see Ortega in this volume). This peripheral urban world also represents not just the present. It has a layered history (Logan, 2021; May, 2020; Van Damme and Oosterlynck in this volume; Shields on settler colonial suburbia in this volume; Robinson on the continuing influence of apartheid in South Africa in this volume). Our book cuts across various themes published in the Global Suburbanisms book series, on governance (Hamel & Keil, 2015, on the theoretical concept; and Hamel, 2021, on in-depth case studies), land (Harris & Lehrer, 2018), and infrastructure (Filion & Pulver, 2019). The current volume also includes cases from variegated geographies of urban peripheries in Europe (Phelps, 2017) and North America (Nijman, 2020a), and many other places in the world (Güney et al., 2019, especially on Istanbul). In the current book, we revisit the concept of suburban and urban peripheries (Harris & Vorms, 2017) and the history and historical continuity (Logan, 2021, on modernist vision).

A View from Planetary Urbanization

Snapshots from Post-Pandemic America

We were finishing this volume as the 2020 American election had just passed and the former president was holding on to his counterintuitions that he won and was cheated of his victory. We now know, many fruitless court battles later and an insurrection in-between, that the former head of state eventually left the White House without conceding and made way for the administration of Joe Biden and Kamala Harris. What will remain with us is the changing sociospatial realities and attached dreamscapes of American suburbia that underlie much of the electoral imaginary and debate in that country. Those profound changes and those altered dreamscapes are the subject of this book. And not just in America. The authors here are concerned with the world of suburbia as we knew it in North America and with a planet on which suburbia is everywhere. Before we get there, let us remain a while longer in the United States.

As it happened in previous elections, the suburbs became once again a major bone of contention for the campaign of 2020. Traditionally seen as "the land of swing voters" and up for grabs by both major parties, this time around the suburbs remained an enigma: many appeals fell flat as the strategists seemed to have lost their grasp on who the suburban voters were (Lung-Amam, 2020a). The incumbent made futile attempts at grasping at outdated ideas of suburban homogeneity and traditional family life behind white picket fences. The outdatedness was twofold: appeals were targeted at an image of suburbs as white, and suburban women's lives and roles as traditional. These entwined *ideas* may have been true (if not unproblematic even then) more than half a century ago but they did not comply with current sociodemographics in the American periphery, and the *politics* associated with white privilege and traditional family values were not sufficiently powerful tropes to attach themselves to a majority of suburban voters in 2020. Instead, as Nijman (2020b) observed for the case of Georgia, rapid urban and suburban demographic change has changed the electoral landscape in fundamental ways. In fact, the dissonance between the incumbent's expressed views on suburbia and the reality were not just demographic, they were political: they handed Joe Biden and Kamala Harris the key to the White House. Lung-Amam (2020b) notes how deep the changes are and what their consequences might be:

> So yes, Black and brown voters deserve credit for evicting a neofascist from the White House. But Democrats cannot, like Trump, stereotype them as "urban" voters when most live in suburbs. Black and brown suburbanites face a range of issues that urban policymakers often miss. Many live in communities with mounting issues of infrastructure decline, underfunded schools, crime and poverty rates, but have a shrinking tax base. Others live in gentrifying neighbourhoods with rising rents, foreclosures and evictions, but lack tenant and affordable housing protections. Many struggle to access public transit, affordable health care and social services that have not moved to suburbs as quickly as those that most need them. Some face fierce anti-immigrant crackdowns and police violence. All struggle with an economic and public health crisis that has hit Black and brown communities the hardest. (n.p.)

While the American suburb continues to incite constant reinterpretations of cultural meaning (Diamond, 2020) and unearthings of alternative histories (Hurley, 2019), such shifts from the centre of cities to the peripheries, and the extended forms of urbanization that now mark our urban regions, point beyond America. We live in a world of suburbs, and urban scholars are beginning to understand that we need to look at what life will be like on a planet where the majority lives beyond the core.

This was also brought home during the unprecedented COVID-19 pandemic that raced around the world in 2020. Scholars had indicated before the pandemic errupted that the spread of infectious disease was linked to massive extended urbanization that character-ized a majority urban world. This meant that (re)emerging infectious diseases now were linked to the geographies of global urban expan-sion in new, diverse, and accelerated ways (Connolly et al., 2020). When COVID-19 began spreading across the planet, it turned out to be mostly a disease of the social, spatial, and institutional periphery of urban society (Biglieri et al., 2020). In a markedly different pattern from SARS in 2003, which had been described as a disease of the global cities network, SARS-CoV-2 (or COVID-19) found the crevices of the extended urban form in which to wreak the havoc it did (Ali & Keil, 2008). While the evidence is still elusive on what role densities and overcrowding played in the narratives of disease transmission, it was front and centre of all responses universally, as communities adapted to a new spatial grammar of lockdowns, quarantines, and social dis-tancing (Acuto et al., 2020).

The snapshots of the American election and the COVID-19 shed light on the increasing significance of the periphery in our urban imagina-tion. In that imagination, suburbs are not what they used to be: linear extensions of the core, passive and dull counterparts to their creative urban cousins. Instead, they are active and dynamic producers of presi-dents and pandemics, arenas of significance for urban life overall. But these are snapshots, and while not anecdotal in the strict sense, they need to be seen in the context of more long-standing tendencies that are the subject of this book.

A Journey to Peri-urban China

Let's take a journey to peri-urban China to illustrate how suburbia needs to be understood in its historical context. Even without a com-plete colonial history, understanding Chinese urban history is impera-tive, as we know from postcolonial studies. In examining the edges of the Kolkata Metropolitan Region in India, Roy (2015) stresses the value of the postcolonial perspective. This is not only because Indian cities have a colonial history and the current condition of "suburbs" needs to be understood with reference to the historical condition in the formerly colonized society but also because the concept of the suburb itself needs to be interrogated through decentralizing the universal theory. Roy illustrates three aspects of "postcolonial suburbs": (1) the making of enclaves as spatial governmentality, for example, eviction as state governance; (2) informality as a modality of governance, involv-ing not only capitalist expansion, accumulation, gentrification, and displacement but also deregulatory experiments and greater flexibility provided to middle-class development, resulting in the proximity of slum and suburbs; and (3) politics of suburban periphery beyond the formal bourgeois civil society but populism and patronage. Her postcolonial critique of suburbia is forceful. Even for those cities without the colonial history, attention must be paid to their particular histories, for example, state socialism in the case of Chinese cities (on the comparative paths of urbanization in China and India, see also Ren, 2020).

Despite the notion of *jiaoqu*, a Chinese word that describes the district outside the city, the suburb had not been a distinctive category of jurisdiction in the socialist history until 1978 when China adopted an open-door policy. In the absence of the concept itself, how might it be possible to examine "suburbanization"? To think about suburbanization in China, we need to ask three questions: Where is the suburb? What is the suburb? What is the suburban question?

For the first question, we tend to answer it in geographical terms. That is, if you go out of the city and before you reach the countryside, you might expect to reach the suburb. However, the question is not that simple: there was not an administrative space of "suburb." In the socialist period, there was a profound urban-rural dualism. A resident, according to the population registration (*hukou*), was either inside or outside the "state system." There was no intermediate space in between state and non-state systems. Despite the pervasive role of the state in economic governance and social lives, the reach of the state system was confined within the bounded area of state workplaces. The state welfare system and its associated collective consumption were organized by "state work-units" and largely provided within the boundary of the city. So, if the "urban question" is linked to the reach of collective consumption, the question is confined within the space of the state system. The vast rural area was left underdeveloped as the self-reliant and self-contained traditional society. Hence, the boundary of urban and rural areas was, in essence, marked by the reach of the state versus the non-state sphere. The modern bureaucratic management system only reached the boundary of the city. There was no "in-between" space between the state and the non-state system. In the periphery of the cities, you might find state-owned farms, state-owned industrial enterprises, and their residential compounds. But they all belonged to the state system, even though they were physically located outside the contiguous built-up area. In other words, even if they were located in peripheral locations, they did not constitute the suburb. Outside the state system was the non-state system. The distinction between them was reinforced by the residential registration system known as *hukou*. In other words, despite being a commonly used word, the suburb was not a meaningful concept but only a reference to the peripheral location. The commonly referred to location of suburb included both space inside and outside the state. In terms of living quality, the peripheral location was inconvenient and less attractive to urban residents (Zhou & Logan, 2008).

In regards to the second question, the urban periphery has been transformed since China became a world factory. The state introduced the peripheral area as a new space to extend capital accumulation (Shen & Wu, 2017). Manufacturing industries have been relocated from the central areas to make space for economic restructuring towards the tertiary economy. Foreign investment has flowed into peri-urban areas, and new development zones have been set up, leading to mixed land uses. Rural migrants have migrated into the city, or more specifically into the peri-urban area as they could not find accommodation in the central city and have had to live in dormitories or mostly in private rental housing provided by farmers in the suburbs. The former rural villages thus have been converted into "urban villages" as the enclave of rural migrants. The urban and rural began to be "integrated" as rural migrants arrived in urban China, while rural China witnessed the development of the city – the urbanization process of the rural, most salient as "urban villages" on the periphery. The suburb thus became an interface and even a battlefield where the state extends its formal governance into traditional leftover rural society, while millions of rural migrants try to make a living through informally building and renting out their extended housing. Still, the suburb is a space of mixed land uses and heterogenous state and non-state fragments. But because of urban sprawl and rural self-construction, the interface between the urban and rural areas, or the so-called suburb, has become a vast area of construction. The urban periphery, with its salient feature of heterogeneity, is becoming a more meaningful category – for rural migrants to find a space to live, for the state to initiate new town development (Shen & Wu, 2020; Shen et al., 2020), and for investors to find land for mega urban projects. The suburb has also imposed a governance challenge for the state. If the urban question in state socialism is a state-organized collective consumption, now along with peripheral development with both state sanctioned and private and self-development, the suburban question becomes an imperative one for both capital accumulation and governance.

Now turning to the third question, we argue that this is a complex one: the role of the state is highly visible but not exclusive; non-state actors are active, and their development practices are widespread. In short, the suburban question is a question of intertwined modality of governance and contested peripheral space (Wu in this volume). Resonating what Roy (2015) described as postcolonial suburbs through an Indian political society in relation to its colonial history, the Chinese suburban area needs to be understood through the histories of state socialism and post-reform market pragmatism.

The suburban sphere is created outside exclusive state control but at the same time suburban development reflects the state developmental intention. This understanding of in-between space of state and non-state adds a new narrative of "intertwined modalities": the underlying governance mode forefronts state strategies, while the state proactively uses market development instruments such as land value capture (Shen & Wu, 2020), quasi governance agencies such as the "development management committee" and local government financial vehicles (LGFVs) (Wu & Phelps, 2011; Wu, 2022; Wu in this volume). The urban periphery is also a space of "assetization" or financialization of housing, where "private" and enclosed residential estates are built (Wu et al., 2020). Instead of seeing suburbanization as residential relocation for a new way of life, the Chinese urban periphery is a space of state entrepreneurialism (Wu, 2018, 2020a, 2020b). That is, the suburban question is more than a question of "urbanism of exception" (Murray, 2017) but more one of "state strategies" and state action through the market (Wu, 2020c).

Planetary Urbanization

Originally conceived by Henri Lefebvre (2003/1970) as a world-wide "urban revolution" and a "complete urban society," "planetary urbanization" has become a salient feature in the twenty-first century (Brenner & Schmid, 2015). The process of urbanization is not confined within the city and becomes pervasive, permeating urbanized areas, peri-urban areas, and the countryside. The statistical classification of urban and rural population is only a technical way of detecting and defining the city as bounded settlements but is insufficient to understand urbanization as a continuing process. The reliance on urban and rural classifications to claim an "urban age" does not reveal interconnection and relatedness of urbanization (Brenner & Schmid, 2014).

This Lefebvrian perspective provides an important intellectual influence on the research on urban peripheries. More people are now living in peripheral urban areas, on the literally suburban planet today (Keil, 2018b). But the distinction between "city" and "countryside" becomes blurred, creating the constellated and fragmented segments at the periphery, or as Keil called these phenomena, "suburban constellations" and "disjunct fragments" (Keil, 2013, 2018a). We have seen rapid and unprecedented urbanization in many previously non-urbanized regions of the world, new forms of urbanization (van Duijne & Nijman, 2019) and the emergence of urban megaformations (Sorensen & Labbé, 2020). In short, urbanization in the twenty-first century is increasingly occurring in the peripheral areas outside the city and in the cities in the Global South where the majority of the urban population already resides (Bhan et al., 2020; Simone, 2019).

Another intellectual influence on peripheral urbanization originates in postcolonial critique, which emphasizes the need to contextualize today's Global South in particular (post)colonial histories. This is a growing frame in the Global North, especially in immigrant countries like the United States, Israel, Canada, and Australia, all of which are settler colonies with the latter two being part of the British Commonwealth (Haas, 2019; Keil, 2018c; Shields in this volume). Here, the postcolonial "boomerang" builds different suburbanisms than those imagined by the common narrative of white-flight suburbia. But also in Europe, where the immigrant (and refugee) experience is now very closely tied to life on the urban periphery and reflects back on all manner of sociospatial relationships in the continent's urbanism (Kipfer & Dikec, 2019; Tchoukaleyska, 2019; Tzaninis, 2017). The postcolonial critique, as will be shown later in this book, also makes the adoption of the reference point of post-war North American suburbs problematic in many cities in the Global South as well as "the multiple and contradictory imageries" of a "South-Eastern" perspective (Yiftachel, 2016, p. 485; Robinson in this volume; Roy, 2016). On the periphery that has distinctive agrarian histories and rural presents, Roy (2016) asks "What is the urban?" The urban question needs to revisit historical administration which classifies the urban and rural and its governmentality. Urbanization is a process beyond the change of economic activities from agricultural to non-agricultural activities and the migration of population from the rural to urban areas. It is also a process of the state exerting its sovereignty in governing its subjects. According to this perspective, the changes in peripheral areas are thus more than economic transition to service industries, rising-edge cities and technological enclaves through formal development but also include "autoconstruction" (Caldeira, 2017) by villagers and real estate developers (Gururani and Kusno in this volume) and greater informality and diverse political mobilization (Roy, 2015).

The pattern of classic suburbanization hence becomes just one particular form among vast possibilities of residential relocation and spatial development. The periphery of cities has seen the arrival of people and activities from both the city centre and outside places. Considering increasing densities (Charmes & Keil, 2015) and retrofitted infrastructure (Phelps, 2015), Phelps and Wu (2011) regard the departure from pure residential suburbs to mixed uses as "post-suburbia." However, this notion of "post" still uses suburbia as a reference point. Although recognizing the difference in terms of economic structure and settlement forms, post-suburbia does not fully appreciate the novel features that might not be related to suburbia. For example, Wu (2020c) argues that peri-urban development in India and China might not be associated with the "suburban utopias" identified by Fishman (1987). As will be seen later when we discuss the Chinese example to reveal the limit of suburbia and postcolonial critique of suburbia as a bourgeois civil society in India, which might be more appropriately characterized as a political society in India (Roy, 2015), the perspective of post-suburbia is still quite restrictive.

Here, we explicitly propose that the research on urban peripheries after suburbia should consider the view from planetary urbanization – thinking peripheral changes not from the city but a worldwide process of contemporary urban changes and should cover aspects of change in land, infrastructure, and governance beyond the post-Fordist transition. These changes include both formal and informal land uses, self-built housing by villagers, and large-scale state-led infrastructure and housing projects, creation of governance of different modalities, and everyday life and practices. As we argue elsewhere, "the notion of suburbanization as dependent on one center has to be discarded as the form and life of the global suburb take shape in a general dynamics of multiple centralities and decentralities" (Keil, 2018a, p. 494). The transformation of urban peripheries should not be understood as suburban changes but rather as a form of planetary and extended urbanization, which we discuss more fully below. This is a view of planetary urbanization, but at the same time it is sensitive to historical temporality and geographical contingencies. The geographical scope of this book illustrates the worldwide nature of urbanization at the periphery. The studies here are from both the Global North and the Global South and provide new narratives, as "it is no accident that those places that experience the major thrust of peripheral urbanization are also the chief source of theoretical innovation: African, Asian, and Latin American scholars not only present the scale and extent to which their urban regions are redefined at the periphery but simultaneously theorize this process in novel ways" (Keil & Lynch, 2019, par. 12).

Extended Urbanization in Peripheries: Land, Infrastructure, and Governance

Now that we have adopted a view of planetary urbanization, we wish to identify more concrete forms of extensive urbanization and peripheral urban changes in this volume's chapters. One important feature is a significant extension of the built-up areas, in either concentrated or fragmented forms (Angel et al. in this volume). The fast extension of urban land use indicates that the change may include not only the classic form of suburbanization but also other self-built or autoconstruction activities in the Global South (Caldeira, 2017). Some of these activities take the form of low-density sprawl, and others purposely seek large-scale housing development through city planning and state-led programs. Through infrastructure development that connects previously discrete settlements, a new city-region has been formed. There are diverse descriptions of this extended form locally, such as urban clusters (Wu, 2016), urban corridors (Sorensen & Labbé, 2020), or "regional urbanization" (Balducci, Fedeli, and Perrone in this volume). Thus, the extended form of urbanization is not limited to the urban edge but extended into a vast hinterland that has been linked into the urban system. The notion of suburbia does not fully recognize more recent polycentric city-region development (e.g., Harrison & Hoyler, 2015) and urbanization of the non-urban areas (Monte-Mór and Castriota in this volume).

However, the extended form of urbanization does not simply mean decentralization and dispersion. As Harris (in this volume) argues, density still matters. From this view, the urban edges are still the most dynamic place connecting the centre with the hinterland. Because of the forces of both concentration and diffusion, the urban periphery does not seem as compact a form as the central areas. Nor does it continue to disperse into low densities as it does in the countryside. More often than not, peripheral densification is a consequence of both the real estate market and deliberate state actions intended to either achieve sustainability objectives, economic development, or affordable housing provision (Robinson et al., 2021). Except to suggest that the periphery is transitional and dynamic, it is difficult to prescribe a defined spatial structure of urbanization. The exact form may also depend upon the imaginary in different political and social contexts – some are more interconnected regional clusters such as the Great Bay Area of Guangdong, Hong Kong, and Macao; others are more centred upon world-class megacities and their immediate edges such as New Delhi (see

Gururani in this volume). But all in all, extended urbanization creates conflicts between localities and raises the need for coordination at a regional scale.

Three aspects – land, infrastructure, and governance – define urban peripheries in a complex way, which are different from the suburbia of low-density urban sprawl, suburban subdivision, formal real estate development, car-dependent infrastructure, and middle-class communities. (Following Harris & Lehrer [2018], who ask "how suburban land markets differ," we define *land* as a key defining dimension. In some respects, the suburban land market is simply a watered-down version of that in and around city centres; that is, land is just less valuable [p. 24]). But considering planetary urbanization, the urban periphery is the frontier of more formal modes of land ownership mixing with informal ones. In post-socialist economies the land is subject to commodification; land ownership shows a strong path dependence. In postcolonial societies, complex tenures persist with the movement to land property right entitling to convert informal into a state recognized form. The development of a land market in urban peripheries is evidence of the expansion of capitalist property rights relations, and, hence, no matter what spatial forms they take, these developments are part of planetary urbanization processes. However, there are significant differences between dispersed suburbanism in the Global North and planning-driven land development that simultaneously removes and recreates informality in China (Wu et al., 2013), or land grab and land speculation, with limited influence of the local government but greater power of regional governance agencies and larger developers in India (Gururani in this volume; Harris & Lehrer, 2018; Ren, 2020). More often than not, the land market has not been converging into a complete urban land market.

Infrastructure has long been under-researched in suburban studies, except for the notion of highway construction leading to car-dependent suburbs. For the Global South, the failing of infrastructure provision to connect residents with municipal services is widely noted. However, infrastructure research now occupies a central position in understanding the development of urban peripheries (Filion & Pulver, 2019). The governments in the Global South now attempt to use infrastructure development to strengthen their positions. In many countries, with infrastructure extension in city-regions, more coordination is required between subregions in the metropolitan area. The collaborative relation between different jurisdictions is required for infrastructure cooperation but it also triggers a process of public participation in more democratic economies because of the promise of infrastructure (Filion in this volume). On the other hand, the development of particular forms of infrastructure such as those for transit-oriented development are considered by state and city planning as a way to promote sustainable lifestyle changes (Filion in this volume). The growth of the urban periphery exerts a strong demand or burden on government finance. The urban periphery in the Global South is where the gap of infrastructure demand and supply is the most significant, yet the government may deliberately pursue the development of one type of infrastructure, for example green infrastructure (Macdonald, 2020), for techno-burbs or development zones while neglecting or insufficiently providing another for everyday needs of poorer residents in the periphery. As the development in the urban periphery shows, infrastructure shapes the life of residents, and infrastructure development and provision are material manifestations of urbanization processes, which raise concerns for social justice and service deficits (Biglieri, 2019; De Vidovich, 2019; Miro, 2020). In the Global South, auto-construction has managed to create informal housing and settlements, but the infrastructure has failed to support these initiatives. Even in the most advanced economies in the Global North, the coordination of infrastructure across different cities within city-region continues to impose political challenges (Addie, 2019; Enright, 2016; Jonas et al., 2014).

The third aspect of global suburbanisms is governance, which defines the new feature of urban peripheries in a significant way. While the real estate market is regarded as the key driving force for urban sprawl, governance in peripheral areas is beyond a single-market mechanism and involves multiple modalities (Ekers et al., 2015; Hamel &Keil, 2015). These modalities include the state, capital accumulation, and private governance, and represent the state, the market, and society, respectively, which co-exist in the periphery. However, how these modalities are combined into an overall governance framework varies across different countries, as shown in case studies (Hamel, 2021). In China, the role of the state is significant and forceful. While market development tools are widely adopted, the state, both central and local, maintain certain "strategies" and development intention. This is framed under "state entrepreneurialism" in which the centrality of governance does not shift to the market (Wu, 2018). Development of new towns in Shanghai and Beijing is strategically defined from a spatial planning perspective. But in other contexts, real estate speculation characterizes peripheral property development (Shatkin, 2017). Beyond a single-city perspective, peripheral development raises a challenge for city planning when comprehensive rational planning, originating at higher levels of government, is locally implemented, as is the case, for example, in Ontario, Canada (Macdonald & Keil, 2012). There is now some tendency towards "governing cities through regions" (Keil et al., 2017) but faced with metropolitianisation (Hamel in this volume), spatial planning for infrastructure provision continues to require complex intercity collaboration (Jonas et al., 2014) and negotiation between localities (Hamel in this volume). In some parts of the world, as Ren (2020) argues in her comparative study of "governing the urban in China and India," territorial and associational forms of governance compete in land development and environmental planning and policy; elsewhere, governance of emerging, and often peripheral, urban spaces occurs in what Simone and Pieterse (2017) call "the lived realms of the 'make+shift' city" (p. xiii). The understanding of governance modalities requires us to pay closer attention to the actors in peripheral development. The actors in peripheral development in the Global South include formal market developers, translocal finance, and local landowners. In India, local landowners inherited their power during colonialism and still influence development in a significant way (Gururani in this volume). The new towns in Jakarta, Indonesia, are developed by developers who originated in the financial sector (Kusno in this volume). The "real estate turn" in East Asia (Shatkin, 2017), largely occurring in the peripheral area, demonstrated variegated state agencies. In Indonesia, the national state actors developed the coalition with larger developers and with transscalar financial actors from Chinese state banks (Shatkin, 2020). In Bangalore, India, the "transscalar territorial network" consists of both global finance capital and local developers (Halbert & Rouanet, 2014).

The Content of the Book

In the remainder of this introductory chapter, we highlight the key arguments and findings of the chapters in this volume. The book contains three sections. In the first section of openings, the chapters provide an overview of research after suburbia, evidence and worldwide geographical descriptions of current changes at the urban periphery, and debates over extended urbanization. In the next chapter, chapter 2, Ute Lehrer reflects on the methodology of using images to combine research and teaching in the peripheral urban areas. The chapter also explains the visual information and research sites covered in others chapters of this book.

In chapter 3, through classifying Landsat imagery and mapping land uses, Solly Angel and colleagues provide a comprehensive picture of peripheral changes in terms of the size of new peripheries, the rate of urban expansion, the difference between more developed and less developed countries, densification versus expansion, and the extent of the urban fabric developed in the peripheral areas. As they point out in their chapter, the "new urban peripheries of cities are growing wild." In contrast to well-regulated suburbia, the challenge is to tame "the way land is subdivided and sold on the urban periphery, with a view of getting rural landowners and informal property developers to improve their land subdivision practices."

In chapter 4, Alessandro Balducci, Valeria Fedeli, and Camilla Perrone discuss Italian cities and how the term "after suburbia" might mean a "regional urbanization," that is, cities lined with the urban conditions but lacking a single urban institution to deal with the interrelated urban development across cities. Balducci and colleagues stress a transscalar "regional imaginary" and think that the regional urban "corridors" might be part of "a complex process of regional urbanization that must be read at a national scale, producing large urban regions." They discuss a dispersed urban pattern across scales: "urban regions, small and medium cities, inner areas, polycentric urban networks, emerging suburban contexts highly interrelated and not as single isolated urban processes." Their idea of a regional dimension of urbanization as "post-metropolitan territories" (Balducci et al., 2017) resonates with various notions of planetary urbanization (Brenner & Schmid, 2014), extended urbanization (Keil, 2018a), and city-region development (e.g., Harrison & Hoyler, 2015; see Wu, 2016, for the Chinese case). This chapter, although focusing on Italian cities, links the agenda of after suburbia with the general body of literature on city-regions and the dimension of regional urbanization.

Instead of thinking suburbanization, Roberto Monte-Mór and Rodrigo Castriota elaborate the concept of "extended urbanization" in chapter 5. They illustrate the reach and nature of extended urbanization in Brazil and in Amazonian territories where urban-industrial actors have exerted their influence over underdeveloped areas with severe ecological implications. Thinking planetary urbanization, the authors argue that there might be an urban utopia outside the inescapable "capitalist urbanization given its planetary scale." Other economies outside the deregulated economies and financial capitalism, and neoliberal logic penetrating all realms of life, are working towards "the urban utopias," which "implies to work outside of capitalism, strengthening other economies, other social relations of production, other forms of producing social (and differential) space that can oppose hegemonic destructive capitalist accumulation." They denote "after suburbia" as a phase after or a space outside "alienation" and the development of everyday life, across different places from metropolitan suburbs to the countryside.

In chapter 6, Richard Harris continues to critique the concept of extended urbanization. In the debate over planetary urbanization and urban theory (Scott & Storper, 2015), Harris raises a key issue about density. Instead of emphasizing an extreme of distinctive suburbs or planetary urbanization, he stresses the importance of density and argues that a "dose of density" makes a difference in finding a distinction between urban, suburban, or non-urban. His historical insights suggest that because of this difference, higher densities accompanied with agglomeration define the urban rather than a vast area influenced by pervasive planetary urbanization. According to Harris, the claim of Lefebvre (2003/1970) for an 'urban revolution'" and its contemporary conceptualization in relation to the urban (Brenner & Schmid, 2014) needs to be carefully re-examined, as it would not wipe out the notion of the "suburb." He challenges the idea that the distinction between the city and suburban or rural areas has almost dissolved and that places everywhere are woven into the urban fabric. Because of variable density, Harris argues that "the transitional spaces called suburbs must have their own peculiar charter ... just as cities are still their own sorts of places." As such, the suburb still persists as a different or varying type, which is much less densely settled.

The second section examines three major themes – land, infrastructure, and governance –and begins with chapter 7. In a different take on the suburban land question, Shubhra Gururani investigates a pastoral stretch in peri-urban New Delhi, a suburban new town known as Gurgaon, India's Millennial City. In this new town, hundreds of developers of different scales engaged in frenetic land development and had to negotiate and compete with the landowners who were villagers. Instead of assuming this after suburbia as an urban question, Gururani argues that at this frontier of India's urbanization we see "an urbanism that does not erase or assimilate the rural but an urbanism in which agrarian and urban dynamics sustain and produce each other." She further suggests that the characteristics of this peri-urban area show that these are a collection of "cities in a world of villages" in that these villages are not part of a complete urbanized fabric. Private rental housing rather than public housing (in the sense of collective consumption, see earlier discussion on China) provides accommodation for thousands of migrant workers and the villages also provide infrastructural services to make urban life possible. The big landowning castes continue to influence land development and thus the villages play a role in property development. Moreover, the division between the villages and the formal planning regime still persists. Subject to this boundary constraint, villages see intensification that demonstrates the caste and class difference. This shows that the development of new towns such as Gurgaon is subject to the agrarian question as supposed to being subsumed entirely under the urban question.

Focusing on suburban infrastructure and its planning, Crystal Legacy compares Toronto and Melbourne in chapter 8. She investigates city-regional infrastructures in relation to the notion of after suburbia. Suburban infrastructure construction is costly but, in order to reduce political contention, decision-making processes have to be streamlined. Legacy looks into the need for formal citizen participation to protect infrastructure planning from "politicization" and the adoption of public-private partnerships. She raises a similar governance issue as Hamel (in chapter 11) who discusses the decision-making process of planning. Suburban infrastructure planning is embedded into a political process in which citizen participation pushes a greater public consciousness about the "promises" of infrastructure and benefits of these projects. Legacy's study provides a portrait of infrastructure development, planning and politics around these large projects in the setting of city-regions, which mismatches the boundary of a single-city administration. She suggests that the political participation of citizens and locally elected officials has "elevated into public consciousness not only the importance of suburban infrastructure and its planning but also the need to take an integrated and regional outlook to understand the politics that suburban infrastructure promises wield."

In chapter 9, continuing along the same theme, Pierre Filion examines the promise of suburban infrastructure as well as the instruments of control and power used in its governance. Such control leads the public to behave according to the norms imposed by administrators or according to commonly accepted standards. Focusing on the region surrounding the City of Toronto, Filion explains the role of infrastructure in the Fordist suburb, the early neoliberal suburb, and the mature neoliberal suburb. He shows that in some mature suburbs new public-transit systems have been developed with linked densification. But in the north-west suburb of Toronto, for example, while immigrants supported the transit project, long-term residents successfully opposed it. Because of suburban diversity and political division, different public domains of transportation infrastructure are created. The lack of transit infrastructure, he writes, "orients behaviour towards the automobile-centric lifestyle." Smart technology and suburban transportation now seem to lean towards the neoliberal and libertarian vision to confine individualistic lifestyles in post-suburbia areas.

Following the analytical framework of governance modalities (Ekers et al., 2015), Fulong Wu reveals the intertwined modalities in post-suburban China in chapter 10. Looking specifically at a new town in Shanghai, he illustrates how "state entrepreneurialism" (Wu, 2018, 2020c) might be an explanatory perspective, different from neither the neoliberal suburbanism (Peck, 2011) nor the authoritarian Chinese state. New towns have been a salient feature of recent peripheral development in China. Recent studies on suburban China reveal a heterogeneous social space on the periphery (Wu & Shen, 2015; Li et al., 2020). Wu stresses the role of the state in governing the peripheral urban areas and developing new towns as new spaces for capital accumulation (Shen & Wu, 2017, 2020). But the geographies of the peri-urban are fragmented as shown in the Lingang new town, involving multiple "development actors" and state agencies, and intertwined modalities of governance in which homebuyers participate and "market instruments" such as financialized development corporations are deployed (Wu et al., 2020). Wu argues that governance after suburbia in China "cannot be simply characterized as state-led or market dominated."

In chapter 11, Pierre Hamel provides a comprehensive review of city planning and its role in the making of global suburbanism. He focuses on governance and its challenges after suburbia for planning. Focusing on recent forms of "metropolitanisation" and the rise of city-regions, he reviews the long tradition of the "postmetropolis" driven by a new mode of space production under post-Fordism and postmodernism (Soja, 2000). Further, he examines the role of planning and planners associated with the change of "planning model" from rational comprehensive planning, activist planning, and the introduction of participatory mechanism. The "new regionalist" approach in the post-suburban era challenges this cityness tradition as multiple actors across scales are involved in city-region planning. Working with the planetary urbanization proposed by Brenner and Schmid (2014), Hamel argues that "the requirements for metropolitan governance are on the rise." The peripheries reinforce economic, social, and cultural diversity. Thus, "fragmented political geographies" after suburbia require a different planning approach to solving conflicts across regions. As such, planners should not regard themselves as the privileged experts to promote an exclusive normative model after suburbia (e.g., the new regionalism) and need to be aware of planning practices as concrete processes of deliberative governance.

The third section examines various pathways into studying urbanization after suburbia. These include the challenges African urbanization poses to the conceptual world of the suburbs, the alienated suburbs, the suburbanization of sex shops, new towns in Indonesia, transactional migration into Manila's peri-urban areas, and revisiting the Indigenous North American suburbs. These studies not only enrich our understanding of different forms of the planet's peripheries but also break out the stereotypes and bring together separated narratives from the Global North and Global South.

Chapters 13 and 14 are both about temporality of after suburbia research. In chapter 13, Rob Shields revisits North American suburbs through the perspective of settler colonial society and postcolonial critique. As part of Indigenous research, Shields examined the diasporic and Indigenous communities in the suburbs and looked at a series of case studies, including Indigenous suburbia, Indigenous veterans, erased suburbs, incorporation, municipal colonialism, land dispute, and settler colonial suburbia. He highlights that "suburbia is a site of both exclusion and the creation of not just a stereotypical 'suburban middle class' but of a domesticated, planned environment and range of modern citizens who are workers in a division of labor that is economic and sexual." The research recovers the "lost history of previous occupation of the land." Just as Harris (2015) uses Toronto to show that the "suburban ideal" does not fit well with the actual North American cities, the postcolonial perspective, when applied to the heartland of the suburban research, reveals quite complex and neglected facets of the peripheral area.

Similarly referring to the time before suburbia, Ilja Van Damme and Stijn Oosterlynck in chapter 14 remind us of the need to understand the history of suburbanization and to compare after suburbia through time. Despite the newest forms of "technoburbs," "edge cities," and urban mutations, they suggest that after suburbia could also mean a reversal of urbanization and a possible "reruralization" as "weakening of operations, contact ties and mental identifications with an urban centre, effectively shrinking or displacing the suburban phenomenon as such." From a historical perspective, they argue that after suburbia "becomes a category of the past, an emergent and unstable phase defining the contact zone between 'city' and 'countryside' throughout history." The darkness of a future after suburbia may mean, for some cities, such as the historical city of Antwerp, for example, rapid depopulation and loss of connections between the city and its hinterland.

The next two chapters are all about geographical contingency, using African urbanization as an example. In chapter 15, Jennifer Robinson thinks, from the perspective of methodological and conceptual innovations, that suburbs "remain active forces in the global urban landscape, shaping emerging configurations of the urban." She thinks the "afterlives" of the suburb in "its US and UK-centric cultural resonances" and the "specific suburban histories continue to be operative in shaping current and future configurations of the urban." Through a comparative urban studies perspective, she urges reconceptualization of the suburbs, which "can begin anywhere"; for Johannesburg, South Africa, for example, continuing high-income suburban developments co-exist with "large-scale state-led initiatives for poor housing located on the peripheries of the city." Robinson suggests, despite its singularity, that this case presents interconnected spaces, as she has found presented elsewhere, specifically London and Shanghai, in her research (Robinson et al., 2021).

In chapter 16, Robin Bloch, Alan Mabin, and Alison Todes provide a comprehensive overview of historical, current, and future (expected) developments on the world's most rapidly urbanizing continent, Africa. Focusing on sub-Saharan Africa, they dispel existing myths about the region's urbanization and demonstrate that its urban regions display a wide and growing variety of urban and suburban forms. Noting the ongoing dialectic of centreand periphery in sorting populations by wealth and origin, they report on the "search for better terms for and understandings of African suburbs and suburbanisms" and place the debate on the continent's remarkable advances in urban forms and lives to extant debates in urban and suburban theory. African cities have demonstrated fast urban expansion, in terms of investment flows into peripheral developments, state-affordable housing, movements of middle-class in inferior locations through self-construction, and coming migrants in the places yet to become urban. The authors note that the suburb in African contexts contains the settlement "before suburbia," as the suburb implies a place that has not been fully urbanized, with the reference to other places than the city. The debate of suburbanism versus peri-urban development reveals the complexity of African urban peripheries.

The next two chapters are back to the heartland of suburbia and reveal the reality of the peripheral areas today. Chapter 17 by Matt Hern continues to pursue the complex character of suburbia, including suburban poverty and different kinds of urban peripheries. As he argues, suburbanization is a substandard and alienated urbanization for newcomers as well as the working-class and poor populations. He highlights that "sub-urbanization 'after suburbia' doesn't look all that suburban anymore," and that a lot of gentrification studies acknowledge displacement but do not trace displaced people to the suburb (see also Lawton, 2020). For some places, "fixing the suburbs" is advocated in a similar vein to aestheticized urbanism in the central city, through densification, transit-oriented development (TOD), and compact communities. But Hern thinks through divergent patterns of peripheralization in the peri-urban Global South and sheds light on some common threads – "how sub-urban places fall off the map, and are rendered outside." Thinking further about dispossession, Hern suggests that the land constitutes the very proprietary model upon which the suburban social relations rely: "new suburban developments transform non-valuable, stagnant rural, agricultural and in-used land immediately into property commodity." Just like settler-colonial logics to use suburban land to delineate the boundary of the suburban community, now low-income residents are displaced out of the city into messy and cheaper urban fringes. As such, Hern sees "urban periphery across the global are becoming increasingly fraught, complex and destabilized."

Focusing on the cultural dimension, Paul J. Maginn and Christine Steinmetz in chapter 18 tell the story of suburban sex shops and present a sharp contrast to the common image of domesticated and conservative suburbs. Through studying the representation of sex toys in films and detailed geographies of sex shops in Perth and Sydney, Australia, they challenge the social and cultural stereotype of the twentieth century suburbia as places of "domesticity, conformity, heteronormativity, and conspicuous consumption." They reveal that the rise of online shopping makes sex toys commonplace in many suburban households. The major wave of mass commodification and consumption associated with adult retailing has made the distinction between the city and suburbs less meaningful. With products on female sexuality and more female entrepreneurs in the industry, the image of suburban male consumers for sexualized spaces in the city does not provide an accurate picture of today's geographies of adult retailing and consumption. Their research is a cultural study of suburban life after suburbia.

The final two chapters in the third section focus on actors and agency in the process of development, and they both interrogate who builds the periphery. In chapter 19, Abidin Kusno examines the new cities being built by a major developer in Indonesia. Through describing the lineage of "city builders" – the gigantic Lippo Group – Kusno reveals the "logic" of business operations behind building the suburbs. Beyond a meta narrative of capital accumulation, Kusno shows that the acquisition of three large sites in the suburban and exurban areas of Jakarta is fairly incidental, and, in fact, are the legacies of property markets because they are the collaterals of another developer who borrowed the money from the Lippo Bank. Subsequently, the restructured business of Lippo gave up the financial sector and concentrated in real estate development itself. The Lippo Group reimagined these sites as the future cities of Indonesia. The building of these future cities is also associated with global connections: the California gated communities, the Japanese real estate developer, and the Chinese style of high rises, but they are Indonesian projects. The Jakarta case reveals the history of post-suburbia but not as a history of residential areas changing into mixed uses in the suburbs; rather, it highlights the operation of real estate and financial business across time and space.

In chapter 20, Andre Arnisson Ortega examines peri-urban Manila and reveals the role of transnational migration in the formation of wealthy enclaves. These "clusters of eclectically styled two- and three-storey mansions loom prominently over the rest of the village." This phenomenon is the development of translocal spaces after suburbia. These rich and often gated estates are built not only by foreign direct investment but also by diasporic capital, namely, remittances from overseas Filipinos. The study enriches the earlier notion of "desakota," a term coined by McGee (1991) that is based on the Indonesian example of describing mixed rural-urban land uses to highlight new transnational residential spaces. Peri-urban Manila cannot be fully explicated through a singular suburban narrative of city-periphery relations. As Ortega argues, the neighbourhoods of transnationalism in the Global South demonstrate the need to go beyond the "usual Anglophone suspects" of middle class suburbs. On the other hand, it is relevant to global suburbanism research as in North American suburbs we also witness racially diverse "ethnoburbs" (Li, 2009; and the chapters by Hern and Keil in this volume).

In the concluding chapter of the book, chapter 21, Roger Keil discusses theoretical implications, methods, and new research areas "after suburbia." He reflects on how research on global suburbanisms can contribute to urban theory and practice. As the principal investigator of the "Major Collaborative Research Initiative (MCRI) Global Suburbanisms" project, Keil (2013) started from the notion of "suburban constellations" and elaborated and developed his focus on governance (Ekers et al., 2015; Hamel & Keil, 2015), and finally proposed the notion of "after suburbia" to study the planet's urban peripheries (Keil, 2018a, 2018b). As a highlight of this post-suburban research agenda, the notion of "after suburbia" captures a diversity of forms on the peripheries, ranging from in-between spaces of various densities, to suburban high-rise estates, for example in China and Turkey (Güney et al., 2019). It also eventually includes the emergence of new research areas linked to extended urbanization such as the relationships of suburbanization and disease (Connolly et al., 2020), the suburban infrastructure fix (Logan, 2021), and suburban political ecologies (Newman, 2015;

Tzaninis et al., 2021). This research agenda of linking across different cities in the world provides insight into particularities of places but also maintains comparable understandings (Peck, 2015). Keil, in this conclusion, points out that by recognizing the periphery has become demographically and socio-economically hyper diverse, the practice of rebuilding suburbs would have to abandon the stereotypes. Focusing on planning, he stresses the need to mobilize the political agendas to tackle social contradictions such as poverty, exclusion, and space and resource allocations, beyond a simple densification fix (what he unpacks as "Vancouverism" in the Ontario city of Brampton, located northwest of Toronto).

The book is rounded out by two visual concepts – through images and text – that highlight important methodological contributions to the MCRI Global Suburbanisms project throughout its duration. These visual concepts provide illustration on the variety of suburbanization processes and suburban ways of life featured in the book. They also articulate important pedagogical and methodological dimensions of the MCRI Global Suburbanisms project. Ute Lehrer in chapter 2 exhibits and explains the results of experiential education events that she spearheaded between 2010 and 2019. The images related to this project provide a portrait of her systematic approach to working with undergraduate and graduate students at multiple locations. They reveal lived experience, on-the-ground perspectives, and perceived-conceived lived spaces of the global suburb. The images represent a perforation of the boundaries between research and teaching, recognizing the students' key roles as being learners and scholars at once. Markus Moos in chapter 12 reveals and recounts work done under his leadership on suburbanisms as ways of life. His team's methodological innovation extends to both data generation and visualization and contributes immensely to our understanding of the shifting categories and adaptive lenses used in global comparative research of this kind. The images from both these projects serve as markers of chapter beginnings, randomly distributed so as to invite the reader to stroll through the various enactments of the suburban planet.

Conclusion: Beyond Stereotypical Suburbia

Urban transformation has been a core topic of urban studies and geographical and social research. Conventional studies on urbanization have paid attention to rural-to-urban migration, while suburbanization research traditionally has not been linked to general processes and conceptualizations of urbanization but largely concerned with "white flight" in the Anglo-American context. The conceptual division of the urban, the rural, and the space in between as the suburban has been provocatively criticized by the concept of "planetary urbanization" (Brenner & Schmid, 2015), as the urban processes are not confined by the arbitrarily defined boundaries. Parallel to suburban studies are a prolific body of studies on "peri-urban areas" in the Global South. In contrast to "suburban community studies," the research on the peripheral areas reveals informality and autoconstruction (Caldeira, 2017). The politics of development in the periphery exceeds the formal and regulated land market.

The urban periphery in the twenty-first century is no longer a replica of post-war Anglo-American suburbia. The dichotomy between the city and countryside does not exist, and the place between them, commonly referred to as the suburban community in studies by American sociologist Herbert Gans (1967), is more diverse and heterogeneous. Suburbia as known in classic suburban studies ceases to have a distinctive referent. The research on urban peripheries requires a more plural approach. Here we emphasize four aspects: historical temporality, geographical contingency, suburban reality, and actors and agencies in peripheral development.

First, a perspective of historical temporality sheds light on the peripheral world before and after suburbia. From a longer timeframe, suburbia is a historical period created by specific conditions. The urban peripheral has a history of rural and Indigenous communities before the middle class came to the suburb on a large scale (Shields in this volume). The development of suburbs is also a process of labour reproduction in a domesticated and planned environment. Before the creation of the suburban residential environment, the urban peripheral areas had unattended wildness. Most suburban studies paid attention to cul-de-sac suburbia but failed to recognize the space outside real estate development and lingering Indigenous communities in the peripheral area. Adopting the view of temporality, we need to pay more attention to preexisting communities in the history of suburban development. Indeed, for some cities, after suburbia may not mean a continuing trend of converging into the city and the development of edge cities and technoburbs. In fact, industrial suburbs, their decline and restructuring are of particular concern in many capitalist and former socialist economies (Bernt, 2019; Wilson et al., 2019). Research on urban shrinkage reveals the trend of depopulation and state withdrawal, declining service provision in industrial cities in Eastern Europe (Bernt et al., 2017). The urban periphery as a zone of transition can experience a reversal process of "re-ruralization" (Van Damme and Oosterlynck in this volume). Through a historical perspective, the periphery after suburbia may mean a detachment from the city and a holding on to rural identities (Van Damme and Oosterlynck in this volume).

Second, the perspective of geographical contingency reveals the diversity of both formal and informal development in urban peripheries. The research on peripheral urbanization in the twenty-first century needs to understand geographical contexts. Adopting an approach that demands an acknowledgment of the changing "geographies of theory" (Roy, 2009) allows us to consciously pursue comparative urban studies through "theorizing from elsewhere" (Robinson, 2016). Robinson in this volume argues for an ex-centric reference point to embrace a wide variety of urban forms. For example, studying a mega housing program (the Corridor of Freedom) in the peripheral area of Johannesburg in post-apartheid South Africa shows that the project reflects a more developmental and redistributive state rather than middle-class suburbanization (see also Robinson et al., 2021). Similarly, the densification in outer suburbs of London (in the area of Old Oak and Park Royal) does not follow the trajectory of "suburbanization" but rather reflects the state's effort to cope with the crisis of housing affordability through promoting higher densities and land value capture. Studies after suburbia described under MCRI Global Suburbanisms include a new look at the North American example, once considered the ideal case of suburbanization (Nijman, 2020a): European perspectives east and west (Phelps, 2017): decolonial and post-colonial perspectives from "the Global South": "Asian" perspectives on "sub/urban theory" (Wu & Keil, 2020): work on African sub/urban constellations (chapters by Bloch, Mabin, and Todes and by Robinson in this volume): and the multiple worlds of "massive suburbanization" (Güney et al., 2019). Although the studies presented in this book might not directly compare different places, they all engage with the process and forms of peripheral changes around the world.

Third, beyond stereotypical suburbia, the real world of urban peripheries reveals significant social heterogeneity and inequalities. For the suburbs in the Global North, poverty becomes a prominent issue (Hern in this volume), and, in terms of consumption and cultures, the suburb is less homogeneous. On the one hand, the suburb today may have a substandard environment and receive the poorer residents displaced from the city centre. On the other hand, the distinction between urban and suburban consumption behaviour becomes meaningless. For example, adult retailing is not limited to the high street but prevails in the suburb, which is supposed to have family-oriented domestic life (Maginn and Steinmetz in this volume). After suburbia research thus reveals great potential to adopt a flexible and realistic approach to recognize the messiness of urban peripheries, not just the slums in the Global South but also the poverty, underdevelopment, and alienation in the Global North. These are comparable, although they are different, originating from different historical and social contexts.

Fourth, similar to studying diverse social realities of urban peripheries, the research on urban peripheries need to pay more attention to actors and agencies in the development process. The question is, who builds the periphery? In suburbia studies, the role of the middle class in pursuing its ownership is highlighted. The political economic perspective emphasizes the real estate developer in the formal land market and the state in relation to capital accumulation (Ekers et al., 2015). In the Global South, the peripheral area is a place largely constructed by its residents through what Caldeira (2017) calls "autoconstruction." In many places, the peripheral development is a mix of smaller builders, larger development corporations, and state agencies. Mega-urban projects are often initiated to peruse global competitiveness and world-class cities. From the Global South, we have seen the role of landowners who were villagers before the real estate boom. In India, big landowning castes are important actors in land development (Gururani in this volume) and large developers created through land redevelopment in Indonesia (Kusno in this volume). Paying closer attention to actors means that we need to trace the actual development processes. These actors are not static and pre-made. They gained their power through and made a fortune in the process of development. They are beyond the urban periphery as a locality but related to those in other places in relational geography. In the case of the outskirts of Jakarta, for example, the Lippo Group development company is connected with experiences and investors in California and Shenzhen (Kusno in this volume). In Manila's peri-urban fringe, diasporic capital is an important source for housing development in gated communities (Ortega in this volume). These gated communities are not associated with centrifugal residential "suburbanization" and bourgeois suburbia but rather transnationalism and the dream of overseas Filipinos. The attention to actors means a further investigation of their motivation and behaviours.

As several of the diverse contributions in this book show, the research from the Global South suggests that the stereotypes of suburbia are no longer valid. Recent studies in after suburbia in the Global North indeed reveal greater heterogeneity and substandard conditions of the suburb. In short, the urban peripheries in the twenty-first century are becoming very different for their counterparts in history. There are similarities and dissimilarities between the urban peripheries across the planet. Through various approaches and testimonies, the chapters in this book demonstrate variegated forms of peripheral urbanization. The periphery after suburbia is not a linear interpolation from its past suburbia (Keil, 2018b). Through three major aspects – land, infrastructure, and governance – this volume collectively contributes to our understanding of the planetary process of urbanization on the periphery. These studies also raise a need for new vocabularies to describe the urban periphery.[1]

NOTE

1 We understand the continuing debate over the concept of planetary urbanization, especially the postcolonial critique; here we use the concept in a more flexible sense, without a more restrictive reference to capital accumulation in capitalism as a core driving force.

REFERENCES

Acuto, M., Larcom, S., Keil, R., Ghojeh, M., Lindsay, T., Camponeschi, C., & Parnell, S. (2020). Seeing COVID-19 through an urban lens. *Nature Sustainability, 3*, 977–8. https://doi.org/10.1038/s41893-020-00620-3.

Addie, J.P. (2019). In what sense suburban infrastructure? In P. Filion & N.M. Pulver (Eds.), *Critical perspectives on suburban infrastructures: Contemporary international cases* (pp. 45–66). University of Toronto Press.

Ali, H., & Keil, R. (2008). *Networked disease: Emerging infections in the global city.* Wiley-Blackwell.

Balducci, A., Fedeli, V., & Curci, F. (2017). *Post-metropolitan territories: Looking for a new urbanity.* Taylor & Francis.

Bernt, M. (2019). Estates under pressure: Financialization, shrinkage, and state restructuring in East Germany. In M. Güney, R. Keil, & M. Üçoğlu (Eds.), *Massive suburbanization: (Re)building the global periphery* (pp. 81–93). University of Toronto Press.

Bernt, M., Colini, L., & Förste, D. (2017). Privatization, financialization and state restructuring in eastern Germany: The case of Am Südpark. *International Journal of Urban and Regional Research, 41*(4), 555–71. doi:10.1111/1468-2427.12521.

Bhan, G., Caldeira, T., Gillespie, K., & Simone, A. (2020, August 3). The pandemic, southern urbanisms and collective life. *Environment and Planning D: Society and Space.* https://www.societyandspace.org/articles/the-pandemic-southern-urbanisms-and-collective-life.

Biglieri, S. (2019). *Planning dementia-inclusive suburban landscapes* [Unpublished doctoral dissertation]. University of Waterloo.

Biglieri, S., De Vidovich, L., & Keil, R. (2020). City as the core of contagion? Repositioning COVID-19 at the social and spatial periphery of urban society. *Cities & Health*. doi:10.1080/23748834.2020.1788320.

Brenner, N., & Katsikis, N. (2020). Operational landscapes: Hinterlands of the Capitalocene. *Architectural Design*/AD 90, no. 1: 22–31. doi:10.1002/ad.2521.

Brenner, N., & Schmid, C. (2014). The "urban age" in question. *International Journal of Urban and Regional Research, 38*(3), 731–55. doi:10.1111/1468-2427.12115.

Brenner, N., & Schmid, C. (2015). Towards a new epistemology of the urban? *City, 19*(2–3), 151–82. doi:10.1080/13604813.2015.1014712.

Caldeira, T.P. (2017). Peripheral urbanization: Autoconstruction, transversal logics, and politics in cities of the global south. *Environment and Planning D: Society and Space, 35*(1), 3–20. doi:10.1177/0263775816658479.

Charmes, E., & Keil, R. (2015). The politics of post-suburban densification in Canada and France. *International Journal of Urban and Regional Research, 39*(3), 581–602. doi:10.1111/1468-2427.12194.

Connolly, C., Keil, R., & Ali, S.H. (2020). Extended urbanisation and the spatialities of infectious disease: Demographic change, infrastructure and governance. *Urban Studies*. https://doi.org/10.1177%2F0042098020910873.

De Vidovich, L. (2019). *Governing local welfare at the urban edges: Issues, challenges and suburban perspectives. A threefold investigation in Italy* [Unpublished doctoral dissertation]. Politecnico di Milano.

Diamond, J. (2020). *The sprawl: Reconsidering the weird American suburbs*. Coffee House Press.

Ekers, M., Hamel, P., & Keil, R. (2015). Governing suburbia: Modalities and mechanisms of suburban governance. In P. Hamel & R. Keil (Eds.), *Suburban governance: A global view* (pp. 19–48). University of Toronto Press.

Enright, T. (2016). *The making of grand Paris: Metropolitan urbanism in the twenty-first century*. MIT Press.

Filion, P., & Pulver, N.M. (Eds.). (2019). *Critical perspectives on suburban infrastructures: Contemporary international cases*. University of Toronto Press.

Fishman, R. (1987). *Bourgeois utopias: The rise and fall of suburbia*. Basic Books.

Gans, H. (1967). *The Levittowners: Ways of life and politics in a new suburban community*. Columbia University Press.

Güney, M., Keil, R., & Üçoğlu, M. (Eds.). (2019). *Massive suburbanization: (Re)building the global periphery*. University of Toronto Press.

Gururani, S. (2020). Cities in a world of villages: agrarian urbanism and the making of India's urbanizing frontiers. *Urban Geography, 14*(7), 971–89. https://doi.org/10.1080/02723638.2019.1670569.

Haas, O. (2019). Suburbanisms of ethnocracy: Building new peripheries in Israel/Palestine. In M. Güney, R. Keil, & M. Üçoğlu (Eds.), *Massive suburbanization: (Re)building the global periphery* (pp. 285–302). University of Toronto Press.

Halbert, L., & Rouanet, H. (2014). Filtering risk away: Global finance capital, transcalar territorial networks and the (un) making of city-regions: an analysis of business property development in Bangalore, India. *Regional Studies, 48*(3), 471–84. doi:10.1080/00343404.2013.779658.

Hamel, P. (Ed.). (2021). *Governing suburbia: Comparing collective action on eight urban peripheries around the world*. University of Toronto Press.

Hamel, P., & Keil, R. (Eds.). (2015). *Suburban governance: A global view*. University of Toronto Press.

Harris, R. (2015). Using Toronto to explore three suburban stereotypes, and vice versa. *Environment and Planning A, 47*(1), 30–49. doi:10.1068/a46298.

Harris, R., & Lehrer, U. (Eds.). (2018). *Suburban land question: A global survey*. University of Toronto Press.

Harris, R., & Vorms, C. (Eds.). (2017). *What's in a name? Talking about urban peripheries*. University of Toronto Press.

Harrison, J., & Hoyler, M. (Eds.) (2015). *Megaregions: Globalization's new urban form?* Edward Elgar Publishing.

Hurley, A.K. (2019). *Radical suburbs: Experimental living on the fringes of the American city*. Belt Publishing.

Jonas, A.E.G., Goetz, A.R., & Bhattacharjee, S. (2014). City-regionalism as a politics of collective provision: Regional transport infrastructure in Denver, USA. *Urban Studies, 51*(11), 2444–65. doi:10.1177/0042098013493480.

Keil, R. (Ed.). (2013). *Suburban constellations: Governance, land and infrastructure in the 21st century*. Jovis verlag.

Keil, R. (2018a). Extended urbanization, "disjunct fragments" and global suburbanisms. *Environment and Planning D: Society and Space, 36*(3), 494–511. doi:10.1177/0263775817749594.

Keil, R. (2018b). *Suburban planet: Making the world urban from the outside in*. Polity Press.

Keil, R. (2018c). Canadian suburbia: From the periphery of empire to the frontier of the sub/urban century. *Zeitschrift für Kanada-Studien, 38*, 47–64. http://www.kanada-studien.org/wp-content/uploads/2020/03/ZKS_2018 -68_3_Keil.pdf.

Keil, R., Hamel, P., Boudreau, J.A., & Kipfer, S. (Eds.). (2017). *Governing cities through regions: Canadian and European perspectives*. Wilfrid Laurier University Press.

Keil, R., & Lynch, L. (2019, October 30). Suburban change is transforming city life around the world. *The Conversations*. https://theconversation.com /suburban-change-is- transforming-city-life-around-the-world-125598.

Kipfer, S., & Dikec, M. (2019). Peripheries against peripheries? Against spatial reification. In M. Güney, R. Keil, & M. Üçoğlu (Eds.), *Massive suburbanization: (Re)building the global periphery* (pp. 35–55). University of Toronto Press.

Kusno, A. (2020). Middling urbanism: The megacity and the kampung. *Urban Geography, 41*(7), 954–70. https://doi.org/10.1080/02723638.2019.1688535.

Lawton, P. (2020). Unbounding gentrification theory: Multidimensional space, networks and relational approaches. *Regional Studies 54*(2), 268–79. doi:10.10 80/00343404.2019.1646902

LeBlanc, P. (2020, October 28). Trump uses outdated thinking in attempt to woo suburban women: I'm "getting your husbands back to work." *CNN Politics.* https://www.cnn.com/2020/10/27/politics/trump-suburban -women-2020-election/index.html.

Lefebvre, H. (2003). *The urban revolution.* University of Minnesota Press. (Original work published 1970)

Li, W. (2009). *Ethnoburbs: The new ethnic community in urban America.* University of Hawaii Press.

Li, Z., Chen, Y., & Wu, R. (2020). The assemblage and making of suburbs in post-reform China: The case of Guangzhou. *Urban Geography, 41*(7), 990–1009. https://doi.org/10.1080/02723638.2019.1598732.

Logan, S. (2021). *In the suburbs of history: Modernist visions of the urban periphery.* University of Toronto Press.

Lung-Amam, W. (2020a, October 5). *Why is Trump obsessed with suburbia?* Brookings Institute. https://www.brookings.edu/blog/how-we -rise/2020/10/05/why-is-trump-obsessed-with-suburbia/.

Lung-Amam, W. (2020b, November 12). Why Trump's suburban strategy failed. *CityLab.* https://www.bloomberg.com/news/articles/2020-11-12 /the-role-of-black-and-brown-suburbs-in-biden-s-win.

Macdonald, S. (2020). *Governing greenbelts in southern Ontario and the Frankfurt Rhine-Main Region: An institutional perspective* [Doctoral dissertation, Utrecht University]. Utrecth Uninversity Repository. https://doi.org /10.33540/126.

Macdonald, S., & Keil, R. (2012). The Ontario greenbelt: Shifting the scales of the sustainability fix? *The Professional Geographer, 64*(1), 125–45. doi:10.1080 /00330124.2011.586874.

May, L. (2020). *Suburban place-making. Political Economic Coalitions and 'Place Distinctiveness' (Antwerp, c.1860-c.1940).* University of Antwerp.

McFarlane, C., & Lancione, M. (Eds.). (2021). *Global urbanism: Knowledge, power and the city.* Routledge.

McGee, T.G. (1991). The emergence of desakota regions in Asia: Expanding a hypothesis. In N.S. Ginsburge, B. Koppel, & T.G. McGee (Eds.), *The extended metropolis: Settlement transition in Asia* (pp. 3–25). University of Hawaii Press.

Miro, J. (2020). *Suburban (dis)advantage: Views of suburban life from low-income immigrants in Surrey, BC* [Unpublished doctoral dissertation]. University of British Columbia.

Murray, M.J. (2017). *The urbanism of exception: The dynamics of global city building in the twenty-first century.* Cambridge University Press.

Newman, A. (2015). *Landscape of discontent: Urban sustainability in immigrant Paris*. University of Minnesota Press.

Nijman, J. (Ed.). (2020a). *The life of North American suburbs: Imagined utopias and transitional spaces*. University of Toronto Press.

Nijman, J. (2020b, November 7). Georgia's political shift – a tale of urban and suburban change. *The Conversation*. https://theconversation.com/georgias -political-shift-a-tale-of-urban-and-suburban-change-149596.

Peck, J. (2011). Neoliberal suburbanism: Frontier space. *Urban Geography, 32*(6), 884–919. doi:10.2747/0272-3638.32.6.884.

Peck, J. (2015). Cities beyond Compare? *Regional Studies, 49*(1), 160–82. doi:10.1 080/00343404.2014.980801.

Phelps, N.A. (2015). *Sequel to suburbia: Glimpses of America's post-suburban future*. MIT Press.

Phelps, N.A. (Ed.). (2017). *Old Europe, new suburbanization? Governance, land, and infrastructure in European suburbanization*. University of Toronto Press.

Phelps, N.A., & Wu, F. (Eds.). (2011). *International perspectives on suburbanization: A post-suburban world?* Palgrave Macmillan.

Ren, X. (2020). *Governing the urban in China and India: Land grabs, slum clearance and the war on air pollution*. Princeton University Press.

Robinson, J. (2016). Thinking cities through elsewhere: Comparative tactics for a more global urban studies. *Progress in Human Geography, 40*(1), 3–29. doi:10.1177/0309132515598025.

Robinson, J., Harrison, P., Shen, J., & Wu, F. (2021). Financing urban development, three business models: Johannesburg, Shanghai and London. *Progress in Planning, 154*, 100513. https://doi.org/10.1016/j .progress.2020.100513.

Roy, A. (2009). The 21st-century metropolis: New geographies of theory. *Regional Studies, 43*(6), 819–30. doi:10.1080/00343400701809665.

Roy, A. (2015). Governing the postcolonial suburbs. In P. Hamel & R. Keil (Eds.), *Suburban governance: A global view* (pp. 337–47). University of Toronto Press.

Roy, A. (2016). What is urban about critical urban theory? *Urban Geography, 37*(6), 810–23. doi:10.1080/02723638.2015.1105485.

Scott, A.J., & Storper, M. (2015). The nature of cities: The scope and limits of urban theory. *International Journal of Urban and Regional Research, 39*(1), 1–15. doi:10.1111/1468-2427.12134.

Shatkin, G. (2017). *Cities for profit: The real estate turn in Asia's urban politics*. Cornell University Press.

Shatkin, G. (2020). Financial sector actors, the state, and the rescaling of Jakarta's extended urban region. *Land Use Policy*, 104159. doi:10.1016/j .landusepol.2019.104159.

Shen, J., & Wu, F. (2017). The suburb as a space of capital accumulation: The development of new towns in Shanghai, China. *Antipode, 49*(3), 761–80. doi:10.1111/anti.12302.

Shen, J., & Wu, F. (2020). Paving the way to growth: transit-oriented development as a financing instrument for Shanghai's post-suburbanization. *Urban Geography, on-line first*. doi:10.1080/02723638.2019.1630209.

Shen, J., Luo, X., & Wu, F. (2020). Assembling mega-urban projects through state-guided governance innovation: the development of Lingang in Shanghai. *Regional Studies, 54*(12), 1644–54. doi:10.1080/00343404.2020.1762853.

Simone, A. (2019). Maximum exposure: Making sense in the background of extensive urbanization. *Environment and Planning D. Society and Space, 37*(6), 990–1006. doi:10.1177/0263775819856351.

Simone, A., & Pieterse, E. (2017). *New urban worlds: Inhabiting dissonant times*. Polity.

Soja, E.W. (2000). *Postmetropolis: Critical studies of cities and regions*. Wiley-Blackwell.

Sorensen, A., & D. Labbé (Eds.). (2020). *International handbook on megacities and megacity-regions*. Edward Elgar Publishing.

Taylor, P. (2021). *Advanced introduction to cities*. Edward Elgar.

Tchoukaleyska, R. (2019). Redeveloping Montpellier's suburban high-rises: National policy meets local activism in the debate over public space. In M. Güney, R. Keil, & M. Üçoğlu (Eds.), *Massive suburbanization: (Re)building the global periphery* (pp. 126–41). University of Toronto Press.

Tzaninis, Y. (2016). *Building utopias on sand: The production of space in Almere and the future of suburbia* [Unpubished doctoral dissertation]. University of Amsterdam.

Tzaninis, Y., Mandler, T., Kaika, M., & Keil, R. (2021). Moving urban political ecology beyond the "urbanization of nature." *Progress in Human Geography, 45* (2), 229–52. doi:10.1177/0309132520903350.

van Duijne, R.J., & Nijman, J. (2019). India's emergent urban formations. *Annals of the American Association of Geographers, 109*(6), 1978–98. doi:10.1080/24694452.2019.1587285.

Walks, A. (2013). Suburbanism as a way of life, slight return. *Urban Studies, 50*(8), 1471–88. doi:10.1177/0042098012462610.

Wilson, D., Boodram, B., & Smith, J. (2019). Public housing, heroin addiction, and America's industrial suburbs: A planetary urbanist perspective. In M. Güney, R. Keil, & M. Üçoğlu (Eds.), *Massive suburbanization: (Re)building the global periphery* (pp. 56–79). University of Toronto Press.

Wu, F. (2016). China's emergent city-region governance: A new form of state spatial selectivity through state-orchestrated rescaling. *International Journal of Urban and Regional Research, 40*(6), 1134–51. doi:10.1111/1468-2427.12437.

Wu, F. (2018). Planning centrality, market instruments: Governing Chinese urban transformation under state entrepreneurialism. *Urban Studies, 55*(7), 1383–99. doi:10.1177/0042098017721828.

Wu, F. (2020a). Adding new narratives to urban imagination: An introduction to new directions of urban studies in China. *Urban Studies, 57,* 459–72. doi: 10.1177/0042098019898137.

Wu, F. (2020b). Scripting Indian and Chinese urban spatial transformation: Adding new narratives to gentrification and suburbanisation research. *Environment and Planning C, 38*(6), 980–97. doi: 10.1177/2399654420912539.

Wu, F. (2020c). The state acts through the market: "State entrepreneurialism" beyond varieties of urban entrepreneurialism. *Dialogues in Human Geography, 10*(3), 326–9. 2043820620921034. doi: 10.1177/2043820620921034.

Wu, F. (2022). Land financialisation and the financing of urban development in China. *Land Use Policy, 112,* 104412. https://www.sciencedirect.com/science/article/pii/S0264837719306313.

Wu, F., & Keil, R. (2020). Changing the geographies of sub/urban theory: Asian perspectives. *Urban Geography, 41*(7), 947–53. https://doi.org/10.1080/02723638.2020.1712115.

Wu, F., & Phelps, N.A. (2011). (Post)suburban development and state entrepreneurialism in Beijing's outer suburbs. *Environment and Planning A, 43*(2), 410–30. doi:10.1068/a43125.

Wu, F., & Shen, J. (2015). Suburban development and governance in China. In P. Hamel & R. Keil (Eds.), *Suburban governance: A global view* (pp. 303–24). University of Toronto Press.

Wu, F., Chen, J., Pan, F., Gallent, N., & Zhang, F. (2020). Assetization: The Chinese path to housing financialization. *Annals of the American Association of Geographers, 110*(5), 1483–99. https://doi.org/10.1080/24694452.2020.1715195.

Yiftachel, O. (2016). The Aleph – Jerusalem as critical learning. *City, 20*(3), 483–94. doi: 10.1080/13604813.2016.1166702.

Zhou, Y., & Logan, J.R. (2008). Growth on the edge: The new Chinese metropolis. In J.R. Logan (Ed.), *Urban China in Transition* (pp. 140–60). Blackwell Publishing.

Image 2. Biking in Florence, Italy

2 The Power of the Image: Integrating Research and Teaching via Experiential Education

UTE LEHRER

One of the key strategies of the MCRI Global Suburbanisms project was an active integration between teaching and research: learning in the field through practice in the far corners of the suburban planet. These days this approach is often referred to as experiential education.

Universities of today believe that learning is not only done in the classroom but also "shaped by instructors' on-the-ground practices for impactful student learning," where "active engagement with teaching as a professional activity" leads to self-reflection, to seeking dialogue, to doing research, and to being open to critique, according to a recent report published by the SubCommittee on Research and Innovation in Teaching and Learning (2019) at York University (p. 2). What has now become part of our own university's agenda was somewhat anticipated by the Global Suburbanisms project's innovative strategy to enhance the relationship of research and pedagogy. The report states that "it may be fruitful to frame teaching as a research-oriented process" (p. 2). Combining teaching and research can be complementary to mindful teaching practices, which are more and more expected in today's world of education (Griffiths, 2007; Rosch, 2015).

In the context of our work, that meant taking students into the field and exposing them not only to the built form of a city but also to the planning legislation, the politics and different forms of investment and disinvestment, as well as the lived experience on the ground. It entailed meeting up with leading scholars from different parts of the world, as well as with professionals, politicians, activists, and journalists. These intentional student learning experiences led to professional reports being produced by the students after they returned home and to proficient presentations to the York University audience as well as to planning departments at the City of Toronto and professional planning organizations.

We applied this method both at the graduate and the undergraduate level, for which two specific courses were developed at York University. Both courses took students into the field and asked them to perceive the surroundings based on a particular topic. At the undergraduate level, students obtained clear instructions to take photographic impressions and documentation of various articulations of the urban infrastructure (which was one of the foundational themes of the research project) while exploring suburban areas in Canadian cities. This course, taught at York University under the directorship of Roger Keil, and with my assistance for the visualization assignments, took students from Montreal (2012), to Toronto (2010), to Waterloo (2015), to Winnipeg (2011), and to Vancouver (2013).

The graduate courses, which were taught under my directorship, brought students first to Leipzig, Germany, in 2011, then to Montpellier, France, in 2012; Shanghai, China, in 2015; Johannesburg, South Africa, in 2016; back to Germany again, this time to Frankfurt, Weimar, and Berlin in the summer of 2019; and finally to New York City in the fall of 2019. In most of these iterations, we integrated the field course directly with our research themes, where we had organized conferences with our international colleagues on our foundational and regional themes – such as governance, land, infrastructure, and European and African suburbanism. Students were able not only to rub elbows with leading intellectuals in the field but also to contribute by giving a group presentation. In addition to attending the symposia, students stayed for about two weeks and learned on the ground to do urban planning in place-specific contexts.

Further, we organized and co-taught a two-week long international summer school in Florence and Milan, Italy, on comparative urbanism, focusing on global suburbanization, land, and infrastructure with students from York University, University of Waterloo, University of Florence, and University of Milan as well as a good number of international students from around the globe. Students had the opportunity to meet with international experts, both formally and informally, to participate in field trips exploring the urban centres and peripheries of Florence and Milan, as well as attending career-building workshops for professionals and early career academics. One of the learning modules was on how to use visual methodology.

Out of this vast array of works, we gathered photographic material and brought it together, showcasing the rich work that had been done by both undergraduate and graduate students. With the help of research assistants Nicole Yang and Kourosh Mahvash from the University of Waterloo, we produced nine panels that where shown at the final Global Suburbanisms conference "After Suburbia," held at York University. These nine human-sized panels have been reduced to book size and are dividing the individual chapters of this collection, together with panels that were produced by Markus Moos (see chapter 12, as well as the visuals throughout the book).

The last two graduate courses that took students abroad were after the large conference and they are represented in this book by stand-alone visual representations, showing the energy that was present during all of those courses, where teaching met research, where urban theory met the lived urban experience.

The concept of the book layout is that each visual impression divides each chapter. These impressions are either based on the student experiences and their photographic articulation as instructed by me or on the work that was done as part of the research project directed by Markus Moos. While the visual doesn't have a direct connection with the chapter that follows, it still speaks to the content in multiple ways. In this sense we follow what John Berger (1972) already knew: "Seeing comes before words. The child looks and recognizes before it can speak" (p. 7). In this sense, it is the knowing that the visual dividers unearth in the reader's mind.

REFERENCES

Berger, J. (1972). *Ways of seeing.* Penguin.

Griffiths, R. (2004). Knowledge production and the research–teaching nexus: The case of the built environment disciplines. *Studies in Higher Education, 29*(6), 709–26.

Rosch, E. (2015). The emperor's clothes: A look behind the Western mindfulness mystique. In B. Ostafin, M. Robinson, & B. Meier (Eds.), *Handbook of mindfulness and self-regulation* (pp. 271–92). Springer.

Sub-Committee on Research and Innovation in Teaching and Learning. (2019). *A framework for engaged teaching at York University: Moving towards evidence-driven practice.* York University. https://www.yorku.ca /teachingcommons/wp-content/uploads/sites/38/2021/01/Updated _Engaged-Teaching_Final.pdf.

Image 3. Greater Toronto, Ontario, Canada (Fall 2010)

3 The New Urban Peripheries, 1990–2014: Selected Findings from a Global Sample of Cities

SHLOMO ANGEL

With contributions by Yang Liu, Alex M. Blei, Patrick Lamson-Hall, Nicolás Galarza Sanchez, and Sara Arango-Franco

Introduction

Cities grow in population and wealth; as they grow, they both densify and expand. As cities expand, they convert new areas on their rural peripheries to urban use and create *new urban peripheries* in the process, newly settled areas or newly annexed villages and towns that were formerly on the outskirts of cities and are now contiguous with or engulfed by the city's expanding urban footprint. These new urban peripheries are the focus of this chapter. Previous research, supported by a rigorous theoretical framework, has established an important empirical regularity that characterizes new urban peripheries: They typically have lower average densities. More generally, it has been shown in numerous studies that population densities decline as distance from the city centre increases and, hence, we can expect that as cities expand outwards and away from their older centres, newly built areas will be further away from areas built earlier and will thus have lower average densities as well. Unfortunately, beyond this important finding, very little is known about urban peripheries and particularly about the new urban peripheries, areas built or incorporated into the urban footprint in recent decades. In this chapter, which is discursive and non-technical in nature, we report on a number of recent findings that characterize the new urban peripheries the world over – in our case, areas added to cities between 1990 and 2014. The report is based on our analysis of satellite imagery and census data in a stratified global sample of 200 cities, a 4.7 per cent sample of the universe of all 4,231 cities and metropolitan areas that had 100,000 people or more in 2010. We also report on a few additional findings from a pilot study now underway in ten cities in ten different world subregions.

We begin by focusing on five areas of concern and answering five important questions about the new urban peripheries that could not be addressed before. We then explore some of the more obvious policy implications of the answers to these questions. The first four questions focus on the more quantitative dimensions of the new urban peripheries while the fifth question focuses on the qualitative differences between the urban fabrics of the areas of cities created before 1990 and those prevailing in new urban peripheries added to cities in the 1990–2014 period. These five questions aim to alert readers – be they municipal officials, policymakers at all levels, civic leaders, and interested citizens – to the vast new territories that were converted from rural to urban use during the 1990–2014 period, with the view to paying more attention and investing more resources in properly preparing new urban peripheries for urban settlement in the years to come before they are occupied. This can be achieved by at the very least identifying the lands needed for public works – arterial road and infrastructure networks – that need to be secured now, protecting areas of high environmental risk from settlement, and improving land subdivision practices.

These are the five areas of concern and associated questions: The first area of concern is the relative size of the new urban peripheries. On average, were the areas of new urban peripheries built between 1990 and 2014 smaller or larger – and by how much – than the areas of cities built before 1990? The second is the rate of urban expansion. During the 1990–2014 period, at what rate were new areas added to urban peripheries of cities compared to the rate of growth of their populations? The third is the urban expansion rates in more-developed and less-developed countries. Were there differences in the rates of urban expansion between cities in more developed and less developed countries? The fourth is densification versus expansion. Of the populations added to cities in the 1990–2014 period, what share densified the areas built before 1990 and what share settled in new urban peripheries? The fifth is the quality of the urban fabric in new urban peripheries. Did the quality of the urban fabric in new urban peripheries built in the 1990–2014 period improve or deteriorate, on average, in comparison to the quality of the urban fabric in areas built before 1990?

We suspect that most urban planners and policymakers – at the local, national, and international level – are not fully aware of the extent to which urban peripheries are now growing and are unsure whether the quality of the urban fabric in these new urban peripheries has improved or deteriorated during the past twenty-five years. The bulk of the chapter reports on these matters. A concluding section outlines a number of simple pragmatic interventions that can assist rapidly growing cities in preparing their urban peripheries to absorb the expected growth of populations, especially in the less-developed countries where most growth is projected to take place in the coming decades.

The chapter is structured as follows: Following a set of short notes on methodology, the subsequent sections provide the answers to the questions posed above and discuss their implications for urban planning and policymaking in general, and for preparing cities for their orderly expansion in particular. The concluding section outlines a pragmatic program of intervention in the urban peripheries of rapidly growing cities.

Notes on Methodology

The methodology employed in seeking answers to the questions posed above proceeded in a discrete set of steps that are briefly described in this section. The outlines below are short summaries of more detailed descriptions of the methodology that are available elsewhere (Blei et al., 2018; Galarza-Sanchez et al., 2018; Lamson-Hall et al., 2018). The reader is referred to these sources for a more elaborate description of this section of the chapter.

Defining Cities by the Limits of Their Built-Up Areas

We defined cities in a manner that blends a number of physical and functional attributes that allowed us to identify and measure the urban extent of these cities using satellite imagery. The urban extent of cities thus defined does not necessarily correspond to the municipal boundary, and may include parts of a large number of municipalities. We identify cities by the edge of their built-up areas – what the ancient Romans used to refer to as the *extrema tectorum* – and their centre as the central business district (CBD) of its main municipal area. Identifying cities as spatial units that are associated with urban labour markets, rather than with administrative boundaries, is in line with our finding that agglomeration economies in cities are metropolitan in scale, grounded in the advantages that metropolitan labour markets offer by making all jobs accessible to all residents.

The 2010 Universe of Cities

Given the above definition of cities, we identified the universe of all 4,231 cities that had 100,000 people or more in 2010. The three main data sources for constructing this universe were the UN Population Division, which provided data for settlements with populations of at least 300,000; the website www.citypopulation.de, which reproduces census data and census maps for all countries; and the Chinese Academy of Sciences, which provided information for Chinese cities and towns.

The Global Sample of Cities

We sampled 200 cities from the 2010 universe of cities based on three sampling stratifications – eight world regions, four city population size ranges, and three categories pertaining to the number of cities in the country.

The first stratification organized cities by eight world regions: (1) East Asia and the Pacific, (2) Southeast Asia, (3) South and Central Asia, (4) Western Asia and North Africa, (5) sub-Saharan Africa, (6) Latin America and the Caribbean, (7) Europe and Japan, and (8) land-rich developed countries (the United States, Canada, Australia, and New Zealand). The second stratification organized cities by city population size, of which there were four ranges – roughly corresponding to small, medium, large, and very large city population sizes – so that the total population of the universe of cities in each of these four ranges was approximately the same, about 622 million: (1) 100,000–427,000, (2) 427,001–1,570,000, (3) 1,570,001–5,715,000, and (4) 5,715,001 and above. An approximately equal number of cities were then sampled from each of the four population size categories. A third stratification was included in the sampling framework so that the sample would contain cities from countries with few cities as well as cities from countries with many cities. The number of cities in the country stratification contained three categories: (1) 1–9 cities, (2) 10–19 cities, and (3) 20 or more cities. Cities were sampled from these categories in proportion to the population of the universe of cities in these categories.

Classifying Urban Extents Using Landsat Imagery and Mapping Urban Peripheries

Using Landsat satellite imagery with a 30–metre pixel resolution, we identified and mapped the urban extents of the 200 cities in the sample in three time periods – circa 1990, circa 2000, and circa 2014. Using GIS mapping software, we calculated a number of metrics pertaining to these maps of the urban extents of cities – population, extent, average density, the saturation of urban extents by built-up areas; the shape compactness of urban extents; the shares of the added extents between two time periods that were infill, extension, leapfrog, or inclusion of older settlements; and the changes in these values over time. We created maps of the urban extents of all 200 cities in the global sample for 1990, 2000, and 2014 (Angel et al., 2016).

Measuring the Attributes of the Urban Fabric with High-Resolution Satellite Imagery

Having identified the new urban peripheries – areas built between 1990 and 2000 and areas built between 2000 and 2014 – in the global sample of cities, we spatially sampled an average of eighty 10-hectare *locales* in every city, both in the areas built before 1990 and in the two new urban peripheries.

Using high-resolution Bing and Google Earth imagery, we digitized several spatial features of these locales to obtain values for a number of metrics that pertain to the quality of the urban fabric – the share of the land in streets, average block size, the share of four–way intersections, and the share of residential land in atomistic development, in informal land subdivisions, in formal land subdivisions, and in housing projects. We also mapped arterial roads in the sample of cities and calculated the density of arterial roads as well as the average beeline distance to arterial roads in these cities. This phase of our investigation allowed us to obtain average values for a number of qualitative attributes of the urban fabric of cities, and to compare these values in areas built before 1990 and in areas built between 1990 and 2000 and between 2000 and 2014.

Densification versus Expansion

Urban densification accommodates urban population growth within neighbourhoods that are already built. Urban expansion accommodates urban population growth in new neighbourhoods built on the periphery of cities. It is generally understood and conceded that cities both densify and expand over time when they grow in population. An evidence-based policy that seeks to balance urban densification and urban expansion should rely on answers to the following questions: (1) What share of the population added to a city between two time periods settled within its existing urban extent and what share occupied new expansion areas? (2) Are recently occupied expansion areas of cities settled at higher or lower urban densities than areas of the cities built earlier? (3) Do newly built areas on the periphery of cities experience densification over time? Given the answers to these three questions, we can project our findings into

the future, seeking answers to the following fourth and fifth questions:
(4) Given existing trends, what share of the added population in coming
decades can be expected to settle within the areas of cities built earlier?
(5) Given existing trends, how much land will cities require to prepare for
their orderly expansion in advance of development in coming decades?
Answers to these questions are of key importance for the future of
evidence-based low-carbon planning of cities.

Mapping a global population grid of one square kilometre, using
WorldPop, in conjunction with other fine grain population data,
allowed us to answer the first three questions for our global sample of
200 cities. More specifically, we have already mapped the expansion
areas of the cities in the sample between 1990 and 2014. We could now
estimate the share of the added population between 1990 and 2014 that
settled in the expansion area and the share that settled in the pre–1990
area. Estimating these shares would allow us to answer the first and
the second questions for the global sample of 200 cities. We could also
calculate the density of areas built before 1990, areas built between
1990 and 2000, and areas built between 2000 and 2014. If densities in
1990–2000 areas were found to be significantly higher in 2014, on average, than they were in 2000, it would prove that densification occurred
naturally as cities expanded.

The Relative Size of the New Urban Peripheries

In 1990, the 4,231 cities in the 2010 universe of cities – all the cities,
metropolitan areas, and urban agglomerations that had 100,000 people
or more in 2010 – occupied a total area of 275,000 square kilometres. A
decade later, these cities occupied an area of 395,000 square kilometres;
and by 2014, these cities occupied an area of 570,000 square kilometres.
To put these numbers in perspective, these cities occupied 0.2 per cent
of the total land area of countries in 1990 and 0.4 per cent in 2014; in
parallel, they occupied 1.8 per cent and 3.6 per cent of the arable land
area in the world at large in 1990 and 2014, respectively. If we refer to
the expansion areas of cities between 1990 and 2014 as the new urban
peripheries, we can conclude that, for the cities in the universe of cities as a whole, *the new urban peripheries were as large as the entire areas
of cities built before 1990.* In Accra, Ghana, for example, the 1991–2014
urban periphery was 5.5 times larger than its 1991 area (see figure 3.1).
In fact, the new urban peripheries in almost two-thirds of the cities in
the global sample (63 per cent) were larger than their areas in 1990. In
twenty cities in the global sample (10 per cent), new urban peripheries
were more than ten times larger than their areas in 1990.

Figure 3.1. In Accra, Ghana, the 1991–2014 urban periphery was 5.5 times larger than its 1991 area. Source: Authors.

From an urban planning perspective, this suggests that, given their relative size – which is considerably larger than expected by most planning professionals – more attention should be given to the new urban peripheries. In other words, had we known in 1990 that cities would, on average, double their areas and that some cities would increase their areas more than tenfold by 2014, we could have better prepared for that expansion, say by laying out arterial road grids and street grids before peripheral lands were occupied. Generally speaking, we did not, and that was a mistake. Needless to say, we do not have to repeat that mistake again.

New Urban Peripheries in More-Developed and Less-Developed Countries

There were substantial differences in the expansion of new urban peripheries in more developed and less developed countries. During the twenty-four-year period between 1990 and 2014, the total area occupied by cities in the universe of cities in the more-developed countries grew by 55 per cent; in other words, the area grew by a half. In parallel, the total urban extents of the cities in the less-developed countries grew by 176 per cent; in other words, the extents almost tripled. *New urban peripheries in cities in less-developed countries were almost twice as large as their areas built before 1990.* Indeed, almost three-quarters (71 per cent) of the total land area of new urban peripheries created in the 1990–2014 period in the world at large was in cities in less-developed countries.

Given current population projections by the United Nations Department of Economic and Social Affairs (2018) – projections that suggest that, between 2015 and 2050, eighteen people will be added to urban populations in less-developed countries for every person added to the urban populations in more-developed countries – focusing on the new urban peripheries of the future will mean focusing more and more on the peripheries of cities in less-developed countries. And the challenge here, it should be noted, is quite different: Preparing new urban peripheries for occupation will often take place in cities with weaker rule of law, weaker adherence to land use and land subdivision regulations, smaller municipal infrastructure budgets and reduced access to infrastructure finance, higher levels of corruption and greater control of private developers over the planning process. More often than not, lessons learned in planning urban peripheries in more-developed countries – relying on elaborate yet enforceable master plans, to take one example – do not travel well to less-developed ones.

The Rate of Urban Expansion Compared to the Rate of Population Growth

During the twenty-four-year period of 1990 to 2014, the total population of the universe of cities grew by 53 per cent – from 1.6 billion to 2.5 billion – while the area occupied by these cities grew by 105 per cent – from 275,000 square kilometres to 570,000 square kilometres. There is no doubt that cities are now expanding at a faster rate than their population growth rate. At current rates, when the population of a city doubles, its urban extent triples. Interestingly enough, the growth rates of both the population and area of cities were found to be statistically independent of city size: On the whole, small, large, and very large cities grew in population and expanded in area at roughly the same rates. On average, the annual population growth rate for cities large and small in the universe of cities was 3.8 per cent; the rate of urban expansion was 5.6 per cent. The difference between them, 1.8 ± 0.4 per cent, was statistically significant. Similar differences were detected in cities in more-developed and less-developed countries. In more-developed countries, the average annual rates of population growth and urban expansion were 1.1 per cent and 2.4 per cent, respectively and the difference between them was statistically significant. In less developed countries, the average annual rates of population growth and urban expansion were 4.7 per cent and 6.7 per cent, respectively, and the difference between them was statistically significant as well.

This amounts to saying that average densities in cities – the ratios of the populations of their urban extents to the areas of their urban extents – are in significant decline. Indeed, between 1990 and 2014, average densities in the universe of cities declined significantly, from 90 ± 11 to 52 ± 5 persons per hectare; in cities in more-developed countries, they declined from 31 ± 5 to 22 ± 3; and in cities in less-developed countries, they declined from 111 ± 13 to 66 ± 6.

These findings suggest that urban planning efforts to contain urban expansion through green belts, urban growth boundaries, and other zoning and land use regulations – efforts that have been in place for the past three decades – have so far not borne fruit. There are anecdotal data to suggest that higher urban densities are associated with lower greenhouse gas emissions, shorter travel distances, and shorter infrastructure network lengths. It stands to reason, therefore, that cities should aim at slowing down their rates of urban expansion – at the very least to match their rates of population growth – so as to keep their average densities from declining, but so far that goal has proven to be quite elusive: Urban expansion continues to outpace urban population growth in the great majority of cities the world over. Refusing to acknowledge the pace of urban expansion and to prepare for it properly – simply because it is perceived as undesirable – is tantamount to shirking our responsibilities for ensuring that cities expand into their peripheries in an orderly manner, one that fosters their productivity, their inclusiveness, their sustainability, and their resilience.

Densification versus Expansion

It is generally understood and conceded that the density of urban neighbourhoods is not simply determined by public policy. Urban densities are oftentimes the outcome of supply-and-demand pressures for residential living space. They may also be the outcome of consumer preferences for larger suburban homes further away or for smaller apartments closer to bustling urban centres. Still, there is an ongoing policy debate on the merits of accommodating urban population growth through urban densification as against through urban expansion. Those engaged in this debate claim that public intervention in the matter is needed to ensure that cities grow in a productive, inclusive, and sustainable manner. Densification is typically the preferred course for those concerned with energy conservation, with global warming, and with excessive public infrastructure costs. It is typically resisted by existing communities that prefer the status quo and by established planning regulations that limit what can be built where. Indeed, community resistance to densification, or the inability to reform planning regulations that prohibit it, may limit densification and accelerate expansion. Expansion is typically the preferred course for those concerned with overcrowding or with land supply bottlenecks that may lead to unaffordable housing. It is typically resisted by homeowners lest it depress property values, by municipal authorities reluctant to extend infrastructure services, and by concerned citizens who want to protect green spaces on the urban periphery. That resistance to urban expansion may compromise preparing for it at the proper scale, by failing to put in place adequate public works in advance of development, for example.

What is typically missing in this debate is the evidence on the relative share of the population added to cities in a given time period that was accommodated through the densification of their existing urban footprints as against the share of the population added to these cities that was accommodated in their new urban peripheries. We have obtained data on these relative shares, as well as additional data on the densification and saturation of existing urban footprints and new urban peripheries, in a representative group of ten cities for the 1990–2014 period. The cities were Bangkok (Thailand), Bogotá (Colombia), Baku (Azerbaijan), Cairo (Egypt), Dhaka (Bangladesh), Hong Kong (China), Wuhan (China), Kinshasa (Democratic Republic of Congo), Madrid (Spain), and Minneapolis (USA).

Before reporting on these results and their implications, let us look at Cairo to get an overall understanding of the parameters used in our analysis (see figure 3.2 below). The map on the left of the figure shows the urban extent of Cairo in 1992 and its new urban periphery – the areas built between 1992 and 2013. The population of Cairo in 1992 was 9.6 million and it grew to 15.7 million by 2013. Of the 6.1 million people added to the city during the 1992–2013 period, 0.7 million (12 per cent) were accommodated in the pre-1992 urban extent, while 5.4 million (88 per cent) were accommodated in the new urban periphery.

The added population to Cairo's 1992 urban extent increased its saturation (saturation = total built-up area ÷ total urban extent) from 72 per cent to 94 per cent. At the same time, despite the increase in its population, its built-up area density (built-up area density = total population ÷ total built-up area) declined from 324 to 268 persons per hectare. The area occupied between 1992 and 2003 had a 44 per cent saturation level by 2003 and this level increased to 72 per cent by 2013. The area occupied between 2003 and 2013 had a saturation level of 47

Figure 3.2. The area of Cairo, Egypt, built before 1992 (dark grey), and its new urban peripheries, areas built during the 1992–2003 period (medium grey), and areas built during the 2003–13 period (light grey).

per cent by 2013. In short, most of the added population to the 1992 urban extent of Cairo was absorbed through increased saturation – occupying urbanized open spaces that were previously vacant – and not through increasing the density of its built-up areas. The new urban periphery became more saturated over time. It also became denser: By 2013, the overall density in the area initially built between 1992 and 2003, for example, increased from 55 to 100 persons per hectare. Most likely, new construction in this expansion area between 2003 and 2013 was added at higher densities than the original construction that took place between 1992 and 2003. We repeated this analysis for the remaining nine cities. We report briefly on these results below.

Figure 3.3. The shares of city populations added between 1990 and 2014 that were accommodated in the pre-1990 urban footprints.

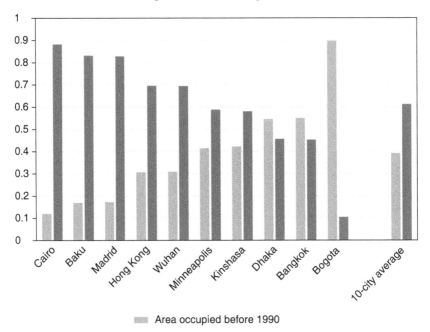

Area occupied before 1990

The shares of city populations added to the ten cities studied between 1990 and 2014 that were accommodated in their pre-1990 areas varied greatly, from 12 per cent in Cairo to 90 per cent in Bogotá (see figure 3.3 below). The very high value for Bogotá can be attributed to the deteriorated security on its suburban fringe and the high threat of kidnapping on suburban roads during this period (Gaviria et al., 2018). While it is difficult to draw conclusions about the universe of cities as a whole from such a small representative group, we note that, on average, 39 ± 17 per cent of the added populations between 1990 and 2014 were settled in the pre-1990 areas of cities, while the rest, 61 ± 17 per cent, were settled in the new urban peripheries of these cities, areas occupied between 1990 and 2014.

If any populations at all were added to the pre-1990 footprints of these ten cities, it clearly densified them because their densities are simply ratios of their populations and their areas, and their areas remained fixed. Even for as small a sample of ten cities, average overall densities in the pre-1990 areas in these ten cities increased by a factor of

Figure 3.4. Average levels of saturation (built-up area ÷ total area) increased significantly between 1990 and 2000 and between 2000 and 2014 in the ten representative cities.

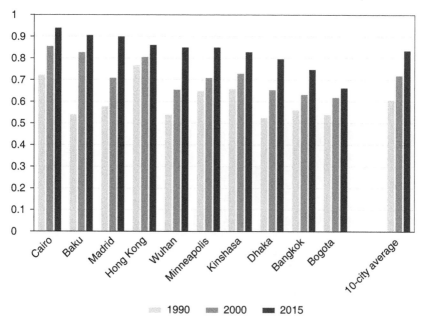

1.5 ± 0.3, a statistically significant increase at the 95 per cent confidence level. That increase in density could take place in two distinct ways: By increasing the saturation of the pre-1990 footprints or by increasing the built-up area density (population ÷ built-up area) within these footprints. Figure 3.4 shows the increase in saturation (built-up area ÷ total area) within the pre-1990 urban footprints of the ten cities. There is no doubt that the added population to the pre-1990 urban footprints of these ten cities significantly increased the level of saturation within these footprints: The ten-city average saturation level increased from 61 ± 6 per cent in 1990 to 83 ± 6 per cent. In three cities – Cairo, Kinshasa, and Wuhan – saturation levels reached 90 per cent or more, leaving little open space – either public or private – within their urban extents. Saturation, no doubt, clearly has its limits.

In contrast to the clear pattern detected in increased saturation levels, the densification of built-up areas within the pre-1990 urban extents of the ten cities studied here presents a more varied pattern. It increased substantially in some cities, for example, Dhaka and Bogotá, and it

Figure 3.5. Average built-up area densities (total population ÷ total built-up area) did not change significantly between 1990 and 2000 and between 2000 and 2014 in the ten representative cities.

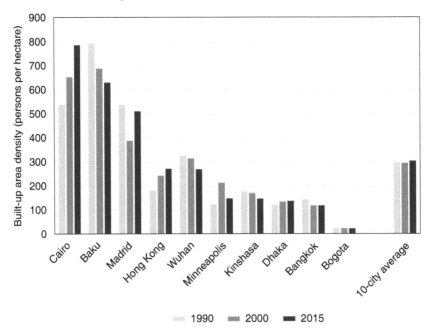

decreased substantially in others, for example, Hong Kong. On average, the built-up area densities in these ten representative cities did not change substantially over the 1990–2014 period (see figure 3.5). There is no natural limit to the densification of the built-up areas of cities, but it is difficult to undertake in practice. It may require smaller apartments, lower vacancy rates, taller buildings, higher plot coverage, or a greater share of land in residential use.

Our position in the debate on the merits of densification versus expansion takes the middle road, seeking proper balance between the two by promoting both acceptable densification and orderly urban expansion. We have no issue with either densification or expansion and we are fully aware that they are substitutes. Both can and do accommodate population growth. It is important to keep in mind, however, that in the presence of inevitable urban population growth one cannot be both against densification and against expansion. Strict adherence to the status quo and uncompromising "not in my backyard" opposition in the presence of urban population growth must be delegitimized and actively resisted. Densification versus expansion, or the relative importance we attach to one vis-à-vis the other, presents a choice we must make, a choice that cannot be avoided. Neither densification nor expansion is simple or easy to implement. Both require strong leadership and public support and, more often than not, regulatory reform. What is of greatest importance in the context of this chapter is that hoping against hope that strict containment of existing urban footprints can relieve cities from their need to expand into new urban peripheries is counterproductive.If successful, containment might lead to land supply bottlenecks and, consequently, to higher land and housing prices; and, if it fails, to unplanned and disorderly expansion.

The densification of the urban extents of cities is easier to accomplish by the infill of urbanized open spaces, many of them still vacant, and much more difficult to accomplish by increasing density in situ in already built-up areas, either by subdividing units into smaller ones, by adding units to existing structures, or by tearing them down and building higher structures in their stead. It should not come as a surprise, therefore, that in the ten representative cities we found significant infill between 1990 and 2014 in the pre-1990 urban extents, but no significant increase in the average densities of built-up areas there. And while densification in situ, at least in theory, has no upper limit, saturation of urbanized open spaces does: Sooner or later the remaining stock of vacant land is exhausted. In the face of population and income pressures on housing demand, limitations on infill or on in situ densification of built-up areas intensify pressures for urban expansion. Lest they fall prey to wishful thinking, urban planners that need to confront these pressures on housing demand, with the objective of ensuring the adequate supply of residential land and housing so as not to render housing unaffordable, must correctly estimate how much of that demand can be accommodated within existing urban footprints and how much will need to be accommodated by urban expansion, while keeping an eye on land price movements on the urban periphery. Anecdotal evidence, from Santiago, Chile, for example, suggests that if land supply bottlenecks – often accentuated by infrastructure shortages on the urban periphery – allow land price to increase, they tend to remain high, particularly when there are no pressures on peripheral landowners to dispose of their lands.

The Quality of the Urban Fabric in the New Urban Peripheries

It is important to know if the quality of the urban fabric – measured, say, by the share of the built-up area occupied by roads, average block size, four-way intersection density, access to arterial roads, or proper residential land subdivision – in new urban peripheries built in the 1990–2014 period improved or deteriorated in comparison to areas built before 1990. The default answer to this question – the null hypothesis, so to speak – is that the quality of the urban fabric in the new urban peripheries of cities remained the same over time. One would hope, however, that in parallel with economic development and social progress along a number of dimensions over the last twenty-five years, it has improved. There would be less cause for worry if the urban fabrics in the new urban peripheries were of similar, or preferably better, quality than those found in areas built before 1990.

Our findings show that the opposite is true, leading us to believe that drawing attention to preparing new urban peripheries for human settlement – the most basic and the most established task of traditional urban planning – is now one of the most critical tasks confronting the rapidly growing cities in less-developed countries. We find that, on the whole, new urban peripheries were less prepared on a number of critical dimensions for urban settlement than areas of cities built before 1990.

That said, along one important dimension – the share of the built-up area occupied by streets and roads – there has been no change, whether for the better or for the worse. The share of the built-up area occupied by roads was 21 ± 1 per cent, on average, in the areas of cities built before 1990; it was 21 ± 1 per cent, on average, in the new urban peripheries of these cities, areas built during the 1990–2014 period. There was no significant statistical difference between the two. Median values in the two areas were not that different either. There were numerous cities, however, both in more-developed and in less-developed countries that had considerably lower values in their expansion areas. For example, the new urban peripheries of London and New York had only 10 per cent and 13 per cent of their built-up areas, on average, dedicated to streets, respectively. Similarly, the new urban peripheries of Kolkata and Dhaka had only 10 per cent and 12 per cent of their built-up areas, on average, dedicated to streets, respectively.

Interestingly, the share of roads that were less than four metres in width increased significantly, from 20 ± 2 per cent in the areas of cities built before 1990 to 28 ± 2 per cent in areas built in the 1990–2018 period. The share of roads less than four metres wide was only 18 per cent and 14 per cent in London and New York, respectively, but as high as 60 per cent and 56 per cent in Kolkata and Dhaka, respectively. There are good reasons to believe that, in the new peripheries of many cities, not enough land was dedicated to roads, but we cannot say that, in general, less area was dedicated to roads in the post-1990 period than in the pre-1990 period.

On a number of other dimensions, instead of seeing progress over time, we saw regress. We can definitely say that about the average block size, a key metric in evaluating the walkability of urban areas, since the larger the block size, the longer the walking distance between two random locations in comparison with their beeline distance. The average block size increased significantly in urban areas built after 1990 in the universe of cities in comparison to areas built before 1990. It increased from an average of 3.4 ± 0.2 hectares in the former to an average of 5.4 ± 0.4 hectares in the latter in the universe of cities as a whole, and the increase was significant in all world regions.

What was true for the average block size was also true for the share of road intersections that were four-way road intersections, another common measure of walkability. Four-way road intersections define blocks; they also make for increasing the number of alternative routes between urban locations, thus reducing the prevalence of bottlenecks and distributing traffic flow more evenly among city streets. The average share of road intersections that were four-way in the pre-1990 areas in the universe of cities was 14 ± 1 per cent; it decreased to 10 ± 1 per cent in the areas of cities built between 1990 and 2014, and it decreased significantly in all world regions except in Europe and Japan. This is indeed a worrisome development: Nine out of ten intersections in new urban peripheries in the universe of cities are, on average, three-way intersections. Indeed, narrow road networks with three-way intersections can, and most often do, ensure road access to every plot – a key requirement for making the plot salable – but that is all they can do. Three-way intersections hamper walkability and they present difficult obstacles for the smooth flow of traffic within as well as across territories where they predominate.

Urban and metropolitan labour markets thrive when all workers have access to all jobs, because that ensures that firms can hire the best workers and workers can find the best jobs. Labour markets are integrated when all locations are connected by intercity arterial roads that allow workers to reach their workplaces all over the urban area rapidly and efficiently. The presence of an arterial road, preferably one that carries public transport, within walking distance of a residence greatly facilitates access to jobs throughout the urban area. We can legitimately inquire, therefore, whether the average walking distance to an arterial road in new urban peripheries built between 1990 and 2014 increased or decreased significantly from the average distance to arterial roads in the areas of cities built before 1990. There is no question that for cities in the universe it increased significantly, from 431 ± 48 metres, on average, in pre-1990 areas to 601 ± 50 metres, on average, in the new urban peripheries developed in the 1990–2014 period. It increased significantly in cities in all world regions, except Europe and Japan, land-rich developed countries (United States, Canada, Australia, and New Zealand), and South and Central Asia.

Figure 3.6a shows the scarcity of arterial roads in a northeastern section of Bangkok in the 1980s. The arterial roads were eight kilometres apart and carried all intracity traffic. As a result people could not walk to arterial roads and arterial roads were highly congested. Figure 3.6b shows the public transit network on the arterial road grid in Toronto circa 2010. The density of the arterial road grid allowed almost everyone to be within walking distance of arterial roads and hence from public transit. The global increase in distance to arterial roads during the 1990–2014 period is also a worrisome development. To function as integral parts of urban labour markets, new urban peripheries must be connected to the overall network of arterial roads that connect workers to jobs throughout cities. The productivity and inclusiveness of cities is impaired when job markets become fragmented and cities cannot function as integrated labour markets.

Anecdotal evidence suggests that laying out residential areas in an orderly manner before they are occupied – that is, subdividing land into regular plots and arranging these plots along a regular network of streets – has economic and social benefits that far exceed its costs. First, regular layouts facilitate the provision of infrastructure and reduce the expenditures associated with it. Second, introducing roads into fully built irregular settlements (see figure 3.7a) has proved both costly and inefficient (Abiko et al., 2007). Third, regular layouts accelerate the transition of informal settlements away from being perceived as "slums," as has been the case of the gridded squatter settlements on the periphery of Lima, Peru (see figure 3.7b). Fourth, settlements with orderly layouts increase in economic value over time more rapidly than atomistic development, as Guy Michaels and his co-authors (2017) describe:

> [We studied] "Sites and Services" projects implemented in seven Tanzanian cities during the 1970s and 1980s, half of which provided infrastructure in previously unpopulated areas (de novo neighborhoods), while the other half upgraded squatter settlements. Using satellite images and surveys from the 2010s, we [found] that de novo neighborhoods developed better housing than adjacent residential areas (control areas) that were also initially unpopulated. Specifically, de novo neighborhoods are more orderly and their buildings have larger footprint areas and are more

Figure 3.6. Arterial roads in a northeastern section of Bangkok (this page) were eight kilometres apart. Most arterial roads in Toronto (following page) carry public transit.

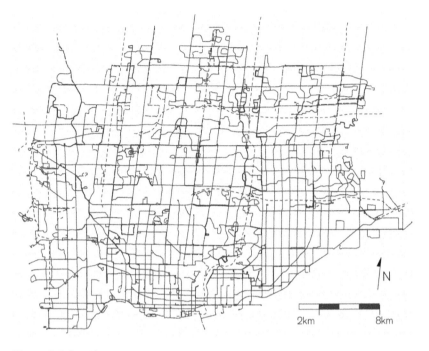

Figure 3.6. (*cont.*)

likely to have multiple stories, as well as connections to electricity and
water, basic sanitation and access to roads … While we have no natural
counterfactual for the upgrading areas, descriptive evidence suggests that
they are if anything worse than the control areas … the mean land value in
[Dar es Salaam's] de novo neighborhoods is in the range of $160–220 per
square meter, while in its upgrading neighborhoods it is about US$30–40
per square meter. (pp. 1, 26)

Given the importance of orderly land subdivision, we can ask whether
land subdivision practices are gaining in importance in the new urban
peripheries or whether they are losing ground. Unfortunately, evidence from the universe of cities is by no means encouraging. We
examined the shares of residential areas within randomly selected ten-
hectare locales in the global sample of cities and divided them into
four categories: (1) atomistic, where houses were added one by one
without any layout; (2) informal land subdivisions, where there was
a more or less regular layout but roads remained unpaved; (3) formal
land subdivisions, where there was a regular layout and roads were

Figure 3.7. The cost of providing residential infrastructure in the Matinha favela in Rio (top) was three to six times the cost of providing it in a new land subdivision. Some houses in the Comas district, a properly laid out former-squatter settlement in Lima, Peru (bottom), are now selling for $180,000.

paved; and (4) housing projects, where there was a regular layout and a repetitive house design. We generalized our findings to the universe of cities as a whole.

In the universe of cities as a whole, the share of residential areas that were not laid out before they were occupied – that is, those with atomistic housing – increased significantly, from 22 ± 4 per cent in the areas of cities built before 1990 to 32 ± 4 per cent in the new urban peripheries built during the 1990–2014 period. That increase was significant in all world regions as well, except in South and Central Asia. Globally, we are seeing a large increase in unplanned urban expansion, an expansion that will make it difficult for many people to obtain infrastructure services and to see their settlements transform from "slums" into regular urban neighbourhoods.

In parallel with the increased share of atomistic layouts, we also observed an increased share of informal land subdivisions coupled with a decreased share of formal ones. The share of informal land subdivisions in the universe of cities increased significantly, from 20 ± 4 per cent in the areas of cities built before 1990 to 32 ± 4 per cent in the new urban peripheries built during the 1990–2014 period. In parallel, the share of formal land subdivisions in the universe of cities decreased significantly, from 47 ± 4 per cent in the areas of cities built before 1990 to 24 ± 4 per cent in the new urban peripheries built during the 1990–2014 period. The share of residential areas in housing projects did not change significantly. The absence of formal layouts is especially critical in sub-Saharan Africa, where the share of atomistic residential development increased from 26 ± 13 per cent in pre-1990 areas to 43 ± 14 per cent in new urban peripheries, while the share of informal land subdivision decreased insignificantly, from 50 ± 15 per cent in pre-1990 areas to 43 ± 14 per cent in the new urban peripheries built during the 1990–2014 period.

Municipalities in rapidly growing cities are often reluctant to engage informal developers because they do not conform to municipal land subdivision regulations, regulations that often require the provision of a full complement of infrastructure services before the construction of houses can begin. Informal developers, on their part, sell unserviced land with the promise of services to be provided later, typically by pressuring municipalities to provide them. Still, engaging informal developers with a view of assisting them or encouraging them to generate better layouts with regular plots and more land allocated to streets, can yield great benefits to all. For example, developers would end up with smaller amounts of salable land, but would sell it at a higher price; buyers would obtain plots that would increase faster in value; and municipalities could upgrade infrastructure at lower cost. Urban planners must work to ensure that new urban neighbourhoods on the urban periphery are laid out in an orderly fashion. That has always been their traditional role and it appears that they have been neglecting it of late, having lost interest in affecting the emerging shape of their cities on the ground. Too little is known about informal land developers on the new peripheries of rapidly growing cities, especially the poorer ones, and too little is being done to engage them.

To conclude, along a number of important dimensions we can determine without a doubt that the urban fabric in the new urban peripheries of cities is of a lower, rather than a higher, quality than the urban fabric in areas of cities built earlier. This suggests a serious failure on the part of urban planners entrusted with ensuring that essential public works, as well as essential public open spaces, are put in place before cities get built. The concluding section of this chapter outlines a pragmatic set of practical interventions that can repair and correct past failures and ensure a modicum of order in new urban peripheries to be built in the decades to come.

Conclusion: Preparing Future Urban Peripheries

The most important conclusion of this chapter is that urban planners and policymakers must stop neglecting the peripheries of cities. Instead, they must direct both attention and resources to learning more about urban peripheries, to understanding the processes now guiding their formation, to becoming more involved in these processes, and to taking a few basic actions that can help shape new urban peripheries so as to make them more productive, more inclusive, more sustainable, and more resilient. All of this can be done at modest cost and with modest changes to current practices, while yielding economic and social benefits that far outweigh these costs.

Particularly, urban planners and policymakers would do well to engage in the long-term planning of new urban peripheries, preferably preparing now for the expansion of cities in the next three decades, with the year 2050 as a realistic time horizon. Preparing for urban expansion in advance indeed requires a long-time horizon. It is considerably cheaper to engage in planning, as well as in land acquisition, in areas of expansion far away from the existing edge of cities than in areas close by. Looking far into the future while acting now reduces present outlays that can vastly multiply future benefits.

We now know the magnitude of the lands that cities will need for their future expansion in the coming thirty years. Each city can now determine its urban extent and use more precise statistical tools to project it into the future. These projections need to be done realistically, shying away from tendencies to underestimate urban expansion in the hope that it never occurs. Projections must take into account both population and income growth, remaining alert to the fact that cities typically expand at a faster rate than the rate of growth of their populations. They must also take into account the potential for the densification of the areas of cities that are already built. In fact, estimating the areas needed for urban expansion must go hand in hand with estimating the shares of the population that will need to be settled in new urban peripheries as against the shares that can be accommodated in already-built areas.

That said, there is good reason to prepare more lands for expansion than what predictions call for. As long as investments in expansion areas are small, there is no great loss if expansion does not occur at full scale. There are heavy penalties, however, for not preparing adequate areas for expansion – cities may then expand in a disorderly and thus costly manner or, if not allowed to expand, create land supply bottlenecks that may result in unaffordable housing. Planners must also understand that the urban peripheries of cities cannot be built at high densities. Density levels typically decline with distance from city centres but increase over time as cities expand outwards. Hoping to densify cities and limit their footprint by building dense settlements on the urban periphery, while certainly possible here and there, is unlikely to change this overall pattern, a pattern that has been observed everywhere.

From the perspective of urban population growth, urban planners and policymakers the world over should come to understand that urbanization – insofar as it entails the growth of urban populations – is new essentially over in the more developed countries; it is now essentially an issue facing the less developed ones. And preparing for urban expansion in less-developed countries – with weaker rule of law and limited human, fiscal, and financial resources – is a task of quite a different order than managing urban expansion in more-developed countries. Emphasis on complex master plans that rely on strong compliance must give way to simpler interventions that can utilize the limited available capacities for making significant changes on the ground.

Finally, the quality of the urban fabric in new urban peripheries can be enhanced with three simple sets of actions: (1) preparing an arterial road grid throughout the expansion area; (2) identifying and securing a hierarchy of public open spaces that need to be protected from development; and (3) improving land subdivision practices on the urban periphery.

The first involves creating an arterial road grid that covers the entire projected urban periphery to 2050 with the view to connecting it to all urban locations, to facilitating public transit provision, to creating natural corridors for transit-oriented development, and to shifting urban development away from lands with high environmental risk. This arterial road grid – say one with thirty-metre-wide road rights-of-way spaced one kilometre apart – can allow the entire expansion area to be within walking distance of arterial roads and hence from public transport. The key to success in creating such a grid hinges on acquiring the rights-of-way for the entire grid now. Our Urban Expansion Program at New York University has been assisting cities in Ethiopia and Colombia, for example, in preparing such grids and some of these cities are now in advanced stages of preparation (Lamson-Hall et al., 2018; Vásconez et al., 2015). Two cities in Colombia – Montería and Valledupar – are now planning to plant trees at ten-metre intervals along the future sidewalks of their future arterial road networks.

The second involves identifying a hierarchy of public open spaces, large and small, that need to be protected from development and instituting a set of pragmatic arrangements that can ensure that they remain open in the face of pressures from formal and informal developers to occupy them. Again, the key issue here is not to create a map of desired open spaces, paint them green, and hang the map in the municipal office to gather dust, but to put together a strategy for protecting a limited hierarchy of open spaces on the ground, be it by creating stronger institutional arrangements, by creating incentives, by organizing communities, or by soliciting funds. The success of such an effort must be measured by individual accomplishments on the ground, accomplished one at a time, rather than by good intentions that are ultimately frustrated.

The third intervention involves the engagement of informal suppliers of residential plots in improving their land subdivision practices, especially in the rapidly growing cities in Sub – Saharan Africa and in South and Southeast Asia, where most urban expansion is expected to take place in coming decades. This will require learning more about these practices, removing the stigma associated with them, modifying regulatory practices where necessary to recognize the value of regular subdivisions that do not offer a full complement of infrastructure services at the outset, and engaging informal developers in planning and disbursing plots in more orderly land subdivisions. Espinoza and Fort (2018), for example, have taken an important step in this direction, creating a smartphone application – "Lotizer" – for generating instant layouts, and testing it in the new urban peripheries of several cities in sub-Saharan Africa.

Needless to say, all three proposed interventions – acquiring the rights-of-way for the arterial grid, protecting a hierarchy of public open spaces from development, and increasing the capacity of informal developers to engage in proper land subdivision – are much simpler, much cheaper, and much easier to implement than the complex master plans that still command the attention of local and national politicians in less-developed countries, and that are still aggressively marketed in these countries by unscrupulous planning consortiums from more-developed countries. Both those selling and those buying these master plans see them as ends in themselves that have no visible effect on the facts being created on the ground. In fact, there are good reasons to believe that most of those engaged in master plan transactions have never been to the urban periphery and have never sought to under-stand how it is now being formed.

At this point in time, as this chapter has shown, the new urban periph-eries of cities are growing wild. They need to be tamed and they can be tamed. They can be tamed by projecting how much land will be needed for their expansion to 2050 and where these lands would be. They can be tamed by an active initiative aimed at conserving public open spaces and protecting them from development. They can be tamed by creating a skel-eton plan in the entire area of expansion, a plan that already determines the location of an arterial road grid now and that secures the right-of-way for that arterial grid, before development takes place. Finally, they can be tamed by focusing new attention on the way land is subdivided and sold on the urban periphery, with a view of getting rural landowners and informal property developers to improve their land subdivision practices. While these minimal interventions by no means address all the myriad problems confronting rapidly growing cities in less-developed countries, they offer real hope in making something of real value happen there.

ACKNOWLEDGMENTS

The authors wish to thank the Lincoln Institute of Land Policy and the Marron Institute of Urban Management at New York University for funding the research leading to this work.

REFERENCES

Abiko, A., Cardoso, L., Rinaldelli, R., & Haga, H. (2007). Basic costs of slum upgrading in Brazil. *Global Urban Development* 3(1). https://www .globalurban.org/GUDMag07Vol3Iss1/Abiko.htm.
Angel, S., Blei, A.M., Parent, J., Lamson-Hall, P., & Galarza-Sánchez, N., with Civco, D.L., Qian Lei, R., & Thom, K. (2016). *Atlas of urban expansion – 2016 edition: Vol. 1: Areas and densities*. NYU Urban Expansion Program

at New York University, UN-Habitat, and the Lincoln Institute of Land Policy. https://www.lincolninst.edu/publications/other/atlas-urban-expansion-2016-edition.

Blei, A., Angel, S., Civco, D.L., Galaraza, N., Kallergis, A., Lamson-Hall, P., Liu, Y., & Parent, J. (2018). *Urban expansion in a global sample of cities: 1990–2014* (Lincoln Institute of Land Policy Working Paper WP18AB2). https://www.lincolninst.edu/sites/default/files/pubfiles/blei_wp18ab2.pdf.

Espinoza, A., & Fort, R. (2018, June). *Planificar la informalidad: herramientas para el desarrollo de mercados de "urbanizaciones informales planificadas"* [Planning informality: Tools for the development of markets for "informal planned urbanizations"]. Grupo de Análisis para el Desarrollo (GRADE), Contribuciones al debate sobre la formulación de políticas públicas, No. 40. https://www.grade.org.pe/wp – content/uploads/GRADEap40.pdf.

Galarza-Sanchez, N., Liu, Y., Angel, S., Blei, A., Kallergis, A., Lamson-Hall, P., Salazar, M., Civco, D.L., & Parent, J. (2018). *The 2010 universe of cities: A new perspective on global urbanization* (Lincoln Institute of Land Policy Working Paper WP18NG1). https://www.lincolninst.edu/sites/default/files/pubfiles/galarza_sanchez_wp18ng1.pdf.

Gaviria, S., Goldwyn, E., Liu, Y., Galarza, N., & Angel, S. (2018). *Invisible walls: Organized violence, densification, and the containment of urban expansion* (Marron Institute of Urban Management Working Paper #41). https://marroninstitute.nyu.edu/uploads/content/Invisible_Walls_Final_Paper(1).pdf.

Lamson-Hall, P., Angel, S., De Groot, D., Martin, R., & Tafesse, T. (2018). A new plan for African cities: The Ethiopia Urban Expansion Initiative. *Urban Studies, 56*(6), 1234–49. doi:10.1177/0042098018757601.

Lamson-Hall, P., Angel, S., & Liu, Y. (2018). *The state of the streets: New findings from the Atlas of urban expansion – 2016 edition* (Lincoln Institute of Land Policy Working Paper WP18PL1). https://www.lincolninst.edu/sites/default/files/pubfiles/lamson-hall_wp18pl1.pdf.

Michaels, G., Nigmatulina, D., Rauch, F., Regan, T., Baruah, N., & Dahlstrand-Rudin, A. (2017). *Planning ahead for better neighborhoods: Long run evidence from Tanzania* (London School of Economics Working Paper IZA DP No. 11036). https://docs.iza.org/dp11036.pdf.

United Nations Department of Economic and Social Affairs. (2018). *World urbanization prospects 2018*. https://esa.un.org/unpd/wup/Download/.

Vásconez, J., Galarza, N., & Angel, S., with Montezuma, R., & Fonseca, S. (2015). Preparando ciudades de rápido crecimiento para su expansión: Informe sobre Valledupar y Montería, Colombia. *Medio Ambiente y Urbanización 83*, 129–54. https://www.ingentaconnect.com/content/iieal/meda/2015/00000083/00000001/art00009?crawler=true.

Image 4. Leipzig, Germany (June 2011)

4 Regional Urbanization Processes in Contemporary Italy: Beyond the City, from the Country of One Hundred Cities

ALESSANDRO BALDUCCI, VALERIA FEDELI, AND CAMILLA PERRONE

Patterns and Pathways of Urban Transformation: The Italian Contribution to the International Debate

Over the last twenty years, a relevant amount of research in Italy has been dedicated to explore urban change, in particular observing the complex interplay between the persistence of the urban territorial patterns inherited from the past and the emergence of new urban phenomena, connected with globalization processes and their impact on the sociospatial fabric. The debate has been rich and challenging (Balducci & Fedeli, 2007; Boeri et al., 1993; Clementi et al., 1996; Indovina, 1990; Secchi, 2005), though it has had quite a limited international resonance. Actually, for, and notwithstanding its specificities, this debate contributes to enriching the international debate, proposing possible alternative interpretations to those offered by international literature, which is questioning by large the nature of the urban, providing reconceptualization which is both inspiring and tricky (Roy, 2009).

The current Italian academic debate has its roots in the studies of the 1960s, when a series of important changes started to affect some of the liveliest urban contexts in Italy. Scholars like Giancarlo De Carlo (1962), Carmelo Samonà (1959), and Ludovico Quaroni (1967) introduced the concept of *città-regione* in order to describe the emergent effects of urban regionalization in contexts such as the Milan urban region. Since then, those original positions have been reused and adapted to discuss more recent processes that occurred during the late 1980s and early 1990s, when in particular the concept of *città diffusa* was introduced (Indovina et al., 1990). This umbrella term has been used to describe the contradictions and potentialities, the threats and opportunities of a kind of urban development with new forms, size, and meaning both in more typical metropolitan conditions (the case of the Milan urban region) and in non-metropolitan ones (the Veneto Region) (Boeri et al., 1993; Indovina et al., 1990; Perulli, 2012; Secchi, 2005; Turri, 2000). In this respect, it has been used to observe the alteration to the traditional Italian urban structure, as well as the diffusion of new urban landscapes and lifestyles, apparently almost invisible to both research and policies. A part of the debate has stigmatized the externalities of a "diffuse urbanization" (Indovina, 1990), though exploring the specific nature of the Italian diffuse city, as a mixed-use city, not necessarily low density and not necessarily rich and suburban – thus different from the concept of sprawl articulated by US literature.

The concept of *città infinita* (Bonomi & Abruzzese, 2004) renewed this debate, anticipating somehow the idea of the *endless city* by Ricky Burdett and Dejan Sudjic (2007) and suggesting the idea of a city, which is infinite under different perspectives. On the one hand, due to the fading of the contrast between urban and non-urban in large conurbations; on the other, due to the infinite complexity that is present in all its components, where problems and opportunities of (traditional) urban areas could be unexpectedly found even in apparently (traditional) suburban or peripheral areas. An important contribution to this debate has been offered by the long tradition of studies on the historical polycentric nature of the Italian context (Magnaghi, 2005; Perrone, 2010) and its interaction with the conceptualization of the Third Italy in the economic sphere, by authors like Giacomo Becattini and others, who have revealed the peculiar interaction between a small and medium sized manufacturing system and the historical urban grid in specific parts of the country (namely in the Lombardia, Veneto, Emilia Romagna, Toscana, and Marche regions; see Bagnasco, 1977; Becattini, 2015).

Nevertheless, despite the width and depth of the debate, which has been sometimes influential in the European context, the last systematic attempts to interpret sociospatial change processes were produced almost twenty years ago, when the ITATEN research project (Clementi et al., 1996) tried to map new emerging and plural "urban forms"/"forms of urbanity" all over the country. After that inspiring experience, very limited empirical research has been produced in order to provide an overall picture of the state of the art of urbanization processes in Italy.

Beyond Metropolis: A New Urban Question and the Need for a New Research Urban Agenda in Contemporary Italy

The research project known as Programma di Ricerca di Interesse Nazionale (PRIN) Post-metropolis moved from these assumptions and in particular from the acknowledgment that the city has become an ambiguous object (Martinotti, 1999), the description (and government) of which is particularly complex because it has become more and more difficult to isolate the contemporary urban fabric in terms of a stable and definitive sociospatial fact that is clearly distinguishable from the non-urban realm.

In particular, when confronted with the the many ways of describing the new characteristics of the urban – conurbations, global city-regions, megalopolis, megacities, polycentric regions (Borja & Castells, 1997; Florida, 2006; Geddes, 1915; Gottmann, 1957; Hall & Pain, 2006; Sassen, 2001; Scott, 2011) – the project confronted several recent interpretations that proved to be particularly stimulating in so far as they allowed us to go beyond a conceptualization of the contemporary urban as a form of crisis/death, decentralization, dispersion, or even "disloyalty" to the constituent characteristics of the historical city.

In particular the idea of the post-metropolis, as formulated by Soja (2011), in terms of "the emergence of a distinctive new urban form, the extensive polynucleated, densely networked, information-intensive and increasingly globalized city region … where relatively high densities are found throughout the urbanized region" (p. 684), was the starting point of the project. This post-metropolis concept was adopted strategically as a "portmanteau" supporting, on the one hand, the exploration of a country in which urbanity seems everywhere, but at the same time is specific and peculiar, where history and physiography matter but seem to be progressively (sometimes suddenly) obliterated by profound changes, with enormous socio-economic impacts. On the other hand, it seemed equally useful to nourish the debate on the nature of the "new urban question," going beyond a simple amplification of the typical problems of the twentieth-century city, such as environmental degradation, social polarization, inequalities in the distribution of and access to resources.

Throughout the research project, the post-metropolis input was hybridized and tested through other emerging theoretical contributions, establishing a dialogue with a number of other conceptualizations confirming or opposing Soja's perspective, all embedded in Henri Lefebvre's "radical hypothesis of the complete urbanisation of society [implosion/explosion], demanding a radical shift in analysis from urban form to the urbanisation process" (as cited in Brenner & Schmid, 2014, p. 12). The first relevant inspiration came from the concept of "suburbanization" formulated by Roger Keil (2018b): The suburban space is seen as an emerging form of the contemporary urban and part of extended urbanisation "but not one that is subordinate to state or city-centric rationalities" (p. 506). In this sense, it pushes for an urban theory able to go beyond the traditional understanding of the city from the inside out and to "challenge, reverse and propose to eventually abolish the centralist bias in urban theory and introduce a new way of looking at what we call 'suburban' in the context of urbanisation processes overall" (Keil, 2018a, p. 41). In this perspective "the urban century is really a suburban century where most activity in terms of expansion and contraction of urban population, built form and economic activity will occur in peripheral area" (p. 15).

The second is the thesis of "planetary urbanization," formulated by Brenner (2014) and by Brenner and Schmid (2014), according to which "under contemporary conditions ... the urban can no longer be understood with reference to a particular 'type' of settlement space, whether defined as a city, a city-region, a metropolis, a metropolitan region, a megalopolis, an edge city, or otherwise" (p. 13). Formulating "the idea of the 'non-urban' as an ideological projection derived from a long dissolved, preindustrial geohistorical formation" (Brenner & Schmid, 2012, p. 12), the authors propose development of "new strategies of concrete research and comparative analysis that transcend the assumptions regarding the appropriate object and parameters for 'urban' research that have long been entrenched and presupposed within the mainstream social sciences and planning/design disciplines" (Brenner & Schmid, 2014, p. 740).

Based on these perspectives, our research project (PRIN 2010 – Post-Metropolitan Territories as Emergent Forms of Urban Space: Coping with Sustainability, Habitability, and Governance)[1] explored the production of urban regionalization or "regional urbanization" (Soja, 2011) of the major urban areas in Italy in an effort to map the emergence of multiscale processes of urbanization, based on a complex interaction between path-dependency and innovation. A large part of our effort was actually concentrated on the development of a research framework useful for producing new analytical and interpretative portraits of the urban in Italy. We adopted a quantitative research approach, deciding to conceptualize and operationalize our main research questions through the production of a set of indicators that measure and describe processes of social, economic, environmental, and political change. On the basis of this research "protocol," we produced an "Atlas of Post-Metropolitan Territories" (www.postmetropoli.it) in the form of an open web resource to be used by researchers, scholars, practitioners, and decision-makers, as well as everyday citizens, which provided the basis for the main research[2] (Balducci et al., 2017a, 2017b, 2017c, 2017d). Drawing from the Atlas, we produced nine Regional Portraits

(Turin, Milan, Venice and Veneto, Florence and Tuscany, Rome, Naples, Palermo, South-Eastern Sicily, and Gallura – the latter two as counter-cases) to represent some of the most interesting urban areas in Italy or unexpected processes of the post-metropolitan nature in non-typical urban contexts.[3] The Atlas, together with the Regional Portraits, provides an image of urban Italy which is quite challenging and can contribute to the thinking about the notion of "after suburbia."

The "Urban" in the Dialectics between Fixity and Motion (Limitation and Facilitation): Additional Elements for Interpreting the Italian Case

The exploration of the Italian case, through the lens of processes of regional urbanization, offered a crucial opportunity to further measure the relevance of some additional contributions to urban theory coming from economic geography and urban studies. Recently, Schmid (2015) addressed the question of specificity as the constitutive feature of cities while exploring its meaning for urbanization. He investigates "how specificity is produced and reproduced, what role it plays in the production of urban spaces, and how it influences the planetary trajectory of urbanization" (p. 286). Schmid describes the interplay between *distensive* and *tectonic* forces as an interaction between *facilitation* and *limitation*: Whatever the territorial material pattern is (whether it be monocentric, polycentric, extensive, or densely interconnected), it incorporates a dual dimension; that is, "it facilitates processes of inter-action, but it also channels them and thus hampers alternative possibilities of development" (p. 294). Schmid sees this conceptual pair as part of the dialectic of *fixity* and *motion* introduced by Harvey (1985) to explain the constitutive and permanent contradiction inherent in the production of the built environment.

In Schmid's (2015) words, the process of facilitation and limitation is then described as "the dialectics of fixity and motion, the contradiction between the dynamics of urbanization and the permanence, the per-sistence of the spatial structures it produces ... This also explains why urbanization manifests such a high degree of path dependency: the built environment cannot be changed overnight, or at least not without caus-ing massive destruction and devaluation of existing investments. Thus, an urban fabric arises that can often barely be fundamentally changed and can only be adjusted with considerable efforts" (pp. 294–5).[4]

During the process of physical transformation of the territories (urbanization), nature (understood as a geo-bio-physical substrate) is both destroyed and recreated as a second nature through society's appropriation and transformation.[5] Despite this, the natural spaces do not disappear but can reappear at any time in a dialectical contraposition that involves two antagonistic "modes" of path-dependency: the resistance of geo-historical matrices and the persistency of urbanization. Within this conceptual frame, "the city, as a second nature, is caught between fixity and motion. Every urban development creates new possibilities, but at the same time also establishes fixed structures, thus limiting the potential for later corrections or changes to the course of development" (Schmid, 2015, p. 295). The dialectics between *facilitation* and *limitation*, *fixity* and *motion*, is therefore recognized as a distinctive character of the process of urbanization, as well as an emerging issue of the post-metropolitan transition. It is then possible to say that the post-metropolitan and post-suburban transition (Keil, 2011, 2013, 2018a, 2018b; Soja, 2011, 2015) takes variegated features. Such features show some dominant tendencies (multiscalar regional urbanization, planetary urbanization), while revealing the role of path-dependence and the resistance of geo-historical background, in creating the specificity of Italian urban landscapes (Paba & Perrone, 2016, 2017, 2018; Perrone et al., 2017).

Is Italy Special? A National and Regional Portrait

A general result of this project is the production of a national and regional interpretation of several relevant urban processes occurring in the Italian context: a multiscale portrait of urban regionalization processes in Italy, where a complex interplay between path-dependence and change is not only conceptualized but also exhibited in detail (thus being a crucial case to the discussion in chapter 14 by Ilja Van Damme and Stijn Oosterlynck in this volume).[6] Beyond any extreme position affirming or denying an epochal transformation of cities, the new pattern of urban spaces results in the coexistence of different urban configurations. Peripheral centres are growing because of their internal vitality and attractiveness, thus creating a new spatial pattern. Cities of different sizes in emerging urban regions are not losing their role as urban communities where residents are also local workers; in fact, they belong to a space based on flows of different kinds that pertain to different scales. Despite this change, the polycentric Italian urban geography articulated in large, small, and medium-sized cities is still evident while responding to variegated sociospatial dynamics (the flattening of the gradient of urban density; the gradual erosion of the boundary between urban and suburban; an increasing specialization of the suburban).

A rewriting of the initial requirements of the urban in Italy and its compliance with the traditional centre/periphery categories is then occurring (with significant inversions). The degree of actual diversification is such that the urban forms anchored to the past codified models (both metropolitan, polycentric, and dispersed) are still recognizable (see the case of Turin and Florence, or the disaggregated models of the urban in the case of Veneto, throughout the northwest and east of Italy, as well as in the centre of the country) in contrast to the emerging trends similar to those described in international literature as regional urbanization processes (Soja, 2011; see the case of Milan, located at the centre of the northern sector, in between Turin and Venice, where multiple hierarchies overlap the transformation dynamics).

Regional urbanization processes are indeed taking place between the local and non-local, which is reshaping the traditional urban structure, and between the urban and non-urban, which has never been simple to conceptualize in an ordinary and linear way. The process therefore emerges as a diversified *in-between* urban dimension that embeds explosive dynamics as well as deep processes of implosion of the concentrated urban, within a thick interplay between path-dependence and change.

This dimension is described by the selection of aggregate socio-economic indicators that detect, in particular in the centre-north of the country, the emergence of what we call an *urbanoid galaxy* (Fregolent et al., 2017, p. 284) as an effort to qualify and develop further the concept of the *infinite city* (Bonomi & Abruzzese, 2004), which was originally introduced by the authors with reference to the Milan urban region and the largest conurbation of which it is a part. The *urbanoid galaxy* is a provisional and explanatory conceptual and methodological proposal to read new urban assemblages and the complexity evolving from the new regional urbanization processes in the whole north-centre of the country, which is highly urbanized and, what is more, highly interrelated (from Piedmont to Veneto, from Lombardy-Switzerland to Emilia Romagna and Tuscany). In fact, we have recognized the trajectories of concentration and extension of the urban, measured both in their physical dimension (which plays an important role in Italy, with reference to the resistance of geographical and landscape features) and in their economic and social dimension. A path dependence emerges of historical patterns and more recent socio-economic models, suggesting a form of diffuse and archipelago urbanization, which is somehow post-metropolitan because of the way regional urbanization is taking place, producing a highly interconnected polynuclear urban formation, whose forms of urbanity can no more be defined in a unitary and holistic way, with reference to the traditional idea of the polycentric city.

The Milan urban region, in particular, appears easier to be conceptualized and observed, because of the historic interplay between polycentrism and metropolization, which has shaped the regional space in the last century. It is an apparent urban *continuum*, where new discontinuities emerge exceeding all traditional urban and metropolitan kinds of boundaries, imaginaries, and shapes (Balducci et al., 2017a, 2017d). But similar dynamics can be observed in the case of Naples, in the south of Italy (Laino, 2017), as the outcome of a clash between the historical monocentrism of the regional central place and the polycentrism of internal areas, towards the formation of an urban continuum constrained only by the complex geomorphology of the area.

As explored by the project, these phenomena are part of a complex process of regional urbanization that must be read at a national scale, producing large urban regions. A new combination of material and immaterial relationships is greatly impacting the use of space. Spatial relationships have not disappeared, but distance relationships are creating new types of "communities" and networks. The dynamism of former peripheral areas has often been interpreted as an alternative and a threat to the role of central cities. However, it is now evident that this is not true. It is not a development against the wealth of central cities; rather, it is a part of the same new form of urban development.

On the contrary, a specific condition seems to characterize the Turin and Rome cases (Caruso & Saccomani, 2017; Cellamare, 2017), which seem to represent more traditional processes of transition from a monocentric to a polynuclear urban structure. Although keeping a metropolitan structure from the spatial point of view, as well as often social and actor-related, with a strong role of the central cities in the interplay between centralization and decentralization, they locally present interesting forms of a post-metropolitan nature: awkward local conflicts, new political-social alliances, unexpected sociospatial assemblages, which are evidence of a post-metropolitan nature. Dynamic phenomena, produced by incomplete, overlapping, and highly interconnected processes of a transcalar nature, are forms of polynuclear urbanization, such as in the Veneto and Tuscany cases (Fregolent et al., 2017; Paba et al., 2017).

Finally, post-metropolitanity appears in different ways in cases such as eastern Sicily and north Sardinia. These are contexts that have never been metropolitan, but where peaks of urbanity, rather than simple urbanization processes, are completely redefining the organization of space: flows of people, connectivity and accessibility, social and demographic dynamism; heterogeneity of people and functions (Decandia, 2017; Lo Piccolo et al., 2017).

Conclusion: A Transcalar Regional Imaginary for a New Urban Agenda

Fragmentation, fractalization, redistribution, and restructuring trajectories of urban and urbanity are at play in the Italian context; in so far as being a crucial stake for a country whose identity is tightly linked with the urban condition (like other European Union countries; see Kazepov, 2005) and at the same time is lacking both a national urban agenda/strategy and any institutional reform able to grasp and deal with the new interrelated nature of processes (Urban@it, 2017). The main outcomes of the research we have conducted provide sufficient evidence to support the development of both robust and flexible new spatial imaginaries that are able to deal with the emergence of the trajectories of change and path-dependence which we have identified and discussed.

New and differentiated demands and offers of urbanity are emerging in the country, so much so that peripheries and centralities cannot be found in the usual places, although accumulation of inequalities and agglomeration effects are still visible in the largest urban areas; spatial distance still matters, but not so consistently as in the past; and marginality or attractiveness are consistently redefined thanks to digitalization, new lifestyles, and economic restructuring processes. Conflicts and unbalances are rising unexpectedly in unusual places; while the urban rural mix is changing forms and prospective roles. In this respect, administrative boundaries are stressed even beyond what we could expect. In particular, the boundary between urban and non-urban fades, in some cases completely, while in others it changes forms, location, and meaning: a new set of boundaries emerges, between dynamic and stagnant places, between attractive and dull ones, between highly homogeneous and highly differentiated societies. So much so that new assemblages, beyond horizontal and vertical forms of cooperation, are emerging, sometimes with contradictory effects.

The interrelated nature of urban problems highlights the necessity to introduce at least a new spatial perspective in order to understand processes and policy issues that could not be detected out of a transcalar – at least regional – perspective. If the project could not necessarily prove that the new regional urban dimension has generated new spatial injustices of its own, new spatial injustices and a new urban question could ensue, due to the lack of conceptualization and acknowledgment.

In this respect our research highlights the need to take the challenge to turn upside down any process of identification of a new urban agenda in Italy, in particular one that introduces differentiated urban-territorial imaginaries. This means the need in the first place to go beyond, both in terms of problem-setting and problem-solving, binary policy models (centre–periphery, urban–non-urban, concentration–decentralization), looking at the coexistence and plurality but interrelatedness of urban forms. In other words, one that considers urban regions, small and medium cities, inner areas, polycentric urban networks, emerging suburban contexts that are highly interrelated and not single isolated urban processes. What is at stake is the very nature of urban policy: the urban, in its different forms. This means, of course, reducing the distance among territories, practices, problems, and institutions; introducing innovative forms of governance capable of building the territory of policies as the outcome of local and interscalar processes; and producing new social, economic, and political efficacy and legitimacy.

Finally, there are also implications in the planning sphere. Planning has always been based on a homogeneous and linear relationship between territory and authority, while according to our descriptions of the Italian urban space it is now being called to work across territories, without any stable reference to defined authorities and boundaries. In the metropolitan phase, planning was mainly concerned with growth at the edge and new developments in uninhabited territories. Whereas in the post-metropolitan phase, planning has to involve the in-between spaces and the reuse and recycling of sites that have traditionally been inhabited by residents.

Assuming this new perspective, the future of urban-regional planning should include strengthening the understanding of the multicentric nature of urban regions; encouraging the recognition of significant intermediate aggregations of former territorial authorities capable of organizing themselves into meaningful urban areas where complementarities and the integration of service systems can be sought; encouraging the construction of coalitions around important territorial projects; rethinking the issue of land consumption within a new transcalar and multiactoral dimension; enhancing agricultural, green, and natural areas as constitutive urban functions; creating new conditions of habitability for resident or temporary populations across the urban space; enhancing the production of new public spaces at the regional scale, a new "urbanity" based upon economic and social complexity; and producing new systems of centrality capable of enhancing the old structure of existing centres as a means to support the emerging urban settlements.

The debate on the negative or positive nature of urban transformation is progressively losing its meaning. We have tried to show what the contemporary urban space is in Italy. It is not the deterioration of the modern city or metropolis; it is something else. Images are persistent, and if we continue to name this new sociospatial phenomenon a city or a metropolis, we may unconsciously incur a number of implicit consequences. If we continue to look for the reconstruction of an order that has been lost forever, we will miss the opportunity to explore the potential and appropriateness of the post-metropolitan space in dealing with emerging societal challenges.

ACKNOWLEDGMENTS

This chapter is an overview of a research project funded by the Italian National Ministry for Education and University (2013–16).

NOTES

1 The project, which was funded and run between 2013 and 2016, involved nine universities in Italy (Politecnico di Milano, Università del Piemonte Orientale, Politecnico di Torino, IUAV Venezia, Università degli Studi di Firenze, Università di Roma La Sapienza, Università di Napoli Federico II, Università di Palermo, Università di Sassari) and other research centres (in particular ISPRA, Istituto Superiore per la Protezione e la Ricerca Ambientale) with a large and interdisciplinary network of scholars and researchers.

2 The Atlas is designed to be implemented over time, also with the contribution of researchers who are not formal members of the original network. During the second and third year, for example, the construction of the Atlas has been extended to other urban areas – Genoa, Bologna, and Bari – with the contribution of other research units with the aim of promoting further information on related cases (see PRIN Postmetropoli, 2015).

3 Each Research Unit contributed to the production of the research framework and to its application in a specific territorial context, basically corresponding to the reference territory of the local research units and teams, with the aim of producing in-depth interpretations of local cases, as well as contributing to a discussion at national level. On this base, every research unit produced a report dedicated to the different *squares* that we decided to call *Regional Portrait*. The idea was to produce a series of interpretative portraits of the urban regions explored, based on a common

research protocol which was not necessarily comparative but could express significant research hypotheses at both local and national levels.

4 It seems to us that Keil and Addie (2016) go in the same direction when they suggest that emergent regionalized topologies and territoriality blur conventional city-suburban dichotomies in extended urban areas that are now characterized by polycentric post-suburban constellations (p. 891).

5 It would be necessary to investigate the term "nature" as a historical and social construct (or historical category) and its implications in the interweaving of path-dependency and sociospatial transformations. A study that we cannot develop in this chapter. For an initial analysis, see Sassen and Dotan (2011), and Perrone et al. (2017).

6 As they write: "Path dependency explicitly puts time central to causal analysis, is sensitive to the contextually specific interaction between multiple causal factors and operates on a notion of eventful temporality by seeing time as fateful, contingent and complex … it is unhelpful to reduce path dependency to the idea of 'lock in', as the latter is predicated on conventional economic notions of 'equilibrium thinking' … To take history seriously implies leaving open the possibility – and even likelihood – of endogenously generated change and continuous evolution and even transformation occurring."

REFERENCES

Bagnasco, A. (1977). *Tre Italie. La problematica territoriale dello sviluppo italiano* [Three Italies: The territorial problem of Italian development]. Il Mulino.

Balducci, A., & Fedeli, V. (Eds.). (2007). *I territori della città in trasformazione: tattiche e percorsi di ricerca* [The territories of the city in transformation: Tactics and research paths]. Franco Angeli.

Balducci, A., Fedeli, V., & Curci, F. (Eds.). (2017a). *Post-metropolitan territories: Looking for a new urbanity.* Routledge.

Balducci, A., Fedeli, V., & Curci, F. (Eds.). (2017b). *Ripensare la questione urbana. Regionalizzazione dell'urbano in Italia e scenari di innovazione* [Rethinking the urban question: Regionalization of the urban in Italy and innovation scenarios]. Guerini.

Balducci, A., Fedeli, V., & Curci, F. (2017c). Milan beyond the metropolis. In A. Balducci, V. Fedeli, & F. Curci (Eds.), *Post-metropolitan territories: Looking for a new urbanity* (pp. 27–52). Routledge.

Balducci, A., Fedeli, V., & F. Curci (Eds.). (2017d). *Metabolismo e regionalizzazione dell'urbano. Esplorazioni nella regione urbana milanese* [Metabolism and regionalization of the urban: Explorations in the urban region of Milan]. Guerini.

Becattini, G. (2009). *Ritorno al territorio* [Back to the territory]. Il Mulino.

Becattini, G. (2015). *La coscienza dei luoghi* [The conscience of places]. Donzelli.

Boeri, S., Lanzani, A., & Marini, E. (1993). *Il territorio che cambia, Ambienti, paesaggi, immagini della regione milanese* [The changing territory: Environments, landscapes, images of the Milanese region]. Aim-Segesta.

Bonomi, A., & Abruzzese, A. (Eds.). (2004). *La città infinita* [The infinite city]. Mondadori.

Borja, V., & Castells, M., with Belil, M., & Benner, C. (1997). *Local and global: The management of cities in the information age.* Earthscan Publications.

Brenner, N. (Ed.). (2014). *Implosions/explosions: Towards a study of planetary urbanization.* Jovis.

Brenner N., & Schmid, C. (2012). Planetary urbanisation. In M. Gandy (Ed.), *Urban constellations* (pp. 10–13). Jovis.

Brenner N., & Schmid, C. (2014). The "Urban Age" in question. *International Journal of Urban and Regional Research, 38*(3), 731–55. https://doi.org/10.1111/1468-2427.12115.

Burdett, R., & Sudjic, D. (Eds.). (2007). *The endless city: The urban age project by the London School of Economics and Deutsche Bank's Alfred Herrhausen Society.* Phaidon.

Caruso, N., & Saccomani, S. (2017). Turin metropolitan region: From path-dependency dynamics to current challenges. In A. Balducci, V. Fedeli, & F. Curci (Eds.), *Post-metropolitan territories: Looking for a new urbanity* (pp. 53–74). Routledge.

Cellamare, C. (2017). Transformations of the "urban" in Rome's post-metropolitan cityscape. In A. Balducci, V. Fedeli, & F. Curci (Eds.), *Post-metropolitan territories: Looking for a new urbanity* (pp. 117–37). Routledge.

Clementi, A., Dematteis, G., & Palermo, P.C. (1996). *Le forme del territorio italiano* [The forms of the Italian territory] (Vol. I–II). Laterza.

Decandia, L. (2017). The territory of the Sardinian province of Olbia-Tempio on the post-metropolitan horizon: From edge area to node of a new city-world. In A. Balducci, V. Fedeli, & F. Curci (Eds.), *Post-metropolitan territories: Looking for a new urbanity* (pp. 205–28). Routledge.

De Carlo, G. (1962). Relazione finale [Final report]. In Istituto Lombardo di Scienze Economiche e Sociali (Ed.), *Relazioni del seminario "La nuova dimensione della città, la città-regione." Stresa 19–21 January 1962.* Ilses.

Florida, R. (2006, July 2). The new megalopolis. Our focus on cities is wrong. Growth and innovation came from new urban corridors. *Newsweek International Edition.* https://www.newsweek.com/new-megalopolis-112699.

Fregolent, L., Vettoretto, L., Bottaro, M., & Curci, F. (2017). Urban typologies within contemporary Italian urbanisanition. In A. Balducci, V. Fedeli, & F. Curci (Eds.), *Post-metropolitan territories: Looking for a new urbanity* (pp. 281–93). Routledge.

Geddes, P. (1915). *Cities in evolution*. Williams and Norgate.

Gottmann, J. (1957). Megalopolis or the urbanization of the northeastern seaboard. *Economic Geography, 33*(3), 189–200. https://doi.org/10.2307/142307.

Hall, P., & Pain, K. (2006). *The polycentric metropolis: Learning from mega-city regions in Europe*. Earthscan.

Harvey, D. (1985). The geopolitics of capitalism. In D. Gregory & J. Urry (Eds.), *Social relations and spatial structures* (pp. 128–63). Macmillan.

Indovina, F. (1990). La città diffusa [The diffuse city]. In F. Indovina, F. Matassoni, M. Savino, M. Torres, & L. Vettoretto (Eds.), *La città diffusa* (pp. 19–45). Iuav-Daest.

Indovina, F., Matassoni, F., Savino, M., Torres, M., & Vettoretto, L. (Eds.). (1990). *La città diffusa* [The diffuse city]. Iuav-Daest.

Kazepov, Y. (Ed.). (2005). *Cities of Europe: Changing contexts, local arrangement and the challenge to urban cohesion*. Wiley-Blackwell.

Keil, R. (2011). Global suburbanization: The challenge of researching cities in the 21st century. *PUBLIC: Art / Culture / Ideas, 43*(1), 54–61.

Keil, R. (2013). *Suburban constellations*. Jovis.

Keil, R. (2018a). *Suburban planet: Making the world urban from the outside in*. Polity.

Keil, R. (2018b). Extended urbanization, "disjunct fragments" and global suburbanisms. *Environment and Planning D: Society and Space, 36*(3), 494–511. https://doi.org/10.1177/0263775817749594.

Keil, R., & Addie, J-P. D. (2016). "It's not going to be suburban, it's going to be all urban": Assembling postsuburbia in the Toronto and Chicago regions. *International Journal of Urban and Regional Research 39*(5), 892–911. https://doi.org/10.1111/1468-2427.12303.

Laino, G. (2017). The neapolitan urban kaleidoscope. In A. Balducci, V. Fedeli, & F. Curci (Eds.), *Post-metropolitan territories: Looking for a new urbanity* (pp. 138–60). Routledge.

Lo Piccolo F., Picone, M., & Todaro, V. (2017). South-eastern Sicily: A counterfactual post-metropolis. In A. Balducci, V. Fedeli, & F. Curci (Eds.), *Post-metropolitan territories: Looking for a new urbanity* (pp. 183–204). Routledge.

Magnaghi, A. (2005). *The urban village: A charter for democracy and sustainable development in the city*. Zed Books.

Martinotti, G. (Ed.). (1999). *La dimensione metropolitana* [The metropolitan dimension]. Il Mulino.

Paba, G., & Perrone, C. (2016). Physicality e path dependence nella transizione post-metropolitana in Toscana [Physicality and path dependence in the post-metropolitcan transition in Tuscany]. *Territorio, 1*, 45–52. https://doi.org/10.3280/TR2016-076007.

Paba G., & Perrone, C. (2017). Place matters: Spatial implications of post-metropolitan transition. In A. Balducci, V. Fedeli, & F. Curci (Eds.), *Post-metropolitan territories: Looking for a new urbanity* (pp. 256–65). Routledge.

Paba, G., & Perrone, C. (Eds.). (2018). *Transizioni urbane Regionalizzazione dell'urbano in Toscana tra storia, innovazione e auto-organizzazione* [Urban transitions: Regionalization of the urban in Tuscany between history, innovation, and self-organization]. Guerini.

Paba, G., Perrone, C., Lucchesi, F., & Zetti, I. (2017). Territory matters: A regional portrait of Florence and Tuscany. In A. Balducci, V. Fedeli, & F. Curci (Eds.), *Post-metropolitan territories: Looking for a new urbanity* (pp. 95–116). Routledge.

Perrone, C. (2011). *Per una pianificazione a misura di territorio. Regole insediative, beni comuni e pratiche interattive* [A "tailored" regional planning: Settlement rules, common goods, and interactive practices]. Firenze University Press.

Perrone, C., Perulli, P., & Paba, G. (2017). Post-metropoli – tra dotazioni e flussi, luoghi e corridoi, fixity and motion [Post-metropolis – between stocks and flows, places and corridors, fixity and motion]. In A. Balducci, V. Fedeli, & F. Curci (Eds.), *Ripensare la questione urbana. Regionalizzazione dell'urbano in Italia e scenari di innovazione* (pp. 23–52). Guerini.

Perulli, P. (Ed.). (2012). *Nord: Una città-regione globale* [North: A global city-region]. Il Mulino.

PRIN Postmetropoli. (2015). *Post-metropolitan territories as emerging urban forms: The challenges of sustainability, habitability, and governability* [Territori post-metropolitani come forme urbane emergenti: le sfide della sostenibilità, abitabilità e governabilità]. Research Project of National Interest (PRIN 2010–11). www.postmetropoli.it.

Quaroni, L. (1967). *La torre di Babele* [The tower of Babel]. Marsilio.

Roy, A. (2009). The 21st century metropolis: New geographies of theory. *Regional Studies, 43*(6), 819–30. https://doi.org/10.1080/00343400701809665.

Samonà, G. (1959). *L'urbanistica e l'avvenire delle città* [Urban planning and the future of cities]. Laterza.

Sassen, S. (2001). Global cities and global city-regions: A comparison. In A.J. Scott (Ed.), *Global city-regions* (pp. 78–95). Oxford University Press.

Sassen, S., & Dotan, N. (2011). Delegating, not returning, to the biosphere: How to use the multi-scalar and ecological properties of cities. *Global Environmental Change, 21*, 823–34. https://doi.org/10.1016/j.gloenvcha.2011.03.018.

Schmid, C. (2015). Specificity and urbanization: A theoretical outlook. In ETH Studio Basel (Ed.), *The inevitable specificity of cities* (pp. 287–309). Lars Muller.

Scott, A.J. (Ed.). (2011). *Global city-regions: Trends, theory, policy.* Oxford University Press.

Secchi, B. (2005). *La città del XX secolo* [The city of XX century]. Laterza.

Soja, E. (2011). Regional urbanization and the end of the metropolis era. In G. Bridge & S. Watson (Eds.), *New companion to the city* (pp. 679–89). Wiley-Blackwell.

Soja, E. (2015). Accentuate the regional. *Journal of Urban and Regional Research, 39*(2), 372–81. https://doi.org/10.1111/1468-2427.12176.

Turri, E. (2000). *La megalopoli padana* [The Po Valley megalopolis]. Marsilio.

Urban@iT. (2017). *Mind the gap. Il distacco tra politiche e città* [Mind the gap: The detachment between policies and cities]. Il Mulino.

Image 5. Winnipeg, Manitoba, Canada (Fall 2011)

5 Extended Urbanization, Urban Utopias, and Other Economies

ROBERTO LUÍS MONTE-MÓR
AND RODRIGO CASTRIOTA

Introduction

Urbanization, utopias, and economies are recurring topics in these early millennium rough days. We live in an urbanized world where both concentrated and extended urbanization play a central role in organizing society and space, from everyday life to the new Anthropocene geophysical conditions. Hundreds of large cities and metropolitan areas proliferate in the world. Lefebvre (1999) used the *urban tissue* to refer to the sociospatial fabric that resulted from the dual process of implosion/explosion of the industrial city, leading to the "critical zone" in which the urban emerges as a new category. The urban tissue reaches virtually every corner of the planet, articulating places and locales under the hegemonic urban-industrial globalized capitalist economy.

Since the late 1980s, Monte-Mór has been using the concept of *extended urbanization* to describe the sociospatial processes and forms that result from the explosion of the city onto the countryside (Lefebvre, 1999), drawing from decades of research in the Brazilian Amazonian frontier (Monte-Mór, 1988, 2014a, 2014b). Nevertheless, the idea of extended urbanization gained momentum and has been appropriated to describe innumerous urbanization processes worldwide in an attempt to make sense of the modernization of everyday life and the extension of the urban industrial conditions of (re)production as capitalist production (and particularly consumption) reach broader geographical scales, being now referred to as "planetary urbanization" (Brenner, 2014; Brenner & Schmid, 2015; Keil, 2018a; Schmid, 2018). This term, however, is not only completely dependent on the concept of extended urbanization but also does not make much sense as a concept describing spatial processes; instead, it refers either to a research agenda grounded in an "ex-centric" perspective (Schmid, 2018) or to the acknowledgment that "there is no escap[ing]" capitalist urbanization given its planetary *scale*, as Castriota and Tonucci (2018, p. 514) have argued. Urbanization has thus come to mean much more than the mere agglomeration of people in cities and towns to describe a worldly process of social production of space in which urban-industrial sociospatial forms and processes stemming from the industrial era present a myriad of manifestations in many corners of the world, from the Amazon rainforest to the European countryside to the backlands of Africa and India.

What about utopias? How can they fit into the dystopian scenarios we see daily in the media? Utopia has been a reference since Thomas More (1516/2017) published his novel in the early sixteenth century as a fierce critique of European societies, particularly England. Although beheaded twenty years later by Henry VIII, More's depiction of an egalitarian society in an island equivalent to the size of England became a cornerstone for the infant Enlightenment. Differently from his contemporary Machiavelli, whose political work inspired reforms and new political practices within rising mercantile capitalism, *Utopia* retained its revolutionary prospect and radical perspective.

In the twentieth century, Karl Mannheim (1972) stressed the difference between ideology and utopia, presenting them as two opposed forms within the social imaginary. While ideology is a system of representation that points towards the maintenance and reproduction of the hegemonic form, utopia is connected to representations of desire, thus, of transformation, experimentation, and diversion. Mannheim also talks of *Utopia*'s *subverting function* towards the rupture of the established order (as cited in Löwy, 2016, p. 23). Henri Lefebvre (2008) refers to the *concrete utopia* as "a possibility that explains the present and which the present shuns away towards the impossible" (p. 103).[1] He also refers to the *experimental utopia* "as the exploitation of the possible human, with the support of the image and the imaginary, followed by a permanent critique and a permanent reference to the problematic given by the 'real.' Experimental utopia overflows the common usage of the hypothesis in social sciences" (Lefebvre, 1976, p. 125).

As for other economies, the quest for overcoming capitalism – today also in an effort to save the planet, among other factors connected to economic and everyday life restructuring under neoliberalism – has brought back into the scene, and thus, into academic debates, those "alternative" modes of economic integration based on other values and forms of social organization differing from the market economy and bourgeois society. Domesticity, reciprocity, redistribution, and commercial exchange (Harvey, 1975; Polanyi, 2011, 2012) are those "other modes of [socio]economic integration" now made visible again as a response to current capitalist socioenvironmental crisis, its unacceptable levels of wealth concentration, and sociospatial inequalities and structural unemployment. Critical times, as currently recognized, produce new perspectives and ask for alternatives, forcing us to rethink old paradigms.

Implications of Extended Urbanization

There are certainly many sides to extended urbanization.[2] On one hand, the extension of the urban tissue carries within it the general (urban-industrial) conditions of (re)production onto the whole territory and produces the capitalist sociospatial processes and forms that are proper to a certain stage of capitalist development. The state certainly played a central role in producing the infrastructure and services necessary for (extended) urbanization to take place. Abstract space reached everywhere, as we could argue, after Lefebvre (1999). As the various waves of modernization met rural or non-capitalist places, social spaces were transformed and infected by the many viruses of modernity, from individualism to economicism, not to forget consumerism. Transformations are taking place at various paces in different social spaces and the old ways and means of (re)production, both in the economies and in everyday life, are substantially changing. These changes tend to replicate the "spirit of capitalism," bringing into those places and communities new (Fordist) commodities that require new technologies, thus implying the existence of what Milton Santos (1979) named the "techno-scientific and informational milieu," that is, the sociospatial forms and processes generated within the technological transformations in social space. In other words, extended urbanization itself, as Lefebvre (1999) points out, reached *virtually* all corners of the world connecting all social spaces under the hegemony of abstract space.

On the other side, the urban tissue carries within it the seeds and viruses of those attributes that were proper to the city, to the polis, meaning citizenship and politics. More specifically, the social mobilization that characterized the 1960s and 1970s attested to the repoliticization of the city that Lefebvre had announced in the late 1960. Appearing originally in major cities around the world, the so-called "urban social movements" were extended along with urbanization to eventually imply the dysfunctionality of the adjective "urban." In the late 1980s, even in developing countries like Brazil, that adjective became meaningless since social movements had reached everywhere, from backlands to the Amazon rainforest, as landless movements, extractivist groups, rubber-tappers, and Indigenous peoples got organized from local to national levels, fighting for their rights in the constitutional political process that led to the 1988 National Constitution (Monte-Mór, 2004, 2014a).

Extended urbanization thus implies a dialectical process in which abstract space is extended (virtually) onto the planet, with all its implications for the strengthening and consolidation of capitalist relations of production as urban-industrial space takes over new spaces, therefore extending the realm of the industrial era, while at the same time extending the politicization of social space and its integration at global scale and producing the conditions for the emergence of differential space beyond centres. In this sense, extended urbanization partially overlaps, conceptually and empirically, the domains of suburbanization, that is, "a combination of non-central population and economic growth with urban spatial extensions" (Ekers et al., 2015, p. 22). Both processes and concepts highlight the importance of "foregrounding the *explosion* antithesis conceptually and recognizing its own differentiated dynamics" (Keil, 2018a, p. 500). In Brazil, the invisibility and representational homogenization of non-central spaces and peoples resembles the suburban dialectics of order and disorder (Keil, 2018b). Particularly in Amazonian territories of extended urbanization, such invisibility and homogenization paves the way for exogenous, often city-centric actors to reproduce urban-industrial patterns of land use and territorial development with severe ecological implications. As in the suburbs, these territories become "some outsider's punch line" (Keil, 2020, p. 2).

On the other hand, suburbanization and extended urbanization are not interchangeable.[3] Although the idea of the "non-central" is extremely encompassing – and maybe too broad – the image evoked by the "suburb" is not too far from the city itself and still operates within the outskirts of global metropolitan centres. Unlike extended urbanization, suburbanization can be analyzed through city lenses – as suburbanization can still mean "cityization" in some contexts, even if it regards the city's edge. If more recent processes of suburbanization, as Keil (2020) argues, were fed through "people and economic activity moving from the center of the city to the suburbs" as well as "from the rural countryside to the periphery of cities" (p. 5) there is still a large conceptual and territorial domain that extends beyond the city and the suburban to which we might pay close attention. That is especially important in the Latin American context in which, contradictorily, as extended urbanization advances, the questions of the peasantry, of Indigenous and so-called "traditional peoples" and of many city "outsides" are strengthened instead of weakened. Moving towards the "non-central," thus, means critically engaging with the city's co-constitutive "Others" in terms of cosmologies, ecological practices, and modes of social reproduction.

This shift from the focus on production itself onto reproduction is a key feature of Lefebvre's (1999) urban era where concerns about life quality and collective subsistence become the centre of societal concerns. In this sense, the emergence of environmental concerns that characterized the turn of the century certainly add to the urban concerns about collective reproduction, at this point bringing to the centre of the scene concerns about survival of humanity as a whole. Contradictorily, as more and more people are brought together in urban agglomerations, their fates lie beyond them.

In such a scenario, Lefebvre's (1999) proposition of the dawn of an urban era, superseding the former industrial period, appears as a concrete utopia to guide us into the future, as opposed to the various dystopian propositions and representations we see today in films, novels, talks, and even advertisings that point towards the crumbling of contemporary industrial society being extended everywhere. Theory must be able to draw from actually existing utopians, non-capitalist everyday struggles, other economies and modes of social and economic integration that currently requalify the experience of urbanites beyond cities. Popular and academic "calls" for more socially just and ecologically viable futures should pay attention to these *opportunities* to "bridge the divide between the precarious now and the possible urban futures" (Simone & Pieterse, 2017, p. 54) if we are to "re-root" the urban in decentralized, suburban, ex-centric perspectives beyond the "ivory tower" (Keil, 2018b, p. 1591). As such, collective reproduction, everyday life and the politics of social space redefine both the production of space and the production of theory.

Exploring Urban Utopias

In general, when talking about urbanization, people are talking about cities. Most people (scholars included) see the world from the perspective of cities, albeit this perspective is already being deconstructed through a systematic critique of "cityism."[4] Most people on earth live today in urban centres, or in places categorized as cities (Ibañez & Katsikis, 2014; United Nations Human Settlements Programme, 2007; see also Brenner & Schmid, 2014)). Even in developing countries like Brazil, more than 80 per cent of the population live in so-called "urban areas," which explains, but does not justify, such a bias. Of course, there are cities, and there are cities. The cityist thinking is mostly grounded in large metropolises, which are in fact urbanized regions where the urban tissue has been extended much beyond the limits of metropolises, such as metropolitan regions, urban agglomerations, and other urbanized areas.

The idea of an extended urbanization that goes much beyond city limits reaching regional, national, and even planetary scales is each day more recognized and accepted. The urban tissue, or simply "the urban,"[5] as the sociospatial form and process resulting from the exploding encounter of the industry and the city, as Lefebvre (1999) described it, appears as a contemporary category that encompasses both the city and the countryside modifying both of them (but not eliminating) through the extended urbanization and their integration in the urban-industrial realm.[6]

The social, cultural, political, and economic synergies that develop in cities, in the country, and in the urban all point to new articulations and to the strengthening of new (often, old) forms of social organization, which I refer to, following Polanyi (2011) and Harvey (1975), as modes of economic integration. It is in that mixed context of rescue and reinvention that we can, and should, talk again about utopias. It is exactly in these periods of crises and deep transformations that the utopia thought imposes itself as necessary, be it as an antidote, be it as an inspiration of new experiments. Therefore, we are talking about concrete utopias (Bloch, 2006; Löwy, 2016; Münster, 1993), about utopia as new experiences that point towards futures that may seem impossible today, but which are already virtually announced as possibilities to be construed (Lefebvre, 1976).

Those utopias have the urban as their privileged territory, not the cities or the countryside, but the new spatial forms and processes that are being developed beyond cities and in old rural areas, particularly the new centralities that surround metropolitan regions and connect to traditional spaces – rural peripheries, savannahs, open fields, forests, and others – where traditional forms of living and survival still exist and thrive more strongly each day. "Peasant communities" of various types, now connected through extended urbanization to the urban-industrial tissue (and society), from native populations to new forms of community organizations, both in the Global North but particularly in the Global South, reinvent their modes of sociopolitical and economic integration, both at local and distant realms, to deal with the profound changes they have been experiencing.

Therefore, these contemporary utopias, which do not necessarily strive for progress but more often strive for ascension[7] and emancipation, can only be built in connection to old peasant utopias, in the fringes of the cities and in the urban. Monte-Mór has named them "peasant urban utopias" in another context, but we could also perhaps call them "suburban utopias." They arise from the transformation of the urban-industrial into its announced contemporary form, the urban-natural,[8] struggling to become the urban-utopia (Monte-Mór, 2018). Why does it happen? It seems that in Lefebvre's urban era, the centrality of life space, and consequently of everyday life, dictate the future to come.

Other Economies Asking to Be Taken Seriously

Since the crises of Fordism and the welfare state, capitalism has not been capable of proposing a solution to the many problems it has created in the previous two centuries during the rise and consolidation of an industrial society. It is widely accepted that the inclusive proposal of an industrial regime based on mass production and mass consumption, with its virtuous cycle of increasing productivity and recurring transfers to real salaries, allowed for significant enhancements in life quality for the working classes in the developed capitalist societies. The articulation of Fordism and the welfare state represented the highest stage of that mode of production, creating in its golden years the well-known win-win condition in those countries where it thrived. Despite its migration to developing countries in the 1970s, when the Fordist crisis began in the capitalist centres, it rarely found the political and sociocultural conditions in the periphery to achieve the same levels of success.[9]

The following flexible regime of accumulation (Boyer & Mistral, 1978; Harvey, 1989), articulated with deregulated economies and financial capitalism, inaugurated the contemporary neoliberal logic that has been penetrating all realms of life (Dardot & Lavall, 2010). Nevertheless, flexible production (post-Fordism) also created the conditions for the emergence of new/old economic (sociocultural and political) processes that have since continuously challenged capitalist hegemony, albeit peripheral amidst globalized financial capitalism.

In the global peripheries, however, those new/old economic processes (informal economy and other economies) have always played a central role in societal reproduction. Several theories and analyses have tried to explain the various forms by which working classes guarantee their reproduction in poor countries' disarticulated economies (De Janvry, 1981). In those countries, the capitalist formal sector does not include the majority of the population, who survive mostly within an "inferior circuit" that relates to the capitalist "superior circuit" in functional ways (Santos, 1979). Those working families subsist through a plurality of non-capitalist relations and practices, many of them involving non-mercantile relationships. Those practices and agents constitute the realm of the "other economies," which contemporarily also produce politicized alternative forms of economic organization, more collective and more democratic, eventually self-governed. We have, thus, in the last decades, seen the emergence and/or strengthening of several "other economies," complementary and/or opposed to the capitalist economy, producing theoretical and practical developments: environmental or green economies; ecological economies; social economies; popular economies; solidarity economies; functional and cooperative economies; sharing and gift economies, among others. They are gaining visibility and experimentation, from the capitalist peripheries where they have been historically endemic, to the developed capitalist countries, where they are (re)emerging in reinvented forms.

Very diverse and with distinct objectives, these "other economies" have several features in common. One of these salient features is the concern about the collective reproduction of life, from the local to the planetary scales, focusing on the environmental question, but also on social and cultural diversity. Beyond that, in addition to the obvious planetary environmental threat, is the crisis of industrial capitalism in its globalized financial phase, aggravating its disembeddedment from concrete reality and the exacerbation of abstract space as the dominant mode of socio-economic articulation, apart from and at the same time penetrating the core of everyday life.[10] We may consider that there is an articulation between the capitalist global economy, the public-sector economy, and a popular economy,[11] taken as an expression of the multiplicity of other economies strengthened in resilient peripheral social spaces and reborn and/or redefined in the centres of capitalism themselves.

A common feature among "other economies" is also the "labour fund"; that is, those economies that are mostly organized around the working capacity of its participants, organized in various forms that might vary from extended family and community units to associations, cooperatives, and others. In all cases, the other economies have the "amplified life reproduction" (Coraggio, 1994) as their central concern, in opposition to the wealth accumulation that is proper to capitalism. In most cases, they are also embedded within everyday life and life space, thus bringing "alienation" back to the debates as a crucial concept, which has since long been central to the political economy, but which is now redefined and enriched by the ideas of territory, life space, and nature itself.[12]

Another common feature of these other economies rests on the sense of collectiveness and collaboration, in opposition to the sharp competition and individualism that characterize capitalism and bourgeois culture. These economies are also bringing back publications by authors who emphasized the gregarious and collaborative practices that marked early capitalist history, making visible the experiences and theories of the nineteenth century and before, which rested on collective propositions and practices, not only in Europe and North America but also in peripheral societies that informed anthropologists, artists, philosophers, and others. Radical community experiments, from the narodnik movement to communal organizations of traditional peoples, are being avidly revisited and researched, many times rescued to inspire other modes of coexistence and subsistence.[13]

In this sense, these other economies go back to the etymological roots of *oekonomia*, to the *nomos* of the *oikós*, that is, the management of the (extended) house, of collective life space itself, which we could eventually extend to include the planet Earth, if we consider the ecological question. That is quite different from what Aristotle called chrematistic, the formation of market prices, which is what economics eventually came to mean. Having everyday life, life space, or the territory as a central concern of the economies imply new ways of economic thinking, moving towards the organization of societies, the strengthening of identities, the guarantee of collective subsistence – not only of humans but also of non-humans in this "economic" concern centred on the management of life space.

Work, in such a context, also gains a different meaning, searching for disalienation from life space and everyday life, in an effort to construe collective forms of social organization in articulation with nature, rescuing and/or inventing new modes of economic integration that surpass the hegemonic capitalist relations of production. Lefebvre (1978) stresses that capitalism survives by producing (abstract) space and reproducing its social relations of production. However, it seems that contemporary capitalism is not capable of, or is no longer interested in, reproducing in ample ways its relations of production, leaving large portions of the territory and of the working people to their own fate, thus creating the conditions for the emergence and strengthening of other relations of production. These "other" new and old relations of production, nevertheless, (re)emerge and/or grow in articulation with and under the realm of hegemonic capitalism, many times establishing "dual circuits," as stressed by Santos (1979), or a "popular economy" connected to capitalist and public sector economies, as proposed by Coraggio (1994). In this context, new social relations, including new relations between the human and the non-human, and with nature itself, can begin to be explored and redefined, opening room for new experimentations.

Conclusion: Searching for Virtual (Sub)urban Utopias

It seems that the revolution we might contemporarily experiment, shyly announced here and there, is centred on everyday life, in the efforts for collective (re)production and appropriation of life space, and in the reproduction of non-capitalist social relations of production.[14] Not centred on industrial processes and forms, but instead, in urban sociospatial processes deeply related to nature. It is no more a matter of taking arms to overthrow constituted power, as in twentieth-century revolutions. Not a sudden change in the capitalist mode of production, or in the social structure or cultural sphere of what is called the West (Europe and industrial colonies). It is about gradual changes that question the hegemony of capitalist exploitation, founded in the dialectical character and the opportunities generated in a societal crisis that surpasses the crisis of the state and the multiple capitalist crisis, without denying them; on the contrary, these gradual changes feed on them and are articulated by them. It is about thinking, understanding, and exploring the planetary crisis which, in its limits, sets human survival on the planet at the centre of our concerns.

In this sense, it can be a more profound revolution, once it implies a radical transformation at a planetary scale imposed by the advent of the Anthropocene.[15] It seems that this can only be achieved through other economies, based on different values and goals, which are emerging within capitalism's interstices, rescuing and (re)inventing other modes of social and economic integration. Capitalist hegemony seems to be challenged by its own impossibility of being inclusive and of reproducing its social relations of production opening room for other epistemologies and ontologies, borrowing from other cosmologies focused on collective reproduction and survival, facing the threat of not having a "world to come."[16]

Therefore, a central option towards the urban utopias implies to work outside of capitalism, strengthening other economies, other social relations of production, other forms of producing social (and differential) space that can oppose hegemonic destructive capitalist accumulation. Progress must gain an oppositional sign, changing into ascension in movements always from within, without abandoning its human emancipation project. In such a context, industrialism and developmentalism are under severe judgment, virtually overcome, while economic growth has no more interest for us, except in specific sectors and under particular conditions that allow it to be socially and democratically generated and appropriated. It implies the overcoming of the illusions of modernity as it presented itself in the past two centuries; we are concerned with "after," "beyond," "post," and "post-isms" of various kinds (Monte-Mór & Ray, 1995). It implies a radical critique preparing for many new forms of living that are asking to be born in intimate relationships with old essential ones. If we accept that the contemporary project of human emancipation implies general conditions for an amplified reproduction of life, and no more capitalist accumulation or economic development (practically, the same thing, today), then some themes become central in our concerns and debates.

The world of differences, the differential spaces proposed by Lefebvre, are surely part of that. It is in the contemporary (extended) urban – or in the suburban (Keil, 2013, 2018a) – in urban centralities of various kinds, from cities to redefined villages, where social and territorial synergies from the urban praxis, the collective sense of belonging, humanism, republicanism, and of the concrete utopia that Bloch (2006) defends as a result of the philosophy of praxis, of hope as human affection and possibility, that these processes are taking place. Necessarily experimental, as Lefebvre would probably add.

Nature and territory also come to the scene as the central themes that imply the inclusion of the human and the non-human as integral parts of everyday life, imperative for their (re)integration, territorial disalienation, and the overcoming of the nature-culture and the city-country dichotomies, both in a process of redefinition through the advent of the (extended) urban. Thus, it is no more a matter of looking at the urban-industrial tissue, but of producing and strengthening an urban-natural tissue, located at the core of the production of space as experimentation of the urban-utopia (Monte-Mór, 2018). An "extended naturalization" is the necessary complement to extended urbanization, redefining life space itself (Monte-Mór, 2014a).

The conscience of the Anthropocene that rapidly disseminates through digital forms and innumerous urban-rural practices highlights the centrality of the environmental crisis in contemporary days, mobilizing towards new values and new sociospatial practices. It is the fear of the "world to come" that contributes to making each day more visible for the traditional peoples, many of them such as the Brazilian Amerindians, who are thought of as belonging to the past and are in a rapid process of extinction. However, they have immensely increased in population and in sociopolitical organization in the last decades, each year more strongly, bringing to societal debates their other cosmologies, more freely manifested beyond the anthropological and academic studies, gaining political space and influencing everyday life in many realms. Although extremely threatened by Brazil's current extreme rightist government, they are politically organized and have entered spaces such as the universities and graduate courses reaching levels of visibility, legitimacy and cultural strength that were unthinkable a few years ago.

Another central theme to discuss in the urban utopias refers to the valorization of everyday life, of work and of life space. Revolution must be centred on everyday life; we know it and many are those who have said it. To Lefebvre, the (re)discovery of everyday life in the twentieth century restated the questions related to exploitation, alienation, and human emancipation in the context of the overcoming of capitalism.[17] The focus on daily life dislocates the centre of concerns from production, where surplus value is more evident, to the sphere of collective reproduction of the working classes, ultimately forgotten and not researched in its multiple contemporary forms. As Lefebvre (2014) makes explicit in the preface of *The Critique of Everyday Life*, his work is guided by the investigation of alienation, beginning with Marx's critique and following an attempt to overcome Marxism's dominant economicism in post-war Europe, in an attempt to rescue his philosophical and sociological dimensions. The implications for today's critique of dominant economicism are quite evident. In addition, Lefebvre states his shift from rural to urban issues, having the city, at first, and then the urban, as one of the focuses of his concerns, eventually leading to his proposition of an "urban era."

In that context, work presents a living dimension, differentiated and proper to a disalienated and revolutionary daily life, and gradually the spatial dimension – life space – emerges in Lefebvre's thinking. Disalienation within everyday life, and a creative commitment to the management of social space (*oekonomia*), inform the dislocation of his focus from industrial production (accumulation) to urban collective reproduction. The implications for public policies and for political actions are many and are vividly expressed in our contemporary (everyday) life, as in the various political struggles around life space in cities, in the countryside and in the urban – the right to the city, to collective wealth, political power and the cultural *fiesta*.

Another central question both implicit and derived from the above is the deepening of democracy and self-management. Central elements and values in the other economies that we are announcing and experimenting, these processes of democratic and participatory management of life are intrinsically contradictory within capitalism's contemporary oligopolistic and financial versions. It has always been so, one could say, but recent political coup d'états in Latin America, and elsewhere, have proved democracy (in its representative bourgeois form) to be less viable with each day. Brazil's juridical and legislative coup d'état in

2016 illustrates quite dramatically the limits of democracy within this highly competitive and globalized capitalism, particularly when the democratic forces focus more directly on the conditions of collective reproduction and set limits to the capitalist national, and more so international, vested interests.

Several of the other economies that are emerging are based on self-management, as is the case of solidarity economies, encouraged by the Brazilian government since 2003.[18] Despite the combat against those economies conducted by the recent extreme rightist governments in Brazil, their strength among social movements is significant and seems to be growing each day.[19] Ecological economies, social and popular economies, among others, search for more participatory forms of management, not only in Brazil, but also in several other Latin American countries, like Uruguay, Bolivia, Ecuador, Mexico, and Venezuela, not to mention Cuba. Experiences like Bolivia's and Ecuador's Buen Vivir, Brazil's Landless Workers Movement, Mexico's Zapatistas, Uruguay's Cooperatives, among others, suggest that in spite of political and economic constraints, these other economies have come to stay as possible solutions to problems of poverty, structural unemployment, and environmental concerns. Local popular banks, local currencies, sharing economies of various forms, and other modes of socio-economic organization among traditional peoples are pointing towards other possibilities in economic integration and concern.

The modes of economic integration that Polanyi described seventy years ago find today more interest and visibility from scholars, researchers, and the social movements themselves, rescuing concepts like "the commons," in opposition to capitalism's central form of private property.[20] Practices linked to domesticity, reciprocity, redistribution, and local markets are found in all economies in the world, at different levels. They are being rescued in Europe, for example, with great visibility, mostly connected to environmental concerns and movements around climate changes, garbage and waste, housing, transportation, and other collective issues. Solidarity forms related to reciprocity; redistribution forms with self-management and internal democracy; exchange non-capitalist markets connected to "fair trade" organizations, among others, gain visibility across digital networks and the internet and even corporative media.

In poor peripheral countries, where capitalism has manifested in incomplete forms, these other modes of economic integration persist as they constitute survival strategies for large portions of the population. At the same time, these economies are making complementary processes to the capitalist sector more viable by making them more competitive for the global and internal markets. These non-capitalist modes of economic integration, now strengthened in the flexible mode of regulation of capitalism itself, lead to growing hybridisms and unimaginable integrations and complementarities potentialized by digital informational technologies that are disseminated by globalization and extended by urbanization itself.

The articulation between capitalist relations of production and other modes of economic integration has existed everywhere and at different levels for some time, as argued before. The novelty of contemporary days is its potential for politicization and the incapacity or lack of interest of fractions of capital in amplifying the reproduction of its relations of production, excluding large portions of the population from the capitalist circuits. In doing so, tthis politicization leads to the strengthening and experimentation of resilient non-capitalist economic forms that are seen as about to be overcome or about to disappear. To the contrary, the other economies are growing and spreading in various contexts and on various scales, particularly in (sub)urbanized areas on the fringes of the cities and in the countryside, and seem to respond more thoroughly to the hopes and values of contemporary societies.

Diversity is a central feature of extended urbanization, from metropolitan suburbs to the countryside, from forests to savannahs, from villages to urban centralities of various kinds; all embrace and contain today, however, the seeds of an urban-natural tissue, of a (re)encounter of society and nature focused on collective reproduction. Although cities will surely go "greener" each day, it is in the *urban* that utopian experiments shall truly embed themselves. Peasant utopias and urban utopias were never so close and interconnected. To the centrifugal forces of extended urbanization correspond the centripetal forces of an extended naturalization. And it is there, in that myriad of diverse manifestations of peripheral urbanized social spaces that this urban-natural encounter shall flourish more intensely. In the postsuburban world, other economies will have to redefine economics, bringing the concept closer to its etymology, that is the management of the home, of our life space; today, more obviously, the Earth. Extended (sub)urbanization is certainly a terrain of possibility for postcapitalist transition as the extended (sub)urban fabric becomes, instead of *the* city, the privileged locus for the consolidation of an urban-natural society.

This way, we come back to the urban utopias, involving both cities and countryside in their diverse manifestations of today. If Lefebvre's urban era surpasses the industrial era at the same time in which the conscience of the Anthropocene imposes itself, the project of human emancipation implies new socio-natural relations, as in Michel Serres's (1991) "natural contract." The urban-natural surpasses the urban-industrial without extinguishing its legacy, pointing towards an urban-utopia in which themes such as nature, territory, work, everyday life, life space, radical democracy, self-management, new socio-economic forms of organization and institutions and the rescuing or (re)inventions of other modes of economic integration beyond or outside capitalism present themselves as possibilities for overcoming the crises and construing emancipation processes.

In other words, what might seem impossible and utopian today might be reread as possible, concrete, and experimental and, if necessary, eventually present itself as an instrumental response to the societal planetary crises we are currently experiencing. Fifty years later and we can once again use phrases from the events of May 1968, when the urban era seemed to announce itself intuitively: let's be realistic, let's ask for the impossible!

NOTES

1 All translations into English from Lefebvre's texts in Portuguese or Spanish, are free translations by the authors.

2 For an account of the implications of extended urbanization for urban and regional theory, see Monte-Mór and Castriota (2018).

3 As Keil (2018a) has pointed out, "suburbanization and suburbanisms cannot and should not be fully subsumed under the notions of planetary urbanization or extended urbanization" (p. 496), just as, for us, the notion of extended urbanization cannot be subsumed under the notion of suburbanization or planetary urbanization.

4 Castriota (2016) and Castriota and Siqueira (2019) discuss the perceived, conceived, and lived dimensions of cityism (citadismo) and its political relevance for social mobilization, struggles, and (not only urban) knowledge production, beyond the dimension of the "methodological" forwarded by Angelo and Wachsmuth (2015).

5 The emphasis on "the urban" as a noun refers to that new reality and not to the urban as an adjective to the city (Monte-Mór, 2007). For a discussion about the emergence of the urban in Brazil, see Monte-Mór (2014b).

6 In a previous paper (Monte-Mór, 2007), I proposed the urban to be taken as the third term in the city-country-urban dialectical triad, or trialectical, alluding to the term proposed by Soja (1996).

7 It is curious enough to notice that in the I Ching's *Book of Change*, the hexagram that follows "progress" is the hexagram "ascension."

8 It may be worth noting that Lefebvre did not discuss specifically the urban-natural dimension, which we nevertheless assume to be contained in his idea of the urban society and urban utopia.

9 Furtado (1978) offers a sharp chronological essay about the emergence of the industrial society and presents the reasons for its failure in dependent countries, while Lipietz (1987) qualifies peripheral Fordism as incomplete (following "bloody Taylorism"), asking what did Fordism want in the periphery?

10 The main references, that will not be discussed here, are Polanyi (2011, 2012) about the disembeddedness of the market economy, and Dardot and Laval (2010) about the roots of neoliberalism.

11 A reference to the economic triad proposed by Coraggio (1994) – the international capitalist economy, the public sector economy, and the urban popular economy – here redefined in broader terms.

12 Ollman (1984) presents a broad discussion on the implications of Marx's concept of alienation in contemporary life, while Lefebvre (1991) discusses it in Marxian terms but relating it to his own concept of abstract space.

13 Polanyi's daughter organized a series of his articles discussing "the subsistence of man" (Polanyi, 2012).

14 This section is partly rewritten from a talk by Monte-Mór at Cedeplar/ UFMG in 2017 and published as an e-book, *Alternativas para uma crise de múltiplas dimensões*, organized by Viegas and Arbuquerque (Cedeplar, 2018), www.cedeplar.ufmg.br.

15 Anthropocene is a neologism already with a history and holds several implications. Here, it is used to refer to this period in which humankind has become a geological agent capable of altering the nature of Earth, its oceans and subsoil, and the atmosphere itself, beyond being able to destroy the planet by fire.

16 "Is there a world to come?" ask Danowsky and Viveiros de Castro (2014), while the Yanomami shaman Daví Kopenawa talks about "the falling sky" and his (cosmo)vision of a contemporary world, both Western and Yanomami (Kopenawa & Albert, 2013).

17 Lefebvre published a trilogy about this theme: *Critique de la vie quotidienne* (1947); *Fondement d'une sociologie de la quotidienneté* (1962); *De la modernité au modernisme (Pour une métaphilosophie du quotidien)* (1981), published in English in a single volume (Lefebvre, 2014). He also published *La vie quotidienne dans le monde moderne* (1968), where he revisited the theme from the perspective of modernity and the state.

18 The Brazilian government created the National Secretariat for Solidarity Economy (SENAES), which existed from 2003 until 2016, when the government arising with the coup began to dissolve it. Nevertheless, the solidarity movement is still quite strong in the country, particularly in peripheral poor areas.

19 In May 2018, during a general transportation strike, more than three thousand participants coming from all over the country reunited in Belo Horizonte during the IV National Meeting of Agroecology.

20 Dardot and Lavall (2010) wrote a treaty on neoliberalism, making explicit its implications for everyday life, and soon wrote another work about the *commun*, as an antidote presented as "the revolution of the twenty-first century" (Dardot & Lavall, 2014).

REFERENCES

Angelo, H., & Wachsmuth, D. (2015). Urbanizing urban political ecology: A critique of methodological cityism. *International Journal of Urban and Regional Research, 39*(1), 16–27. https://doi.org/10.1111/1468-2427.12105.

Bloch, E. (2006). *O Princípio Esperança III* [The hope principal, vol. III] (N. Schneider, Trans.). UERJ, Contraponto.

Boyer, R., & Mistral, J. (1978). *Accumulation, inflation, crises.* Presses Universitaires de France.

Brenner, N. (Ed.). (2014). *Implosions/explosions towards a study of planetary urbanization.* Jovis.

Brenner, N., & Schmid, C. (2014). The urban age in question. *International Journal of Urban and Regional Research, 38*(3), 731–55. https://doi.org /10.1111/1468-2427.12115.

Brenner, N., & Schmid, C. (2015). Towards a new epistemology of the urban? *City, 19*(2–3), 151–82. https://doi.org/10.1080/13604813.2015.1014712.

Castriota, R. (2016, October 20). *Reassessing the blind field: Cityism and industrialism at the periphery of capitalism* [Conference presentation]. After Suburbia Conference, York University, Toronto, ON.

Castriota, R., & Siqueira, B. (2019). *Estudos urbanos ou estudos da cidade? Notas sobre o citadismo ou cidade-centrismo* [Urban studies or city studes? Notes on cityism or city-centrism] [Conference presentation]. Annals of the XVIII ANPUR Meeting, Natal, Brazil. http://anpur.org.br/xviiienanpur /anais/.

Castriota, R., & Tonucci, J. (2018). Extended urbanization in and from Brazil. *Environment and Planning D: Society and Space, 36*(3), 512–28. https://doi.org /10.1177/0263775818775426.

Coraggio, J.L. (1994). *Economia Urbana: la perspectiva popular* [Urban economy: The popular perspective]. Instituto Fronesis.

Danowsky, D., & Viveiros de Castro, E. (2014). *Há mundo por vir? Ensaio sobre os medos e os fins* [Is there a world to come? Essays on fears and endings]. Cultura e Barbárie, Instituto Socioambiental, Desterro.

Dardot, P., & Laval, C. (2010). *La nouvelle raison du monde. Essai sur la société néolibérale* [The new reason of the world: Essays on neoliberal society]. Éditions La Découverte/Poche.

Dardot, P., & Laval, C. (2014). *Commun. Essai sur la révolutions au XXIe siécle* [Common: Essay on the revolutions in the 21st century]. Éditions La Découverte/Poche.

De Janvry, A. (1981). *The agrarian question and reformism in Latin America*. Johns Hopkins University Press.

Ekers, M., Hamel, P., & Keil, R. (2015). Governing suburbia: Modalities and mechanisms of suburban governance. In P. Hamel & R. Keil (Eds.), *Suburban governance: A global view* (pp. 19–48). University of Toronto Press.

Furtado, C. (1978). *Criatividade e Dependência na Civilização Industrial* [Creativity and dependency in the industrial civilization]. Paz e Terra.

Harvey, D. (1975). *Social justice and the city*. Edward Arnold.

Harvey, D. (1989). *The condition of postmodernity: An enquiry into the origins of cultural change*. Blackwell.

Ibañez, D., & Katsikis, N. (Eds.) (2014). *New geographies 06: Grounding metabolism*. Harvard University Press.

Keil, R. (Ed.). (2013). *Suburban constellations: Governance, land and infrastructure in the twenty-first century*. Jovis.

Keil, R. (2018a). Extended urbanization, "disjunct fragments" and global suburbanisms. *Environment and Planning D: Society and Space, 36*(3), 494–511. https://doi.org/10.1177/0263775817749594.

Keil, R. (2018b). The empty shell of the planetary: Re-rooting the urban in the experience of the urbanites. *Urban Geography, 39*(10), 1589–602. https://doi.org/10.1080/02723638.2018.1451018.

Keil, R. (2020). After Suburbia: Research and action in the suburban century. *Urban Geography, 41*(1), 1–20. doi.org/10.1080/02723638.2018.1548828.

Kopenawa, D., & Albert, B. (2013). *The falling sky: Words of a Yanomami shaman* (N. Elliott & A. Dundy, Trans.). Belknap Press of Harvard University Press. (Original work published 2010)

Lefebvre, H. (1971). *Everyday life in the modern world* (S. Rabinovitch, Trans.). Harper & Row.

Lefebvre, H. (1976). *De lo Rural a lo Urbano* [From rual to urban]. Lotus Mare.

Lefebvre, H. (1978). *The survival of capitalism: Reproduction of the relations of production*. Allison & Busby.

Lefebvre, H. (1991). *The production of space*. Blackwell.

Lefebvre, H. (1999). *A Revolução Urbana* [The urban revolution] (S. Martins, Trans.). Editora UFMG.

Lefebvre, H. (2008). *Espaço e Política* [Space and politics] (M.M. de Andrade & S. Martins, Trans.). Editora UFMG.

Lefebvre, H. (2014). *Critique of everyday life. The one-volume edition* (J. Moore, Trans.). Verso.

Lipietz, A. (1987). *Mirages and miracles: The crisis in global Fordism*. Verso.

Löwy, M. (2016). *Utopias. Ensaios sobre Política, História e Religião* [Utopias: Essays on politics, history and religion]. Unipop e Ler Devagar.

Mannheim, M. (1972). *Ideologia e Utopia* [Ideology and utopia] (S.M. Santeiro, Trans.). Zahar Editores.

Monte-Mór, R.L. (1988, August). *Urbanization, colonization and the production of regional space in the Brazilian Amazon* [Conference presentation]. 16th Interamerican Congress of Planning (SIAP – Sociedad Interamericana de Planificación), San Juan, Puerto Rico.

Monte-Mór, R.L. (2004). *Modernities in the jungle: Extended urbanization in the Brazilian Amazonia* [Unpublished doctoral dissertation]. University of California, Los Angeles (UCLA).

Monte-Mór, R.L. (2007). Cidade e campo, urbano e rural: o substantivo e o adjetivo [City and countryside, urban and rural: The noun and the adjective]. In S. Feldman & A. Fernandes (Eds.), *O urbano e o regional no Brasil contemporâneo: mutações, tensões, desafios* (pp. 93–114). EDUFBA.

Monte-Mór, R.L. (2014a). Extended urbanization and settlement patterns in Brazil: An environmental approach. In N. Brenner (Ed.), *Implosions/explosions: Towards a study of planetary urbanization* (pp. 109–20). Jovis.

Monte-Mór, R.L. (2014b). What is the urban in the contemporary world? In N. Brenner (Ed.), *Implosions/explosions: Towards a study of planetary urbanization* (pp. 260–7). Jovis.

Monte-Mór, R.L. (2018). Urbanisation, sustainability, development: Contemporary complexities and diversities in the production of urban space. In P. Horn, P. Alfaro d'Alençon, & A.C. Cardoso (Eds.), *Emerging urban spaces: A planetary perspective* (pp. 201–15). Springer eBook.

Monte-Mór, R.L., & Castriota, R. (2018). Extended urbanization: Implications for urban and regional theory. In A. Paasi, J. Harrison, & M. Jones (Eds.), *Handbook on the geographies of regions and territories* (pp. 332–45). Edward Elgar.

Monte-Mór, R.L., & Ray, S. (1995). Post-*.ism & the third world: A theoretical reassessment and fragments from Brazil and India. *Nova Economia 5*(1), 177–208.

More, T. (2017). *Utopia* (M.J. Lourenço, Trans.).Clube do Autor. (Original work published in 1516)

Münster, A. (1993). *Ernst Bloch. Filosofia da práxis e utopia concreta* [Ernst Bloch. Philosophy of praxis and concrete utopia]. Editora UNESP

Ollman, Bertell. (1976). *Alienation: Marx's concept of man in capitalist society.* Cambridge University Press.

Polanyi, K. (2011). *A grande transformação* [The great transformation] (F. Wrobel, Trans.). Elsevier.

Polanyi, K. (2012). *A subsistência do homem e ensaios correlatos* [The livelihood of man] (V. Ribeiro, Trans.). Contraponto.

Santos, M. (1979). *O espaço dividido: os dois circuitos da economia urbana dos países subdesenvolvidos* [Divided space: The two circuits of urban economy in underdeveloped countries]. Francisco Alves.

Schmid, C. (2018). Journeys through planetary urbanization: Decentering perspectives on the urban. *Environment and Planning D: Society and Space, 36*(3), 591–610. https://doi.org/10.1177/0263775818765476.

Serres, M. (1991). *O Contrato Natural* [The natural contract] (E. MacArthur, Trans.). Ed. Nova Fronteira.

Simone, A., & Pieterse, E. (2018). *New urban worlds: Inhabiting dissonant times.* Polity Press.

Soja, E.W. (1996). *Thirdspace: Journeys to Los Angeles and other real-and-imagined places.* Blackwell.

United Nations Human Settlements Programme. (2007). *State of the world's cities 2006/7.* Earthscan.

Image 6. Montpellier, France (October 2012)

6 A Dose of Density: The Urban Counter-Revolution

RICHARD HARRIS

The dose makes the poison.

– Paracelsus

Over the past generation or so, "northern" urban theory has faced two major geographical challenges (see Connell, 2007). The first has come from what we now sometimes call the Global South. There, urbanization – embracing the growth of urban areas and a rising proportion of people living in urban areas – has proceeded on an unprecedented scale and at a bewildering pace. This development has raised questions about the causes, character, and consequences of the urban trend. The second challenge has come from the urban periphery, everywhere: the explosive and discontinuous sprawl of fringe development across the landscape. To an increasing number of observers, sprawl has raised questions of identity, not only about what may be called "the suburbs" but also about the city itself. As Keil (2017) has observed, "much of the urban age is, at closer inspection, rather a suburban age" (p. 4). He is right, and this chapter tries to make sense of that fact.

The suburban challenge can be met in three ways. The first and by far the most common is to duck the issue – to proceed as if nothing had happened, or to acknowledge the fact but avoid definitions and barrel on regardless. I will say no more about this. Drawing above all on the ideas of Henri Lefebvre, the second is to argue that the continuing spread of urbanization has dissolved boundaries between city, suburb, and indeed the regions beyond, so that sharp conceptual boundaries have become unhelpful. Urbanization, then, is seen to be – or at least to be becoming – present in some form everywhere. Along these lines, for example, Keil (2017) argues that "as Lefebvre's revolution turns, the suburbanization of the city region takes over the planet" (p. 12). The third response is to argue that what Lefebvre (2003) called the "urban revolution" is not on the horizon and that it is still meaningful to make qualitative distinctions between types of settlements, although these are more difficult to identify because geographical boundaries have become so blurred. Through my involvement in a multi-year research project directed by Roger Keil, I have come to believe that such a response is defensible. I was responsible for coordinating research on the subject of land, one of the three main project themes. Among other things, this required me to read widely about developments in all parts of the world. With Ute Lehrer, I have indicated some elements of the results elsewhere (Harris & Lehrer, 2018), but I now welcome the opportunity to outline a more complete case.

The point of departure has to be with a critique of Lefebvre's claims. His argument about the urban revolution has recently been sustained by a number of writers, notably Brenner, Schmid, and Monte-Mór (Brenner, 2014; Brenner & Schmid, 2015; Monte-Mór, 2014; Schmid, 2016). In part, they have done so through a challenge to the "urban age" account that speaks of the city's exceptional qualities (Brenner & Schmid, 2014a). More generally, Lefebvre is viewed as a touchstone or logical extreme by many others (e.g., Iossifova et al., 2018; Merrifield, 2013). My argument in the first section is that this line of thinking is unpersuasive because it is in part illogical, contains an unresolved tension, and is incomplete. Instead, referring to diverse and often unrelated literatures, I suggest in the second section that what we normally refer to as urban places are indeed distinctive, both in terms of their character and also their effects. This is an old argument, the elements of which I survey in a recent work (Harris, 2021). Their qualitative difference is highlighted by the unique character of urban land: beyond a certain point, more density – a larger dose – changes the impact. In this context, it becomes possible to argue that the urban periphery, which for convenience I will in the third section refer to as "the suburbs," also has a distinctive character, again in terms of land. The peculiar qualities and significance of cities and suburbs define an agenda for counter-revolutionary work.

A Critique of Revolutionary Theory

As many writers have observed, Henri Lefebvre was a prolific writer who did little empirical research but who had a fertile mind and was capable – like William Blake – of cheerfully contradicting himself (Merrifield, 2002, 2013). Various ideas of Lefebvre have been picked up by others, and the present discussion focuses on the strand of his thinking that he developed most fully in *The Urban Revolution* (2003). It has two main elements, related but separable. The first is the claim that, as he states in the first sentence of the book, "society has been completely urbanized" (p. 1). The second, which he often implied but which has been made more explicit by some of his interpreters, is that planetary urbanization is now, increasingly, shaping our lives. As Brenner and Schmid (2014a,) put it, "urbanization is a process that affects the entire world, and not just isolated parts of it ... there is ... no longer any outside to the urban world" (p. 333).

Let us take the second claim first. It must be said that on this point Lefebvre (2003) is inconsistent. At the outset, he identifies urbanization as a significant but secondary factor in shaping our lives. He defines "urban society" as "that [which] results from industrialization, which is a process that absorbs agricultural production" (p. 2). He then notes that "urban reality modifies the relations of production without being sufficient to transform them" (p. 15). But then, looking past the present to an emergent future, he asserts that "the urban problematic becomes predominant" (p. 5), and it is this statement that becomes his springboard: references to industrialization fall away, and discussion shifts to the character and effects of "urban reality." He says little or nothing about suburbs – which of course had different connotations in France than in Anglo-America – except to comment that they are "growths of dubious value" (p. 4).

It was the overall thrust of Lefebvre's argument, rather than his opening statements, that immediately attracted criticism from an ex-student. Responding also to Lefebvre's *Le droit à la ville* (1968), Castells (1977, pp. 86–95) argued that his work gave too much weight to the effects of urbanism, thereby deviating from a defensible position, Marxist or otherwise (see also Merrifield, 2002, pp. 91, 117). Although he has emphasized his debt to Lefebvre, Harvey has consistently articulated a similar view. In early commentary, Harvey (1973) regarded Lefebvre's claim that the urban phenomenon now dominated industrial society was "a hypothesis not a proof" (p. 311). Then, in later work, he elaborated an argument that the driving force is capitalism: modern urbanization is above all "a built environment ... supportive of capitalist production, consumption and exchange" (Harvey, 1985, p. 192).

It would be possible to argue that this is what Lefebvre really meant, but this would be a stretch. Certainly, it is not how he has generally been read and interpreted. In the writings of Brenner and Schmid (2014), for example, there are only generalized references to capitalism or industrial society; the emphasis is on the causal efficacy of urbanization, and of an "urban process." This begs the question as to what causes urbanization and, most importantly, what other forces are shaping our lives.

It is unreasonable to expect extended treatments of either of those very large questions, but without clues and elaborations – such as those that Harvey (1973, 1985), for example, has provided – it is impossible to say how much the urban revolution matters, assuming of course that it exists.

This brings us to the first and most basic claim: that the world has become, or is rapidly becoming, urban. This points to an incontrovertible fact. As observers of the American scene have commented for more than half a century, urban areas have sprawled while the densities of central-cities have generally declined. At the extreme, in Detroit, densities may even be higher in some suburbs than in the central city. The same is true, for different reasons, in a few other places, including urban centres in what was the Soviet bloc, where high rises were erected near the urban fringe (Keil, 2015). Then, too, even in prospering cities that have seen redevelopment, a decline in household size has worked to depress residential densities. As a result, as Shlomo Angel's co-authored chapter in this collection shows, even in cities that have seen new developments, residential densities have not changed much. And so, except in terms of political jurisdictions, the distinctions between city and suburb have become blurred; urban development surrounding hitherto separate cities has coalesced; new Edge City subcentres have sprung up. In a vivid phrase, Friedmann (2002) suggests that the city has no more reality any more than "the smile of the Cheshire cat" (p. xii). But although some of the historical associations of "city" and "suburb" are no longer valid, and despite (or perhaps because of) numerous competing suggestions, no alternative language has won a consensus (see also Phelps et al., 2010). On those grounds, certainly, it is reasonable to raise questions about whether "suburb" has much useful meaning, a point of view that is probably shared by many writers, including those who do not draw on Lefebvre. Lewis (2017), for example, has suggested that "suburb" may now be "chaotic concept" (p. 142).

But Lefebvrians go further, both geographically and conceptually, challenging the idea that only particular places should be called "urban." Their core argument hinges on two related propositions. The first is that the distinction between rural areas and all types of urban territory – city or suburban – has almost dissolved, so that everywhere is woven into "the urban fabric." Lefebvre (2003) suggests that "this expression, 'urban fabric,' does not narrowly define the built world of cities but all manifestations of the dominance of the city over the country" (pp. 1–2), thereby obliterating distinctions; as Merrifield (2014) says, it "stretches to envelop everywhere" (p. x). To be sure, densities vary, but these have ceased to be useful in differentiating one type of place from another. As Monte-Mór (2014) observes, speaking about a remote region in the Brazilian Amazon, "urbanization processes differ only in degrees of intensity relative to trends in more centrally-located Brazilian cities and in metropolitan areas elsewhere in the industrialized world" (p. 112). There is a continuum, and so Brenner and Schmid (2014b) challenge the "urban age" assumption that there is a "type of territory that [is] qualitatively specific" (p. 161). They conclude that, although there may be differences between places it is "no longer plausible" to distinguish urban from rural (Brenner & Schmid, 2013, p. 162).

The second proposition is that what are usually spoken of as urban and rural places are now intimately, indissolubly related, and on a global scale. It is, of course, a cliché that all urban places are now part of a "global urban network" (Friedmann, 2002, p. xv). They are also tied more intimately than ever to less-dense places. As Brenner (2014, p. 20) and Schmid (2016, p. 31) both note, cities have always depended on hinterlands. Lately, however, the spread of commercial agriculture and of tourism into ever more remote locations, including the Amazon basin, means that the corollary is true: almost all rural areas now have ties to urban places. Accordingly, they believe that lately the ante has been upped so far that the game itself has changed, noting that "even spaces that lie well beyond the traditional city cores and suburban peripheries ... have become integral parts of the worldwide *urban fabric*" [emphasis added] (Brenner & Schmid 2014b, p. 162).

The premise of the second proposition is correct, but the inference is illogical. Places can be interdependent and yet distinct. The early twentieth-century central business district and its surrounding residential districts were utterly dependent on one another, but they were not the same in kind. Likewise, Sao Paulo and Amazonia, Zurich and the Alpine ski resorts. It is possible to be mutually dependent and yet fundamentally different.

The first proposition, that differences are at most matters of degree, coexists with an acknowledged fact that sets up an unresolved tension. Lefebvre and his interpreters observe that cities, as that term is normally understood, not only exist but are growing and remain significant. As Brenner and Schmid (2014b) note, "the process of agglomeration remains essential" (p. 162; see also Brenner & Schmid, 2015, p. 170); more substantially, Monte-Mór (2014) observes that "of course there are important differences among the regional examples [in Brazil] – among other elements, in the intensity of market relations, in the character of urban and territorial networks, and in the quality of social and natural space" (p. 114); and Brenner (2014) observes that "population density, inter-firm clustering, agglomeration effects or infrastructural concentration" (p. 19) still matter. But for agglomeration to be "essential," it must have some particular attraction, serving some unique functions, whether for the city's residents or for the companies and institutions that locate there. And, if it has its own essential qualities, then that implies it is a distinctive sort of place, meriting its own label. How can this be reconciled with the claim that it is no longer useful to make a distinction between the city and what lies beyond? To my knowledge, neither Lefebvre nor his interpreters have addressed this question.

To answer it, we need to pay close attention to the purposes, calculations, and actions of people, businesses, and governments. Lefebvre and his interpreters have not done that, empirically or conceptually; instead, they have focused on the built environment. Schmid (2016) acknowledges that there are "social modalities" of urbanization (p. 38), but, with Brenner, he prioritizes studying "the various processes through which urban configurations are produced, contested and transformed" (Brenner & Schmid, 2015, p. 165). "Configuration" implies a physical form, and this is confirmed by their suggestion that, although urbanization is "linked to" capital expansion and globalization, "its specificity lies precisely in materializing the latter" (Brenner & Schmid, 2015, p. 172). It follows that the way to substantiate claims about the ubiquity of the urban process is to trace the penetration of "settlements, transport routes ... pipelines and high voltage lines, logistics facilities, and many other elements" – into once-rural areas around the world (Schmid, 2016, p. 29). That is what Monte-Mór in Brazil and what Schmid and his associates at the ETH Studio in Basel have done, with impressive visual results (Diener, 2015; ETH Studio Basel, 2016; Monte-Mór, 2014). They have shown that infrastructure has penetrated very widely. But there are, nonetheless, marked concentrations, which exist for particular reasons and with distinctive effects. To interpret the underlying logic of concentration and spread – what Lefebvre called implosion/explosion – we must also examine those reasons. When we do so, the familiar features of an urban counter-revolution reappear, "counter" in the sense that it is a reaction to the arguments that Lefebvre styled as revolutionary.

An Urban Counter-Revolution

Many of those who write about cities have no interest in identifying a specifically "urban" aspect to their subject (see Brenner & Schmid, 2014a, p. 323). Many others assume, or have some idea, that urban areas are distinctive sorts of places, but they do not aim to disentangle their elements, or to make them explicit. Tacit treatments are common in at least one of the subfields that I know best, urban history (Harris, 2021; Jensen, 1996; Kwak, 2017; Lewis, 2017; see also Clark, 2013). But over the decades, in all of the major urban disciplines, "urban age" writers have articulated powerful arguments that density and agglomeration matter. In so doing, they illustrate Granovetter's (1978) argument that, crossing thresholds, quantitative change can have qualitative effects (see also Gladwell, 2000). This well-established argument has continuing relevance in the era of sprawl, and implies the usefulness of paying attention to the character of urban land (see Scott, 1980).

There is wide agreement that a major function of cities, and a major raison d'être, is economic. Many writers have shown that the concentration of businesses and labour promotes business efficiency and innovation (Glaeser, 2011; Jacobs, 1970; Lampard, 1955; Scott, 2017; Scott & Storper, 2015; Storper, 2013). Concentration enables specialization among firms and workers alike, a process that depends on tight interlinkages and face-to-face communication. Urban concentration was key to the rise of manufacturing, its elaboration into corporate ownership and management, and remains no less important today. It is a truism that agglomeration has supported the specialization of consumer businesses, including niche restaurants, clothing stores, and personal services. For even longer, around ports and lately airport hubs, it has articulated regional and long-distance trade (Braudel, 1981–4). This has supported specialization, efficiency, innovation, and growth at still larger scales. The social side-effects of urbanization have not always been beneficial, but its economic significance remains huge.

The concentration of people and firms in urban centres poses extreme problems of governance that are unique. Many of these arise from the distinctive character of urban land, which derives its value from its location, not its intrinsic quality (Harris & Lehrer, 2018; Scott, 1980, 2017). Despite sprawl and polycentric development, land value gradients indicate that the greater physical accessibility of urban nodes still matters. Even so, at the urban periphery in regions where land use is regulated, sites zoned for urban use are worth many times more than those devoted to agriculture. That is why the owners of urban sites care far more about the precise delineation of property boundaries. Then, too, because cities are inherently dynamic environments, the pressures for land use change are frequent and strong. Accordingly, there are large incentives to ensure that land becomes a fungible commodity whose ownership is individualized, unambiguous, and readily transferable. What we now call urban "real estate" supports the markets in labour and other commodities, not least by modelling the character and value of market ways of thinking. In that sense, it can be argued that cities are the natural home base of capitalism.

The distinctiveness of urban land does not end there. Those who put it to use generate externalities – uncompensated costs and benefits – whose effects are felt most strongly by immediate neighbours in the form of congestion, noise, air pollution, fire, and other health risks. According to Scott and Storper (2015), these effects are sufficiently strong as to make urban land "simultaneously private and public" (p. 8). This can mean various things in widely different historical and cultural contexts, and certainly researchers need to be alert to this (see Robinson, 2015). But concentrated externalities demand, and have received, exceptional levels of regulation, whether of construction or use. For that reason, and although it may seem counter-intuitive, this means that the *potential* incidence of informal development – understood as activity that violates government regulation – is higher in cities than in rural areas (Harris, 2018a). To manage urban congestion, as well as the needs for concentrated power, water provision, and waste disposal, cities also demand a relatively sophisticated infrastructure of pipes, wiring, trucks, and roads. In sum, the character of urban land defines many of the functions of municipal government in a context that favours capitalist development.

The social dimensions of urban concentration are also significant. Those who have examined this issue still often take the arguments of the American sociologist Louis Wirth (1938) as a point of departure. Unfortunately, these are often simplified. True, Wirth argued that city life offered personal freedom and anonymity, but he also suggested that it offered the incentive and the opportunity for residents to create new forms of community. This is obviously true for migrants and immigrants, or for minority groups such as the LGBTQ+ or BIPOC communities, who congregate in order to build social support networks. It is no less true for unions, masonic orders, business associations, and the myriad other interest groups that cities support. Arguments about the way cities foster community have been elaborated and substantiated by many researchers, notably Fischer (1976, 1995; see also Ioannides, 2013). Significantly, research suggests that the advent of the internet has not displaced face-to-face contact as the basis for maintaining friendships, as well as social and business networks (Mok et al., 2010; see also Storper, 2013). Even in the age of Facebook, the social life of the city is still not merely richer but qualitatively different from that of the village or rural district.

These economic activities, governance practices, and social communities, working through and upon the urban land market, combine to create an urban culture. To some extent each city has its own, but some elements are shared by all cities in a particular nation so that it becomes possible to speak about the urban culture, for example, of American cities. In such terms, Barth (1980) discusses the role of city newspapers, professional sports teams, and department stores in the nineteenth century, while, more recently, Gilfoyle (2016) has reviewed some twentieth-century examples. It is also meaningful to speak about the distinctive role of cities in shaping the national culture, as Schlesinger Jr.'s work has shown for the history of the United States (see, for example, Schlesinger, 1973). Although many writers have suggested that American cities, and culture, have been especially innovative – of "new ideas, new art, new tools and products, new manners and morals" (Warner, 1971, p. 5) – cities in general have served as centres of cultural change. And so writers such as Mumford (1961) have argued that cities have shaped world history. There is room for debate about how far such interpretations can be pushed. But, insofar as they rest on well-established facts and plausible arguments about the continuing significance of urban living for the economy, social life, and politics, they must be taken seriously.

In society as in medicine, quantities matter. A complex, built-up urban place is not just a denser built environment than a distant district with, say, "a second home, a highway, [and] a supermarket in the middle of the countryside" (Lefebvre, 2003, p. 17); it is a qualitatively different sort of place, with a unique mix of economic possibilities, social opportunities, and political challenges. But, even if we accept that argument, where does that leave the suburbs, those "growths of dubious value"?

A Suburban Counter-Revolution

To answer this question, we must clarify what "suburb" means. Of course, there are many nations and cities where this word (or a synonym) is not used, except perhaps by urban experts; other places have no single word for, or even a concept of, the urban periphery (Harris & Vorms, 2017). Even where "suburb" is employed, its meaning can vary, and carry connotations – stereotypes – that are no longer relevant (Harris, 2018b). For present purposes, "suburb" refers to the urban periphery, that ill-defined zone lying at or near the edge of a built-up area. Thin or thick, suburban territory may surround an urban centre on all sides, or not; it may be roughly circular – the outer ring in a donut – but more often not. Regardless of shape, it includes fully built districts, residential and otherwise, that shade in the outer fringes into partially developed transitional territory that is often known as exurban or peri-urban.

Elsewhere, with Ute Lehrer, I have argued that the nature of this zone arises from transitional character of suburban land (Harris & Lehrer, 2018). It is transitional geographically, lying between the city and what lies beyond, whether it be farms, desert, or mountains. Typically, although not always, it is intermediate in density, and almost invariably so in terms of accessibility: compared with the city, it is deficient in transit, compelling people to rely on individual modes of transportation, commonly the automobile. Because of simple topographic geometry, even high-density suburbs and edge cities are less accessible to other metropolitan residents than is the main urban centre. And, because suburbanites rely on cars, while much suburban land is devoted to uses that also serve the city (highways, rail lines, high-voltage lines, freight terminals, airports), a relatively high proportion of suburban land is devoted to infrastructure.

Suburban land is also transitional historically, an aspect which is often neglected, as Ilja Van Damme and Stijn Oosterlynck discuss elsewhere in this volume. Over decades, suburban land use changes and intensifies. Of course, that is also true of centrally located sites, but suburbs are transitional in distinctive ways. The shift from rural to urban uses creates particular juxtapositions: fields, copses, and piggeries intermingle with junk yards, subdivisions, and quarries. As land is converted to urban use its value balloons – often by a factor of ten or more – encouraging speculative investment. In time, as the urban area expands, fringe districts become relatively more central, and absolutely more valuable, so that vacant parcels are filled in and the pressures for redevelopment grow, along with the probability that public transit will be provided. In time, but at no particular moment, suburban territory becomes urban.

It may be objected that this account assumes that urban growth involves accretion from existing nodes, whereas much recent development appears to follow a metro-wide or regional pattern that is multi- or decentred. There is some truth to this. But the research of Angel (2012) has shown that annular growth is still the norm, albeit in irregular forms and on an ever-widening scale. It also shows that, in general, densities do decline away from centres. It is more difficult than ever to identify suburban territory, but the effort is still meaningful.

It can also be argued that suburbs are transitional in cultural terms. Here, especially, stereotypes can mislead. Some suburbs are residential and middle class, consisting of owner-occupied, single-family homes, but such places were only ever one type of suburb and they do not define suburban culture. Walks (2015) has argued that suburban reliance on the automobile, an individualistic mode, helps shape a distinctive outlook which may be characterized as conservative, in current terms neoliberal or populist. This subculture is complicated in exurban and peri-urban territory, which juxtaposes residents whose lives have different daily and seasonal rhythms. Especially in Africa and much of Asia, it also counterposes two or more forms of land tenure, legal systems, and values (Harris & Lehrer, 2018). In so doing, because land is vital in all societies, it can bring whole cultures into direct contact, whether in negotiation or conflict. As capitalism expands and strengthens the private land market, the suburbs are the thin end of a very large cultural wedge.

Conclusion

It may be that the difference between the revolutionary and counter-revolutionary positions is more apparent than real. Both acknowledge that almost all parts of the world are connected, that some parts are more densely occupied than others and that, at least for some purposes, agglomeration matters. Perhaps the disagreement is one of emphasis, or even semantics. Many Lefebvrians acknowledge that clusters of people and economic activity are different, but not different enough, it seems, to deserve their own label. Given that the boundaries of urban areas are fuzzy, and that any lines we draw must be somewhat arbitrary, part of the debate is about whether we need more than one word to describe differences across space.

But more seems to be at stake. Lefebvrians commonly sound like urban homeopaths, claiming that tiny amounts can have significant effects, whereas "urban agers" believe that the dose matters. Size and density build up to thresholds beyond which the quality of social life changes, economic possibilities are enlarged, and new challenges of governance arise. The argument is not simple. McFarlane (2016), for example, has emphasized how complex and context-dependent the causes and consequences of density can be. But, the associated thresholds combine to define cities, shape nations, and now the world.

To be sure, putting numbers on these thresholds is a huge challenge. Many are intrinsically hard to quantify, and all vary by time and place. Everywhere, they are multiple. As they pile up, they create a sort of irregular stairway to … no, not heaven but, as Beauregard (2018) has lately insisted, a distinctive sort of place with its own potentialities and problems. And there is another type of challenge, that of keeping up with what "we" know. So much is now published that to survey the research even in a specific subfield – urban history, sociology, economics, political science, geography, design, ecology, and so on – is a huge task, and keeping abreast of all seems impossible. But the effort is surely worthwhile if we want to understand the world we live in.

If that is true then suburbs, those "in-between" spaces, still have a distinctive, if fuzzy and fragile, identity. In some ways simply the city thinned out, in other respects they are unique in having their own transitional characteristics, geographically, historically, and culturally. These can be elusive. The widening scatter of fringe growth, and the varied forms of suburban development across the major world regions, make it difficult to see commonalities. But if, as I have argued, cities are still their own sorts of places, connected with much less densely settled areas but creating their own force fields, then the transitional spaces called suburbs must have their own peculiar character. The suburbs are dead; long live the suburbs!

ACKNOWLEDGMENTS

I would like to thank Neil Brenner for comments on an earlier draft. He responded to this critique in an exceptionally constructive way and we found areas of agreement, but it is a fair bet that he would not agree with everything written here. I would also like to thank Roger Keil, Fulong Wu, Bob Beauregard, and Rebecca Madgin for their thoughtful comments.

REFERENCES

Angel, S. (2012). *Planet of cities.* Lincoln Institute of Land Policy.
Barth, G. (1980). *City people: The rise of modern city culture in nineteenth-century America.* Oxford University Press.
Beauregard, R.A. (2018). *Cities in the urban age.* University of Chicago Press.
Braudel, F. (1981–4). *Civilization and capitalism, 15th–18th century* (3 Vols.). (S. Reynolds, Trans.). Harper and Row.
Brenner, N. (2014). Introduction: Urban theory without an outside. In N. Brenner (Ed.), *Implosions/explosions: Towards a study of planetary urbanization* (pp. 14–30). Jovis.
Brenner, N., & Schmid, C. (2014a). The "Urban Age" in question. In N. Brenner (Ed.), *Implosions/explosions: Towards a study of planetary urbanization* (pp. 310–37). Jovis.
Brenner, N., & Schmid, C. (2014b). Planetary urbanization. In N. Brenner (Ed.), *Implosions/explosions: Towards a study of planetary urbanization* (pp. 160–3). Jovis.

Brenner, N., & Schmid, C. (2015). Towards a new epistemology of the urban. *City, 19*(2–3), 151–87. https://doi.org/10.1080/13604813.2015.101471.

Castells, M. (1977). *The urban question.* Arnold.

Clark, P. (Ed.). (2013). *The oxford handbook of cities in world history.* Oxford University Press.

Connell, R. (2007). The "northern" theory of globalization. *Sociological Theory, 25*(4), 368–85. https://doi.org/10.1111/j.1467-9558.2007.00314.x.

Diener, R. (Ed) (2015). *The inevitable specificity of cities.* Lars Müller.

ETH Studio Basel. (2016). *Territory: On the development of landscape and city.* Park.

Fischer, C. (1976). *The urban experience.* Harcourt Brace Jovanovich.

Fischer, C. (1995). The subcultural theory of urbanism: A twentieth-year assessment. *American Journal of Sociology, 101*(3), 543–77.

Friedmann, J. (2002). *The prospect of cities.* University of Minnesota Press.

Gilfoyle, T. (2016). Michael Katz on place and space in urban history. *Journal of Urban History, 41*(4), 572–84. https://doi.org/10.1177/0096144215579381.

Gladwell, M. (2000). *The tipping point: How little things can make a big difference.* Little Brown.

Glaeser, E. (2011). *Triumph of the city: How our greatest invention makes us richer, smarter, greener, healthier and happier.* Penguin.

Granovetter, M. (1978). Threshold models of collective behavior. *American Journal of Sociology, 83*(6), 1420–43.

Harris, R. (2018a). Modes of informal urban development: A global phenomenon. *Journal of Planning Literature, 33*(3), 267–86. https://doi.org/10.1177/0885412217737340.

Harris, R. (2018b). Suburban stereotypes. In B. Hanlon & T. Vicino (Eds.), *The Routledge companion to the suburbs* (pp. 29–38). Routledge.

Harris, R. (2021). *How cities matter.* Cambridge University Press.

Harris, R., & Lehrer, U. (Eds.) (2018). *The suburban land question: A global survey.* University of Toronto Press.

Harris, R., & Vorms, C. (Eds.). (2017). *What's in a name? Talking about urban peripheries.* University of Toronto Press.

Harvey, D. (1973). *Social justice and the city.* Arnold.

Harvey, D. (1985). *The urbanization of capital: Studies in the history and theory of capitalist development.* Johns Hopkins University Press.

Ioannides, Y. (2013). *From neighborhoods to nations: The economics of social interactions.* Princeton University Press.

Iossifova, D., Doll, C.N.H., & Gasparatos, A. (Eds.). (2017). *Defining the urban: Interdisciplinary and professional perspectives.* Routledge.

Jacobs, J. (1970). *The economy of cities.* Vintage.

Jensen, H.S. (1996). Wrestling with the angel: Problems of definition in urban historiography. *Urban History, 23*(3), 277–99.

Keil, R. (2015). Towers in the park, bungalows in the garden: Peripheral densities, metropolitan scales and the political cultures of post-suburbia. *Built Environment, 41*(4), 579–96. https://doi.org/10.2148/benv.41.4.579.

Keil, R. (2017). *Suburban planet: Making the world urban from the outside in.* Polity.

Kwak, N. (2017). History: Understanding "urban" from the disciplinary viewpoint of history. In D. Iossifova, C.N.H. Doll, & A. Gasparatos (Eds.), *Defining the urban: Interdisciplinary and professional perspectives* (pp. 53–62). Routledge.

Lampard, E.E. (1955). The history of cities in the economically advanced areas. *Economic Development and Cultural Change, 3*(2), 81–136. https://doi.org/10.1086/449680.

Lefebvre, H. (1968). *Le droit à la ville* [Right to the city]. Anthropos.

Lefebvre, H. (2003). *The urban revolution* (R. Bononno, Trans.). University of Minnesota Press.

Lewis, R. (2017). Comments on urban agency: Relational space and Intentionality. *Urban History, 44*(1), 137–44. https://doi.org/10.1017/S096392681600033X.

McFarlane, C. (2016). The geographies of urban density: Topology, politics and the city. *Progress in Human Geography, 40*(5), 629–48. https://doi.org/10.1177/0309132515608694.

Merrifield, A. (2002). *Metromarxism: A Marxist tale of the city.* Routledge.

Merrifield, A. (2013). *The politics of the encounter: Urban theory and protest under planetary urbanization.* University of Georgia Press.

Merrifield, A. (2014). *The new urban question.* Pluto.

Mok, D., Wellman, B., & Carrasca, J. (2010). Does distance matter in the age of the internet?" *Urban Studies, 47*(13), 2747–83. https://doi.org/10.1177/0042098010377363.

Monte-Mór, R.L. (2014). Extended urbanization and settlement patterns in Brazil: An environmental approach. In N. Brenner (Ed.), *Implosions/explosions. Towards a study of planetary urbanization* (pp. 109–22). Jovis.

Mumford, L. (1961). *The city in history: Its origins, its transformations, and its prospects.* Harcourt, Brace and World.

Phelps, N.A., Wood, A.M., & Valler, D.C. (2010). A post-suburban world? An outline of a research agenda. *Environment and Planning A, 42*(2), 366–83. https://doi.org/10.1068/a427.

Robinson, J. (2015). Comparative urbanism: New geographies and cultures of theorizing the urban. *International Journal of Urban and Regional Research, 40*(1), 187–99. https://doi.org/10.1111/1468-2427.12273.

Schlesinger, A. (1973). The city in American civilization. In J.B. Callow (Ed.), *American urban history: An interpretative reader with commentaries* (pp. 35–51). Oxford University Press.

Schmid, C. (2016). The urbanization of the territory. On the research approach of ETH Studio Basel. In ETH Studio Basel (Ed.), *Territory: On the development of landscape and city*. Park.

Scott, A. (1980). *The urban land nexus and the state*. Pion.

Scott, A. (2017). *The constitution of the city: Economy, society, and urbanization in the capitalist era*. Palgrave Macmillan.

Scott, A., & Storper, M. (2015). The nature of cities: The scope and limits of urban theory. *International Journal of Urban and Regional Research, 39*(1), 1–15. https://doi.org/10.1111/1468-2427.12134.

Storper, M. (2013). *Keys to the city: How economics, institutions, social interactions, and politics shape development*. Princeton University Press.

Walks, A. (Ed.) (2015). *The urban political economy and ecology of automobility: Driving cities, driving inequality, driving politics*. Routledge.

Warner, S.B. (1971). *The urban wilderness. A history of the American city*. Harper & Row.

Wirth, L. (1938). Urbanism as a way of life. *American Journal of Sociology, 44*(1), 1–24.

Land | Infrastructure | Governance

Image 7. Greater Montreal, Quebec, Canada (Summer 2012)

7 Cities in a World of Villages: Agrarian Urbanism and the Making of India's Urbanizing Frontiers

SHUBHRA GURURANI

Introduction

In 1985, the National Capital Region Planning Board Act was passed, and the territory of India's national capital of New Delhi was extended from 1,484 square kilometres to 33,578 square kilometres to create a National Capital Region (NCR). The NCR, which has since been further extended to 55,083 square kilometres, incorporated large swathes of agrarian and pastoral hinterlands in the neighbouring states of Haryana, Uttar Pradesh, and Rajasthan to make room for the urban transformation to come.[1] In anticipation of the new economic policy, which was formally adopted in 1991, urbanization of agrarian lands was identified as one of the key strategies to bolster economic growth and attract foreign direct investment. To accommodate corporate head offices, special economic zones, infrastructural corridors, and housing for a growing urban population, peripheries of all metropolitan cities like Mumbai, Kolkata, Chennai, and Bangalore underwent significant land assetization and de-agrarianization in the 1980s and 1990s. To this end, eminent domain laws were creatively interpreted and several land acquisition rules and land-use practices manipulated to urbanize rural/agrarian land (Gururani, 2013).[2] The acquisition, conversion, and assetization of rural land unleashed an unprecedented land rush and marked a critical turning point in the local and regional political economy of land and land-based power geometries.

In this conjuncture, Gurgaon,[3] an agro-pastoral stretch in the southwestern edge of New Delhi, was one of the first urban peripheries that embraced the dreams and possibilities of a new urbanizing India. In a matter of a few decades, the land-use of hundreds of acres of rural land was changed, and from a cluster of villages, Gurgaon emerged as India's "Millennial City." Due, in part, to its proximity to New

Delhi's international airport, Gurgaon soon became a prime destination for multinational companies that include Google, HSBC, Nokia, Intel, and many more. While the entire Gurgaon district witnessed a significant increase in population, up by almost 75 per cent between 1991 and 2001, the population of the new city increased almost twentyfold. Meanwhile, the area used for agriculture dropped from 80 per cent to 26 per cent of the total land, and the built-up area increased from 9 per cent to 66 per cent (Government of India, 2001).[4] Today, Gurgaon, with its name officially changed to Gurugram, is a bustling city with dozens of shopping malls, golf courses, corporate head offices, and countless gated-housing enclaves that accommodate India's burgeoning middle and upper classes.

As hundreds of acres of agricultural land were acquired, the social, political, and legal boundaries that separated the urban from the rural came to be negotiated and recast in multiple ways. In a country where, according to official statistics, 70 per cent of its population is rural, the reconfiguration of rural and agrarian spaces raises a range of questions and urges us to once again turn our attention to the familiar question of the rural and the urban, the city and its countryside. For example, what do the socio-spatial categories of rural and urban mean in this moment of extended urbanization? How do they intersect with each other, and what can their entanglements tell us about heterogeneous sub/urban forms and processes that the changing relations of urban and rural produce?

In this chapter,[5] I focus on Gurgaon not only to consider suburbanization in India alongside other parts of Asia but also to comparatively consider the landscape of peripheral urbanization or suburbanization and suburbanisms globally in order to take stock after a decade of research in the context of the Global Suburbanisms project, in which I have been a core researcher. By focusing precisely on sedimented histories of land and everyday practices of work, livelihood, mobility, exclusion, and access that make the urban peripheries in India, I work with and against the conceptual categories that have tended to populate the annals of urban theory. Instead of seeing rural, urban, city, or suburb as a thing or a stable entity that is configured by regimes of governance, the goal of the Global Suburbanisms project has been to map how diverse and complex processes of socio-spatial production and reproduction shape the contours of sub/urbanization in different locations (Keil, 2018). It is in this spirit that I turn to the making of Gurgaon and locate its conceptual and ethnographic anchors not in the city, but in its peripheries – peripheries that are simultaneously entangled with agrarian regimes of land and of global capitalism, and that urge us to attend to the enduring colonial and developmentalist legacies and how they have, in this moment of capitalist urbanization, produced a highly fragmented and uneven social-spatial landscape of class, caste, and power.

Straddling the rural–urban divide, the urban peripheries of most metropolises in India are dotted with villages. These social-spatial constellations, often referred to as peri-urban or transitional areas or urban villages, are typically characterized by unpaved roads; self-built houses; and poor water, sewage, and electrical connections. Often hidden behind glassy towers and shopping malls, these villages at first glance appear incommensurate with the dominant regimes of urban planning, aesthetics, and housing, and even considered "slum-like." But these villages are neither slums nor informal settlements that quietly encroach on the city, even if they operate like them, nor are they part of a "completely urbanized" fabric (Lefebvre, 2003, p. 3). Instead, they register an uneven geography of spatial value that is produced out of interconnected processes of agrarian change, colonial policies of land, flexible planning, entrenched caste-land politics, and global capitalism. With dense populations, these villages have come to support and sustain the city in significant ways. Not only does the "city" stand on their land, but in the absence of public housing, these villages have come to provide rental housing for hundreds of thousands of migrant workers who come to work and live in the new city. They also support countless small manufacturing, repair, and infrastructural services that metaphorically and materially make the new city possible. The villages are part and parcel of a metropolitan system in which the city and villages mutually sustain each other in multiple ways and make the frontiers of agrarian urbanism.

In what follows, in tracking the trajectories of urbanization beyond the city, I argue that only by charting how the rural and the urban are co-produced in predominantly agrarian countries like India can we begin to understand the complex processes of sub/urbanization and recognize that *the urban question is simultanesouly an agrarian question.* The emerging urbanism of the peripheries is as equally intertwined with the urban as it is with agrarian dynamics of land, property, and tenure, and thus I suggest may appropriately be called *agrarian urbanism.*[6] I find the framework of *agrarian urbanism* generative, as it adds an important dimension to the map of global suburbanisms and disrupts the standard conceptual terrain of urban theory that tends to be largely city-centric or even urban-centric. It challenges the "intellectual imperialism of the urban" (Krause, 2013, p. 234) and illustrates how, as we work towards a new epistemology of urbanization and consider the forms and processes of sub/urbanization, an engagement with colonial histories and agrarian political economy of land and property remains critical. Robinson (2011) has argued that in order to revisit the analytics of urban studies, we need to locate "cities in a world of cities," and I would suggest that as we try to stretch the comparative arc further and attend to urbanisms in agrarian societies, we must also attend to diverse historical trajectories and relations of land, and locate cities in a world of villages.

The contours and content of agrarian urbanism are far from uniform. It does not unravel along an inevitable history of accumulation and dispossession (see Chakrabarti et al., 2017), but instead, mired in the thick cultural politics of caste-class-land-place-ecology, the contested terrain of agrarian urbanism as I describe below is shaped by everyday practices of speculation, compensation, capture, profit, and collusion. It is this story of agrarian urbanisms that I wish to turn to, but, before I do so, in the next section I briefly discuss the agrarian–urban question in India and then turn to the agro-ecological transformations that took place in colonial Punjab and present-day Haryana, in which the Millennial City is located. In tracing how the changing regimes of land, property, and caste paved the way for a highly differentiated social geography of rural and urban spaces, I wish to show how land and its attendant relations continue to contour the contemporary urban landscape. In the last section, I draw from fieldwork to present an ethnographic snapshot of the agrarian urbanism that is taking shape in Gurgaon and offer a glimpse of why and how, in this moment of global suburbanization, peripheries are the locus of such social and political unevenness.

The Agrarian–Urban Question in the Twenty-First Century

In a provocative essay titled "Is there an Agrarian Question in the 21st Century?" Bernstein (2006) argued that in the wake of varied shifts, including land reforms, capitalist restructuring, accumulation in the post-war years, and, more recently, global capitalism, the agrarian question in its classic sense has lost its analytical purchase. According to Bernstein, the "classic" agrarian question can better be identified as the agrarian question of capital, and, in the context of globalization, the agrarian question has been reconstituted as the agrarian question of labour.

In contrast, other scholars – with whom I am inclined to agree – have argued that despite changes in the agricultural sector, the agrarian question, in the face of neoliberal globalization, remains not only relevant but has even gained renewed importance (see Akram-Lodhi & Kay, 2010a, 2010b; Harriss, 2005; Lerche, 2013). With the entry of new corporate actors, new political forces (local and transnational), new technologies, and new regimes of governance and governmentality (of which urbanization of land and real estate speculation stand out), the agrarian question of land and peasant differentiation remains front and centre (see Gururani & Dasgupta, 2018).

Despite the centrality of rural land for urban expansion, agrarian studies has not, barring a few exceptions, directly tackled the urban question. Likewise, until recently urban studies has also tended to shy away from the agrarian question of land (see Roy, 2016). There is now, however, a growing body of scholarship that focuses on urban peripheries and examines the complex formulation of urban–rural intersection and has engaged with "peri-urban dynamics," "urban villages," "urban peripheries," and takes into account "the greater heterogeneity and segmentation of peri-urban spaces" (Dupont, 2005, p. 17) to make legible the everyday politics of evictions and exclusions that usually mark such urban peripheries (see Balakrishnan & Gururani, 2021; Gururani et al., 2021). In conversation with this body of work, I describe the trajectories of uneven peripheral urbanization and how colonial regimes of land and property continue to undergird the everyday negotiations and contestations that shape agrarian urbanism in this part of the Global South.

Entanglements of Land, Place, and Power: The Making of Gurgaon

In the 1990s, while the peripheries of all megacities underwent rapid de-agrarianization, Haryana was the first state to open its doors to private developers like Delhi Land and Finance (DLF), whose italicized signature now adorns countless buildings and towers in and around Gurgaon. Through their close connections with the then ruling Congress party throughout the 1970s and especially in the 1980s, DLF acquired thousands of acres of land, changed its land use, consolidated it, developed it, and pursued its dream to "build India" (Gururani, 2013). While in the neighbouring township of Noida in Uttar Pradesh the state brokered land deals with developers, in Haryana, the powerful class of agrarian capitalists, mostly from landowning castes of Jats and Yadavs, lobbied local politicians, sold their land directly to private developers, and made significant gains. In his autobiography, K.P. Singh (2011), the founder and CEO of DLF, tells the story of Gurgaon as a tale of personal charisma, wit, care, and determination – a tale of "the mysteries of fate," of serendipity, where things just fell into place for him (p. 180). But the story of Gurgaon that I present is far from magical.

In the uneven terrain of property, ecology, and caste that has shaped Gurgaon historically (see Gururani, 2021 for more detail), the Nehruvian dreams of agro-ecological modernization came to be realized in post-independent India and paved the way for the city's transformation. The Intensive Agricultural Development Program (IADP) was implemented in the mid-1960s amid concerns of food security, drought, and famine in five Indian states, including Haryana. As part of the program, a new agricultural strategy involving the adoption of new high-yielding varieties of crops, modern technology, chemical fertilizers, and irrigation fertilizers was adopted. This strategy came to be euphemistically known as the Green Revolution and was initially hailed as a success. But, it was soon pointed out that the Green Revolution impacted social relations of production and exacerbated inequalities between small- and large-scale farmers and between ecologically diverse regions. For instance, in Haryana, the district of Karnal, most of Hissar, and parts of Rohtak gained from Green Revolution policies, but the district of Ambala and the southern part of Gurgaon district were bypassed and, until recently, the district of Gurgaon and the area covered by the city of Gurgaon were faced with recurrent famines and drought-like conditions, remaining one of the poorest areas in the state. Even within the district of Gurgaon, the uneven trajectories of Faridabad and Gurgaon can be traced back to these agrarian developments. While Faridabad "gained" from a system of tubewells and irrigation that enhanced its already high agricultural productivity and later emerged as a planned urban-industrial complex of Nehru's dream, Gurgaon, with its poor irrigation and fertility, was rendered unfit for development and left out of regional planning. In the 1990s, however, it was precisely Gurgaon's poor agricultural land that made it easier and cheaper to acquire and urbanize.

The landowning Jats, who own close to 90 per cent of the land in Haryana and have larger landholdings than any other caste, were not surprisingly the main beneficiaries of agricultural modernization and soon emerged as the new regional and political elites. In this conjuncture of agrarian change, a segment of Yadavs, a landowning middle-caste of cultivators, also gained from Green Revolution policies and were able to mobilize politically and consolidate their social and political dominance. Comparable to Gupta's (1998) account of agrarian populism in Uttar Pradesh, in Haryana, too, the Green Revolution produced a class of capitalist landowners who came to subsequently play an important role in the evolving land market and real estate speculation. While the agrarian castes of Jats and Yadavs were able to access governmental subsidies and lift-irrigation schemes (Bhalla, 1976), the Gujar pastoralists, who had access mostly to arid land, along with smallholders and landless farm workers, lost out once again in the Green Revolution (Patnaik, 1987), a fate that was to change, for some, with the coming of the urban revolution in the 1990s.

In 1966, in the middle of agricultural intensification, the state of Haryana was carved out of Punjab. The creation of the new state of Haryana took place just a few years after Delhi's first Master Plan was passed in 1962. In 1962, the Delhi Development Authority (DDA) was set up and pushed private developers, like the DLF, out of New Delhi. At the time, Haryana was the only state in the country that allowed private developers to buy land directly from farmers, and since Haryana was a new state and had few established urban administrative bodies, it was easier to manipulate rules and buy agricultural land (see Gururani, 2013). It is against such a backdrop that the passing of the National Capital Regional Act in 1985 proved to be a decisive moment.

While land was acquired in other urban peripheries too, Gurgaon was particularly desirable for several reasons. First, its arid agro-pastoral land with its low land valuation made it easier and cheaper to acquire. In most instances, big and small landholders parted with their land for very little cash and with little resistance. Second, the legislative and political architecture facilitated and even paved the way for a flexible planning regime (Gururani, 2013, p. 119). By the mid-1990s, the entire district of Gurgaon came to be identified as a controlled area, opening up the door for intense real estate speculation, land acquisition, and land sales. Third, the history of uneven development had produced a class of Jat and Yadav farmers who entered the property business and actively participated in the commercialization of land and intensified the process of de-agrarianization in which land served as a critical asset (Paul, 1990). In an interesting twist, amidst the changing economy of land valuation, Gujars, who were once described as an "unsatisfactory race of people" (Darling, 1928, p. 98) and like other pastoralists were considered to be dangerous and unreliable, emerged as important players as the arid/wasteland land they controlled became highly desirable.

In the 1980s, Gujars started to mobilize. They procured property deeds and titles and, along with the Jats and Yadavs, entered real estate speculation. They have since consolidated their political clout and have come to assert a social and economic presence in a city where they have previously been treated with suspicion and even fear. In awe and disbelief, print and news media have covered lavish Gujar weddings (Dash, 2011), their luxury cars, and their changing lifestyle. The shift in caste dynamics has meant that the Jats, who have long enjoyed cultural and political domination, and the Yadavs, who have also established themselves in the real estate business, are now forced to contend with the growing political ascendancy of the Gujars and a few other non-agricultural castes like the Sainis and Gosains.[7] Moreover, the declining role of the agricultural sector in the national economy and the influx of migrants into the villages and the city has meant that the dominant castes, like the Jats, are experiencing a loss of traditional authority and economic status. They have not been able to secure jobs and have in the last few years come out to agitate and demand reservation as Other Backward Classes (OBCs) (Jaffrelot & Kalaiyarasan, 2017). It is in this milieu of flux and uncertainty that the space of the village has come to serve as an important venue for simultaneously supporting traditional caste-based agrarian authority and making room for an unfolding agrarian urbanism of compensation and rent.

Frontiers of Capture (*Kabza*), Compensation (*Muawaza*), and Rent (*Kiraya*)

When I started my first leg of fieldwork in 2007, the first Draft Master Plan 2021 had just been released in February, and the Haryana Urban Development Authority (HUDA), an arm of the provincial state, had entered into extensive land acquisition and development. The city was literally abuzz with land talk. In the next six years, between 2007 and 2013, two more plans – Master Plan 2025 and Master Plan 2031 – were released by the then chief minister B.S. Hooda who, in a masterful stroke of master planning, made 54,000 acres of agricultural land available in Haryana for commercial, residential, and industrial use (Singh, 2013). As more and more land was pulled out of agriculture, the number of private developers went up. By some estimates, close to 500 developers and builders came to acquire land during this period, and they developed over 90 per cent of Gurgaon, including hundreds of gated-housing enclaves that offer state-of-the-art, modern, green, and luxurious living. In this land mania, almost everyone I met was involved in some capacity in the business of land and property. As property dealers, brokers, middlemen, consultants, contractors, suppliers, or land recorders, everyone in and outside of the village was doing what they described as property work (*property ka kaam*). In a very short period, land had transformed from a source of livelihood into a fungible entity that could now be assetized for real estate profit, and village-based relations of caste and authority were significant in making this transformation possible.

I conducted sporadic fieldwork that stretched over a period of a decade from 2007 to 2017 in some of Gurgaon's villages – Chakkarpur, Nathupur, Ghata, Jharsa, Kanhai, Samaspur, Tigra, and Wazirabad. During this period, I had several conversations with villagers who had sold their land to private developers, or whose land had been acquired by HUDA, or who were in the middle of buying and selling land. In 2007, farmer protests against land acquisition in West Bengal were making headlines, and I was expecting to encounter some resistance to land conversion and acquisition in Gurgaon too. But, apart from a few individuals whose land was acquired for the Special Economic Zone (SEZ), most villagers in Gurgaon I met with were by and large not resentful of having lost their land to the developers. If there was any resentment, it was about the trivial compensation villagers received for their land and that said compensation was given in small installments, making it difficult for them to reinvest elsewhere (see Majumder, 2018; Majumder & Gururani, 2021).

Kana Devi,[8] a seventy-year-old Gujar widow in Nathupur, walked me through the village to show me her newly built three-storey house. She lived in a small corner of the house with her unmarried son and rented rooms to young men and women who had come to work in Gurgaon. In describing the early days when land came to be acquired, she said,

> We are Gujars, we have always tended cattle. Our main source of livelihood was selling milk. We had around thirty to forty cows and we sold milk to nearby areas … When developers started buying land, there were middlemen everywhere. They were usually outsiders but they came into our houses, talked to us, and they scared us that if we did not sell it to the developers, HUDA would acquire our land and we would not get any money. We sold in haste and barely received anything. We sold it for 5 or 6 lakhs/killa[9] and the others sold it for 1 lakh/gaj.[10] We got some money, but it was given to us in installments, very small sums, which we could not invest. We felt good then; some bought cars, some spent it on weddings, and others just drank it away. We didn't have the wherewithal to deal with money. We built this house from the small amount we received and now we have the house, we get some money from renting it but other than that we have no land, my sons have no work, and we are left with little. (personal communication, 25 February 2015)

Kana Devi's account was not uncommon, and landowners in almost all the villages were embroiled in long court cases about the compensation they received for their land. In Kanhai, a Yadav-dominated village that is tucked in-between Sector 45 and Sushant Lok, I met Dilip Yadav, whose land was partially acquired by HUDA and partially purchased by the private developer, Ansals, in 1991. Dilip animatedly described how HUDA acquired close to 300 acres of agricultural and pastoral land in the village at a poor compensation rate of Rs. 147/gaj. The villagers were appalled by the paltry rate at which the compensation was calculated, and forty or so families filed a case against HUDA in the High Court to seek better compensation. Having argued that they would not be able to purchase land outside the village at the given rate of compensation, the families were first awarded a better rate of Rs. 247/gaj, which the High Court then raised to Rs. 265/gaj. When the villagers wanted an even better rate, however, the court decided to go back to Rs. 212/gaj. Angered by the court's decision to claw back compensation, several families, in protest, refused to collect the compensation amount, and, as of 2017, there were twenty-five families who still had not done so. A few did, and a few others, pooling other sources of income, purchased land beyond Gurgaon, but according to the group I was speaking with, most were waiting for a new court date.

All my interlocutors, mostly men, were very well versed in the transactions of land and property. In several instances, Dilip and others shared their court documents in the hope that I could assist them or take their case to the media, but in no interaction did I ever hear anyone express resistance to land acquisition. There was indeed some remorse that the village had changed or that the milk from the market did not taste as good, but it was more about the compensation money. Even though the compensation was small compared to the price of land, it was far more than the villagers ever expected and, as a result, most landowners were willing to part with their land or enter into deals with the developers. For instance, in Jharsa, S.S. Thakran, an elderly Jat gentlemen who was fighting twenty-four court cases said, "Land is no longer for agriculture. We did not want to sell land first and now we cannot stop ourselves from selling it. Land has become a big business here" (personal communication, 21 February 2017). Land has indeed become a business, but the story of Gurgaon is not one of dispossession or resistance (Majumder, 2018). As S.S. Thakran explaind, "Here it is all about compensation (*muawaza*) and capture (*kabza*) and rents (*kiraya*). People fight court cases for better compensation, they encroach unnamed land, and they lead a life they never imagined" (personal communication, 21 February 2017). In this conjuncture, the former landowners, mostly from landowning castes such as Jat and Yadav, have emerged as willing stakeholders, and the space of the village has become central to sustaining and facilitating the urbanization in and of the peripheries. For instance, in 2011, the Municipal Corporation of Gurgaon was formed,[11] and village councils (*panchayats*) were dismantled, erasing the formal boundaries between the rural and urban, but despite this governmental shift, the space of the traditional village has persisted. In many ways, the villages have been reconstituted and revitalized along caste–class gradients to allow the dominant caste-groups to once again mobilize their traditional caste-based authority and networks and lay claim to, or "capture," land, including unmarked/common land.

The spatial organization of the village makes this evident. The villages in north India are typically planned around a settlement or residential area, referred to as the *abaadi deh*, which is marked by a colonial-era revenue instrument called the *lal dora* (red tape). The boundary of the village is marked by a spatial ring called the *phirni* or *gher*. In Haryana, unlike Delhi, the *lal dora* area has never been extended, even though the population of its villages has expanded over the years. As a result, the core of the villages has increasingly become densely populated, with very narrow and dark lanes, and with characteristically old, rundown housing structures.

The *phirni* land was traditionally used for pasture or sometimes for storing agricultural tools or grain by all members of the village, including small landholders, and was a commons of sorts. It constituted the material and symbolic interface between the village and the agricultural fields, but the *phirni*, which is now marked by a paved road, has become highly prized real estate asset for both residential and commercial purposes. Based on village authority, the dominant caste of landowners has surreptitiously come to lay claim on this *phirni* land. The Jats, Yadav, and landowning Gujars have moved out from the congested *lal dora* area and have not only built very large houses in the *phirni* but have also built rental housing, from which some earn sizeable incomes. They have informally secured connections to water, electricity, telephone, Wi-Fi, and so on. As a result, the *phirni* has emerged as one of the fraught sites where the micropolitics of caste and power are played out virtually every day. In this changing geography of spatial value, the social-spatial dynamics of the village have provided a strong and reliable network of kin and caste connections that in part explain why, despite having gained from the real estate boom, the upwardly mobile Yadavs and Jats, like Mr. Thakran, continue to live in the villages.

While the landowners have negotiated reasonable compensation, reinvested in land further away, and claimed the *phirni*, for the rest, it is a different story. The majority of those among the lower castes, who are designated as "Other Backward Classes" (OBCs), and the *dalits* (formerly called "untouchables") never had land to sell and have gained hardly anything from the property boom. They continue to live in small, dilapidated structures in the inner core of the village, often with poor or no water and sewage connections. Not only are the landless households pushed into the dilapidated corners of the village, but most of these families have also lost their meager livelihoods. They used to work for the landowning families and supplemented their livelihoods by doing small jobs as stone collectors, thrashers, carpenters, cobblers, and ironsmiths. While some of them in recent years have found employment in government offices, many claim that they are without work because the construction contractors and factories prefer to employ migrants, who are less likely to ask for better wages.

In the course of all these changes, the influx of up to a million migrant workers into Gurgaon has significantly contributed to rearranging the spatiality of the village. In the last two decades, migrants from Bihar, Orissa, Uttarakhand, Uttar Pradesh, and West Bengal have come to Gurgaon to work in factories, in the construction industries, as domestic servants, and as security guards. With no provision of housing for them, the villages serve as cheap dormitories. While some have access to infrastructure, the services for the poorer villagers are at the bare minimum.

Amid these social-spatial changes, the villages have unquestionably been transformed and have become increasingly imbricated with changing political economies of land, housing, and labour. But, instead of being assimilated into the urbanizing landscape, the villages embedded in agrarian and urban regimes of land and property have come to sustain and negotiate the contours of emerging urbanisms – an urbanism I have tried to argue can best be described as agrarian urbanism.

Concluding Notes

In this chapter, I have pursued two interrelated questions. First, I have described how the forms and processes of sub/urbanism that unfold in this political economic conjuncture not only exceed the confines of the city but also the confines of the urban. I have argued that with the entry of new political and economic actors, the questions concerning land and the peasantry have indeed changed in India, but the agrarian question remains at the heart of contemporary urban transformation. Through an examination of colonial and postcolonial regimes of land, revenue, irrigation, and agricultural improvement, I have shown how the complex imbrications of the urban with the rural and agrarian dynamics are constitutive of urbanisms today and how the politics of caste and class continues to be of salience. In doing so, I have suggested that in order to understand trajectories of urbanization that depart from the Euro-American experience, it is critical that we step outside the standard repertoire of urban theory and seek new analytical frameworks. To that end, I have attempted a dialogue between urban and agrarian studies and have outlined the contours of an urbanism in which agrarian regimes of land and property endure and co-produce the urban. An urbanism that can explain how changing relations of caste and land produce an uneven geography of spatial value and how the everyday practices of compensation, capture, and rent play out. An urbanism that does not erase or assimilate the rural but in which agrarian and urban dynamics sustain and produce each other. I have identified such an urbanism as *agrarian urbanism*. Second, in presenting an ethnographic snapshot of Gurgaon and of the many villages that dot its urbanizing landscape, I have drawn attention to the highly fraught and contested terrain that is constitutive of India's agrarian–urban terrain. At a time when there is significant faith among policy gurus and political leaders that urbanization of agrarian hinterlands holds the key to India's neoliberal success, the accounts of uneven urbanization I present not only disrupt the grand narrative of a smart and shining India, but they also highlight how entrenched relations of caste and class endure, and how they continue to configure social geographies of unevenness and difference in globalizing India.

ACKNOWLEDGMENTS

I would like to thank Rajarshi Dasgupta, Will Glover, Roger Keil, Radhika Mongia, Xuefei Ren, and Fulong Wu for their very useful comments on an earlier version of this chapter. The chapter also benefited immensely from the comments of three anonymous reviewers, as well as from the discussion and interventions at the conference "Frontier Urbanism: Tracking Transformation in Agrarian-Urban Hinterlands of South Asia" held at Jawaharlal Nehru University, New Delhi, India, 24–25 February 2017. I remain grateful to Rabab and Harsh Lohit for hosting me and for making fieldwork possible in multiple ways.

NOTES

1 For more on the extent of the NCR, see the National Capital Region Planning Board website: http://ncrpb.nic.in/ncrconstituent.html.

2 I am using the terms *agrarian, village,* and *rural* rather loosely to refer to the social-spatial dynamics that are entangled with agrarian relations of land and property.
3 On 5 November 2016, the city of Gurgaon rebranded itself by officially changing its name to Gurugram, replacing the colloquial "gaon" with the more Sanskritized/civilized "gram," which also means "the village," but a well-disciplined and civil one! I continue to use the old name as that was what was colloquially used in my conversations and interviews during fieldwork from 2008 to 2017.
4 In 1991, the population of the city was pegged at 125,000, which increased to 876,000 within a decade. In 2011, Gurgaon's population crossed the 1-million mark and, today, its estimated population is said to be between 2 and 2.5 million.
5 This chapter is an abridged and modified version of Gururani (2020).
6 Charles Waldheim (2010), a landscape architect, has used the term *agrarian urbanism* to draw attention to the ways in which urban farming is integrated in design and planning in North America. My usage of the term departs significantly from his coinage.
7 The limits of space do not permit a full discussion here, but in the context of the rise of the Hindu right in India, the changes in agrarian political economy, especially the declining role of the agricultural sector in the national economy, have significant analytical and political salience (see Jaffrelot & Kalaiyarasan, 2019).
8 In order to maintain anonymity, the names of all individuals have been changed.
9 4 beegha = 1 killa, lakh = 100,000, so at the current conversion rate, CAD$2,000 for 4 beegha approximately.
10 1 gaj = 9 square feet.
11 See Gururani and Kose (2015) and Pradhan (2013) for discussions of how the Indian census tabulates rural and urban populations.

REFERENCES

Akram-Lodhi, H.A., & Kay, C. (2010a). Surveying the agrarian question (part 1): Unearthing foundations, exploring diversity. *The Journal of Peasant Studies, 37*(1), 177–202. https://doi.org/10.1080/03066150903498838.

Akram-Lodhi, H.A., & Kay, C. (2010b). Surveying the agrarian question (part 2): Current debates and beyond. *The Journal of Peasant Studies, 37*(2), 255–84. https://doi.org/10.1080/03066151003594906.

Balakrishnan, S., & Gururani, S. (2021). New terrains of agrarian-urban studies: Limits and possibilities. *Urbanisation, 6*(1), 7–15. https://doi.org/10.1177/24557471211020849.

Bernstein, H. (2006). Is there an agrarian question in the 21st century? *Canadian Journal of Development Studies, 27*(4), 449–60. https://doi.org/10.1080/02255189.2006.9669166.

Bhalla, S. (1976). New relations of production in Haryana agriculture. *Economic and Political Weekly, 11*(13), A23–A30. https://www.jstor.org/stable/4364490.

Chakrabarti, A., Stephen C., & Dhar, A. (2017). Primitive accumulation and historical inevitability: A postcolonial critique. In T. Burczak, R. Garnett Jr., & R. McIntyre (Eds.), *Knowledge, class, and economics: Marxism without guarantees* (pp. 288–306). Routledge.

Darling, M.L. (1928). *Punjab peasant in prosperity and debt.* Oxford University Press.

Dash, D. (2011, March 13). Big, fat Gujjar weddings: Easy come, easy go. *Times of India.* https://timesofindia.indiatimes.com/home/sunday-times/deep-focus/Big-fat-Gujjar-weddings-Easy-come-easy-go/articleshow/7689163.cms.

Dupont, V. (2005). Peri-urban dynamics: Population, habitat and environment on the peripheries of large Inidian metropolosis: An introduction. In V. Dupont (Ed.), *Peri-urban dynamics: Population, habitat and environment on the peripheries of large Indian metropolises: A review of concepts and general issues* (pp. 3–20). Centre de Sciences Humaines.

Government of India. (2001). *Census of India.* Office of the Registrar General and Census Commissioner. https://censusindia.gov.in/DigitalLibrary/reports.aspx.

Gupta, A. (1998). *Postcolonial developments: Agriculture in the making of modern India.* Duke University Press.

Gururani, S. (2013). Flexible planning: The making of India's "Millennium City," Gurgaon. In A. Rademacher & K. Sivaramakrishnan (Eds.), *Ecologies of urbanism in India: Metropolitan civility and sustainability* (pp. 119–43). Hong Kong University Press.

Gururani, S. (2020). Cities in a world of villages: Agrarian urbanism and the making of India's urbanizing frontiers. *Urban Geography, 41*(7), 971–89. https://doi.org/10.1080/02723638.2019.167056.

Gururani, S. (2021). Making land ouf of water: Ecologies of urbanism, property, and loss. In A. Rademacher & K. Sivaramakrishnan (Eds.), *Death and life of nature in Asian cities* (pp. 138–58). Hong Kong University Press.

Gururani, S., & Dasgupta, R. (2018). Frontier urbanism: Urbanisation beyond cities in South Asia. *Economic and Political Weekly, 53*(12), 41–5.

Gururani, S., Kennedy, L., & Sood, A. (Eds.). (2021). Engaging the urban from the periphery. *South Asian Multidisciplinary Academic Journal, 26.* https://doi.org/10.4000/samaj.7131.

Gururani, S., & Kose, B. (2015). Shifting terrain: Questions of governance in India's cities and their peripheries. In P. Hamel & R. Keil (Eds.), *Suburban governance: A global view* (pp. 278–302). University of Toronto Press.

Harriss, J. (2005, April 1). *"Politics is a dirty river": But is there a "new politics" of civil society? Perspective from global cities of India and Latin America* [Paper presentation]. Conference on International Civil Society, Global Governance and the State, New York, NY, United States. http://eprints.lse .ac.uk/487/.

Jaffrelot, C., & Kalaiyarasan, A. (2017, March 10). Jats in wonderlessland. *Indian Express.* https://indianexpress.com/article/opinion/columns /jats-in-wonderlessland-quote-stir-jat-agitation-obc-status-haryana -employment-4562573/.

Jaffrelot, C., & Kalaiyarasan, A. (2019). The political economy of the Jat agitation for Other Backward Class status. *Economic and Political Weekly, 54*(7), 29–36.

Keil, R. (2018). *Suburban planet: Making the world urban from the outside in.* Polity Press.

Krause, M. (2013). The ruralization of the world. *Public Culture, 25*(2), 233–48. https://doi.org/10.1215/08992363-2020575.

Lefebvre, H. (2003). *The urban revolution.* University of Minnesota Press.

Lerche, J. (2013). The agrarian question in neoliberal India: Agrarian transition bypassed? *Journal of Agrarian Change, 13*(3), 382–404. https:// doi.org/10.1111/joac.12026.

Majumder, S. (2018). *People's car: Industrial India and the riddles of populism.* Fordham University Press.

Majumder, S., & Gururani, S. (2021). Land as an intermittent commodity: Ethnographic insights from India's urban–agrarian frontiers. *Urbanisation, 6*(1), 49–63. https://doi.org/10.1177/24557471211021507.

Patnaik, U. (1987). *Peasant class differentiation: A study in method with reference to Haryana.* Oxford University Press.

Paul, S. (1990). Green Revolution and poverty among farm families in Haryana, 1969/70–1982/83. *Economic and Political Weekly, 25*(39), A105–10.

Pradhan, K.C. (2013). Unacknowledged urbanisation: New census towns of India. *Economic and Political Weekly, 48*(36), 43–51.

Robinson, J. (2011). Cities in a world of cities: The comparative gesture. *International Journal of Urban and Regional Research, 35*(1), 1–23. https://doi.org/10.1111/j.1468-2427.2010.00982.x.

Roy, A. (2016). What is urban about critical urban theory? *Urban Geography, 37*(6), 810–23. https://doi.org/10.1080/02723638.2015.1105485.

Singh, K.P. (2011). *Whatever the odds: The incredible story behind DLF.* HarperCollins.

Singh, S. (2013, May 27). Builder profits soar as master plans proliferate in Gurgaon. *The Hindu*. https://www.thehindu.com/news/national/builder-profits-soar-as-master-plans-proliferate-in-gurgaon/article4753735.ece.

Waldheim, C. (2010, November). Notes toward a history of agrarian urbanism: Architects like Frank Lloyd Wright, Ludwig Hilberseimer, and Andrea Branzi anticipated today's interest in urban farming. *Places Journal*. https://placesjournal.org/article/history-of-agrarian-urbanism/.

Image 8. Metro Vancouver, British Columbia, Canada (Summer 2013)

8 The "Publicness" of Suburban Infrastructure Planning: Cases from Toronto and Melbourne

CRYSTAL LEGACY

Introduction

Public things [such as infrastructure] are part of the "holding environment" of democratic citizenship; they furnish the world of democratic life ... They also constitute us, complement us, limit us, and interpellate us into democratic citizenship.

– Honig, *Public Things*

Bonnie Honig, from whose book *Public Things: Democracy in Disrepair* (2017) this opening quote was taken, posits that infrastructure possesses a "publicness" that connects it to the inhabitants these infrastructures serve, the geographies they exclude, and the neighbourhoods they physically impact. Examining the relationship between infrastructure and suburbs, Addie (2016) takes on the additional conceptual project of describing modalities of infrastructure, through which he develops the idea of suburbanism rendering visible the lived experience and way of being of suburban inhabitants. Similar to Honig's analysis of infrastructure as a public thing, Addie situates infrastructure as a set of political, economic, and geographical relations and, in so doing, exposes how new and, by extension, promised infrastructure can also generate new governance forms, knowledge, and subjectivity. As infrastructure shapes suburban lives, it also ignites into being a political subjectivity in citizens who are affected by infrastructure's material impacts. The ways this materiality can shape those lives and daily experiences (Bissell, 2018) renders infrastructure inherently public because of the influences, both positive and negative, it has on people living in urban and suburban environments. This publicness thereby connects the planning of suburban transport infrastructure – in this chapter, both roads and light rail transit – to participatory action by inhabitants in favour of or in contention with the futures this infrastructure promises or evokes.

Relatedly, in the introduction of their book *The Promise of Infrastructure*, Appel et al. (2018) ask their readers to consider how particular infrastructure can "signal the desires, hopes, and aspirations of a society, or of its leaders" (p. 19). Proposed infrastructure can signify a way of thinking about the future (Larkin, 2013). Examining the convergence of suburban transport planning and inhabitants' participation in this planning raises two questions that this chapter will address. First, in what ways are suburban inhabitants asserting greater levels of political participation in the planning of this infrastructure? And second, how is this participation shaped by concerns about what this infrastructure "promises" about the future of suburban transport infrastructure as it manifests in the socio-technical assemblages connecting the technical fix of different transport technologies with the complex spatial politics of city-regions? (See also Filion & Keil, 2017.)

The burgeoning literature on post-political governance settings (Allmendinger & Haughton, 2012) and important interventions about the different forms the politicization of transport might take (e.g., Henderson, 2013) have created the space to consider how inhabitants might assert their own desires about their mobility in complex city-regions. In this chapter, I consider how the publicness of suburban transport infrastructure may extend to the planning for this infrastructure – the *publicness of suburban infrastructure planning* – illuminating important democratic moments in planning that enliven debates about promises and desires about the future of suburbs.

To do this, I draw insights from empirical research conducted in two city-regions between 2014 and 2018: the Greater Toronto and Hamilton Area (GTHA), and metropolitan Melbourne. The discussion in this chapter is drawn from data collected from semi-structured interviews with transport planners, city officials and councillors, state/provincial politicians and policymakers, and active citizens and community-based groups living in two distinct suburban regions. In the GTHA, I focused on residents living in both the city centre and suburbs of the City of Hamilton, and in metropolitan Melbourne, I focused on citizens living in the middle suburbs west of the city centre including Yarraville, Seddon, Brooklyn, and South Kingsville. The fieldwork was conducted during a period when suburban transport infrastructure planning attracted significant contestation over the future these infrastructure investments might evoke for suburban inhabitants. To capture this turbulent period, my empirical work also draws from an extensive media analysis over this period, as well as participant observation. The latter occurred in Melbourne, and in the GTHA I was able to plan my field-work to correspond with several key decision moments in the rollout of a light rail transit (LRT) line in Hamilton.

The Publicness of Suburban Transport Infrastructure

There has been considerable attention cast upon the processes of sub-urban infrastructure planning by policymakers and politicians alike in recent years. For instance, the global financial crisis of 2007–8, and the economic stimulus programs it precipitated in Western democratic countries including Canada, Australia, the US, and the UK, strength-ened the rhetorical connections between infrastructure planning, job creation, and economic growth (Addie, 2013; Legacy, 2017a). Those connections have long been apparent (O'Neill, 2010), but it was in the wake of this concentrated period of suburban infrastructure invest-ment that the rhetoric of transport infrastructure construction became attached to promises of new jobs and economic opportunities. This was particularly the case for those inhabitants living in suburbs where the impacts of post-industrial processes were resulting in significant job loses, as seen in the western suburbs of Melbourne, or in a transforma-tion of the local economic base, as seen in the case of Hamilton. Infra-structure promises were particularly salient in those cities experiencing structural economic changes. As Boyer (2018) writes, "It is striking that the conceptual rise to intuitiveness of infrastructure roughly parallels the crisis and stasis of neoliberal governance since 2008" (p. 223). For those suburbs experiencing the strain from years of underinvestment in suburban transport infrastructure, and as population growth in the suburbs continued to rise, infrastructure planning elevated in the pub-lic consciousness because of its association with jobs, but also because there was an increasing need for infrastructural responses to popula-tion pressures (Legacy, 2017a).

The significant costs associated with suburban infrastructure con-struction, operation, and maintenance are another reason for greater public scrutiny directed at infrastructure planning. Impacts from sub-urban infrastructure development, particularly in the inner and middle suburbs where an existing built environment will experience significant disruption, can result in financial strain on project proponents and gov-ernments, and result in emotional stress for citizens. When past infra-structure has been poorly costed, or the benefits are misrepresented due to optimistic forecasting, this casts a spotlight on the economic risks of building future large-scale suburban transport infrastructure (Flyvbjerg, 2008). Public and private partnerships may be introduced to assist with the management of economic risks in a way that helps create certainty for corporate investors. Yet, when the proposed infrastructure is controversial – and these projects can be controversial for a range of reasons including the anticipated impacts on built and natural environments – the economic risks can be further compounded by political risk. The latter includes costs introduced from project delays, legal challenges, and sometimes the cancellation of projects. Taken together, these risks may place the infrastructure being promised at risk of ever being built, or the initial promises of a project laid out at its inception eventually being short-circuited during the implementation phase to appease political concerns.

To control these risks, planners and decision-makers may design infrastructure planning processes in such a way as to minimize delays or cancellations borne from politicization, with the view to protect a project from such politically induced risk. Decision-making processes might be streamlined, resulting in the signing of project contracts that lock the project into a construction phase. This may also include project proponents holding governments to account for sunk costs (see Cantarelli & Flyvbjerg, 2013), or governments wanting to protect their projects by inundating public discourse with messages of projects being "done deals," which aims to render any mounted resistance as futile (Legacy, 2017b). Governments may also introduce participatory processes "early" in the process of project planning, oftentimes when key decisions have already been made, to create the perception that infrastructure is being planned in a consultative manner. But too often participatory techniques are only offered as narrow channels for citizens to engage, thus reducing how much citizen-led resistance might render visible the underlying politics of suburban infrastructure planning, including the production of unequal levels of access, which is generated from the structured inequality created from uneven suburban development (Addie, 2013; Filion & Keil, 2017).

The focus of formal citizen participatory processes is on protecting infrastructure planning from politicization that might extend beyond these formal processes (see Legacy 2017a, 2017b). This is partially achieved by veiling the publicness of infrastructure through exclusionary processes of project planning, often fashioned through public–private partnerships. These managerial styles of infrastructure planning corral citizens into consensus-based and outcome-oriented decision-making processes, limiting opportunities to contest the very basis of such proposals and the promises they represent (see Mäntysalo et al., 2011). This is a form of participation utilized in controversial transport infrastructure settings, as seen in Australian cities where participatory processes can be antagonistic and structured to limit engagement, and so frustrate citizen-led efforts to contest projects (Legacy et al., 2017).

If, following Honig (2017), suburban transport infrastructure has an inherent publicness about it, then the promises articulated in its planning (e.g., that it will facilitate greater accessibility, reduce travel times, and attract investment to the suburbs) must be viewed through the domain of spatial politics. These promises present as empty signifiers (Swyngedouw, 2018), which are ultimately detached from the disparate material realities experienced by suburban inhabitants in different parts of the city-region. Suburban transport politics by necessity need to be viewed as part of the "ongoing work of democratic citizenship" that joins together the public things of suburban life with suburban transport infrastructure – as the "truly *public* things – the transitional objects – of democratic life" (Honig, 2017, p. 11).

There are many different ways to understand and to examine the politics of transport. Henderson (2013), for example, describes the politics of mobility as an ideological tension with progressive forms of mobility supporting active, public, and collective forms of transport set against conservative conceptions of mobility, whereby inhabitants might demand unrestricted forms of transport, preferring individual instead of collective transport options. Moving away from an ideological interpretation of mobility politics towards a more socio-spatial examination, Walks (2015), in his study of Toronto, links the politics of mobility to a spatial and political-economic analysis of automobility. Walks brings together geographies, economies, and history to forge a deeper understanding of different publics' reactions to transport infrastructure and planning processes, and to their different transport fortunes (see also Baeten, 2000). The politics of transport can also be described through institutional and social-political perspectives, which highlight how the different politics might be performed through the formation of coalition groups to challenge dominant practices and thinking about transport planning (Stone, 2014; Vigar, 2013).

These politics can be illuminated in the promises governments and project proponents make about infrastructure, particularly the form of politics that might be mobilizing elected leaders into action. This point is underlined by Gupta (2018) when he says that "infrastructures tell us about aspirations, anticipations, and imaginations of the future: what people think their society should be like, what they want it to be like, and what kind of statement they wish to make about that vision of the future" (p. 63). In this context, suburban transport infrastructure planning has a symbolic power (von Schnitzler, 2018); this power raises questions about what new (or reinforced old) trajectories of behaviour, land use patterns, and ambitions it will forge.

An immediate concern is that infrastructure governance can reduce and confine the public's participation in infrastructure planning. As briefly noted above, this is achieved through consensus politics, whereby the contentious aspects of suburban infrastructure, including questions of who will benefit and for whom this infrastructure is being built, are rendered invisible. How the publicnesss of suburban infrastructure is made invisible through planning is an area requiring further attention. To offer an illustration of the publicness and promise of transport infrastructure, I turn to two case study regions – metropolitan Melbourne and the Greater Toronto and Hamilton Area – where the promise of new suburban transport infrastructure has led to significant citizen-led politicization.

Transport Infrastructure Planning in Melbourne and Toronto Regions

Metropolitan Melbourne and the Greater Toronto and Hamilton Area are two regions coming to grips with the challenges of transport infrastructure planning as these regions continue to grow and as their transport debates become attached to discussions about their respective futures. But realizing these futures has not been an easy process.

Local and state/provincial elections are one site where infrastructure contests are waged. They are also sites where the publicness of urban transport infrastructure is most intensely illuminated. In Canadian and Australian cities, local and regional (provincial/state) elections have become events where a vision for the future is attached to certain infrastructure investments shaping suburban localities in particular ways. In 2010 and 2014, the mayoral race for the City of Toronto became a contest of competing transport visions (Alcoba, 2010; Moore, 2014). Would the city commit to the implementation of light rail investment along key corridors in marginalized inner suburbs that are home to some low-income and visible minority groups (Addie, 2013), or would the city prioritize investment in outer suburban subways? The municipal elections of 2014 also saw a mayoral race in neighbouring Hamilton, a medium-sized city compared to Toronto, with candidates proposing divergent views on the future of their city's transport. The issue of transport infrastructure also featured in the 2018 municipal and provincial elections, with questions about whether a Conservative government led by Doug Ford would continue to commit money to support the construction of light rail transit (LRT) in Hamilton. Ford prevailed, and upon the re-election of the pro-LRT mayor in Hamilton, Ford was reported as saying, "If [Hamilton's mayor] wants an LRT, he's gonna get an LRT" (as cited in Craggs, 2018).

In Hamilton, the LRT project forms part of a wider provincial policy ambition to connect the city to the regional transit network and is one of a number of transport projects that would assist in delineating the region "as a spatial frame in which transportation policy would be formulated" (Addie, 2013, p. 205). The project also serves to strengthen Hamilton's local job market and tax base by attracting growth along the corridor of the proposed LRT line. This project precipitated a passionate debate about the future shape of Hamilton. Questions about what might be lost and gained with the construction of a new LRT line sparked divisions predicated on conflicting desires for the future city, particularly the future of Hamilton's downtown and the relationship the city would have to the rest of the region. Reflecting on these divisions, one interviewee, who lives in Hamilton and is also a representative of the local business community, remarked,

> To me, more than anything, LRT is about [asking], what type of city do we want to be? I would put in parenthesis "again" in that question, but I think what is exciting about it is we don't have to look necessarily to the past. We can think about the fact that we have now, at this point of time, a lot of surface parking lots downtown. So, we have lost a lot of heritage, and it's easy to lament that, and as somebody that loves architecture, I do. But you have got to see it as an opportunity, and every surface parking lot that I see in downtown, I envisage some sort of development or a park instead of a place to stash personal modes of transportation. (Interviewee 101)

Looking to Melbourne, Australia, proposals to build new inner-city motorways have been met with considerable resistance in recent years. In 2014, the proposed AUD$6 billion inner-city underground tollway called the East West Link was cancelled following a hotly contested state election that saw a one-term centre-right Liberal government unseated, resulting in the election of the left-centre Labor Party. The cancellation was costly, and it attracted considerable scorn from multiple media outlets and some key industry bodies (see Legacy 2016, 2017b). However, others viewed this payout as necessary to avoid the long-term social and ecological damage that such a project would inflict, including increased car dependency for people living in the suburbs, both east and west of the city centre, and damages to a much-loved inner-city park. Despite this division, or perhaps because of it, the jubilation that was expressed by the anti–toll-road campaigners was quickly dampened when a few short weeks later the Labor government announced the now AUD$6.7 billion West Gate Tunnel, a market-led project proposed by the toll-road operator Transurban.

This project and what it represented, which was the continued prioritization of road-based infrastructure connecting the suburbs to the urban centre, quickly became politicized. This politicization extended to the 2018 November state election, which became a contest between the two dominant parties, the opposition Liberal Party and the Labor government. This contest was centred on proposed large-scale suburban transport infrastructure, including a ninety kilometre, AUD$50 billion suburban rail loop proposed by the Labor Party, and a high-speed rail to regional cities stopping through suburban localities supported by the Liberal Party. Both are long-term aspirational projects with big promises of reducing further reliance on the car and also moving towards a kind of regionalism that would call on existing towns to grow in an effort to de-centralize population away from Melbourne. But these projects sit in tension with shorter-term projects including the AUD$16 billion North East Link and the AUD$6.7 billion West Gate Tunnel projects, which both parties committed to building, and which would lock suburban residents into ongoing car dependency as investments in public transport infrastructure remain as second-order, longer-term priorities.

These two regions – the Greater Toronto and Hamilton Region (GTHA) and metropolitan Melbourne – provide useful illustrations of how the publicness of suburban transport infrastructure planning can present sites of contestation. Moreover, these infrastructure proposals are large and will transform (or lock in) these urban environments into two different trajectories of mobility and land use patterns. The GTHA proposes light rail infrastructure, symbolizing a return to collective mobility and infrastructure that will aid in the intensification of adjacent land uses. Melbourne, on the other hand, is proposing tollways that will offer exclusionary access to jobs and services in the central business district for those who are willing to pay without offering public transport alternatives, thereby continuing the dominance of private automobiles in the suburban parts of the metropolitan area.

The Promise of Light Rail in Hamilton

Harvey (2018) defines promise as "that which affords a ground of expectation of something to come" (p. 80). In the case of the Hamilton LRT, the promise was to "offer 14-kilometres of safe, rapid and reliable transit from McMaster [University] in the west to Eastgate in the east, connecting communities and protecting for future growth through segregated LRT lines" (Metrolinx, n.d.). In comparison, the promise in the case of the West Gate Tunnel in Melbourne was to "end Melbourne's reliance on the West Gate Bridge by building a new tunnel and new links to the port, CityLink and City" and remove 9,000 trucks off local streets (State Government of Victoria, n.d.). In both cases, the promises were the focus of much debate and sparked broad participation by inhabitants in thinking about the futures these proposals could create.

In the case of Hamilton, the proposed LRT became the focal point of a wider contest between two competing visions and trajectories. In contrast to Melbourne, key decisions about the implementation of the Hamilton LRT, including an environmental assessment process, are in the hands of locally elected politicians, even though the project's funding is to be provided by the province. The councillors in Hamilton are elected at a ward level, meaning that a project like the LRT, where investment is concentrated in a particular area of the city, can ignite tensions between those homeowners and larger businesses who seem to gain by living in or near the city centre, and those who do not. For some, including the anchor institutions in Hamilton represented by big business, university, and medical research institutions, the announcement of the funding commitment was welcomed (McGreal, 2018). The LRT was seen to incentivize investment into a downtown area, including the proposed redevelopment of underutilized land currently being used as large service parking areas. Some city councillors representing wards in the suburbs, however, felt that this money could be better spent upgrading ageing infrastructure or expanding the bus network beyond the city centre (Werner, 2017). There were also concerns that the LRT is "a disaster-in-the-making," and that it will "wreck traffic on one of the key roads through the core and suck up scarce tax revenue to run it" (Moore, 2017).

As seen in recent decades, Hamilton is no stranger to divisive capital works projects or transport projects, including the construction of a controversial city bypass motorway in its western suburbs (Craggs, 2013). The promise of the LRT ignited Hamiltonians into yet another debate about the future of their city, and the connection between the downtown and the suburbs of Hamilton brought residents from different parts of city out of their respective corners to claim a stake in those futures. As Harvey (2018) explains, what the LRT presented for some of the residents of Hamilton was "a sense of hope and expectation" that "never erases the uncertainty, and the promise of infrastructural provision" that may excite "both desire and fear of unpredictable and negative consequences" (p. 82).

Since the announcement of the LRT, an extended debate over the possible impacts of the project has ensued in Hamilton. Residents participated in the planning of the Hamilton LRT, as the promise of the LRT "gestur[ed] toward uncertain futures, even as they attempt[ed] to stabilize and channel the grounds of possibility for futures lives" (Harvey, 2018, p. 99). The debate about the LRT in Hamilton rested as much on the details of the project and its impact on immediate geographies as it did on the role the LRT can play in setting the city on a path towards a twenty-first-century vision of a rejuvenated, bustling manufacturing city. At the centre of this vision are investments in a regional public transport system that would draw on the LRT to connect to existing regional railways. This connection would help facilitate greater interconnections between Hamilton's developing urban centre and the wider region of Toronto (see Addie & Keil, 2015), thereby fulfilling the province's ambitions of establishing an interconnected region with the construction of light rails and increased services from the regional heavy rail network.

The challenge for Hamiltonians was on display at an emotionally charged city council meeting in April 2017. The debate that ensued was over a motion to approve an environmental assessment process for the LRT. While the vote eventually passed 10–5, the emotion over the course of the two-and-a-half-hour meeting was palatable as city councillors took turns to express their commitment to making the right decision for their residents and for the city at large, be that to support the LRT or not. The questions of what was considered "right" for some councillors was not clear, and some noted that they would continue to reserve the right to change their mind about the project in the future, which was a position that continued to be held by some councillors in the lead up to the 2018 municipal election (Bay Observer Staff, 2018). Coverage in the local media the following day highlighted some of the key statements made at the meeting as councillors spoke passionately about their city and critically about the level of division this project was creating by pitting urban and suburban inhabitants against each other. In this light, statements about how much the vote "weighed heavily on every councillor" and about how this project "continues to divide the community, giving the false hope [that the project can now be stopped]" (Craggs, 2017) captured the anxiety Hamiltonians had about their future, which manifested as uncertainty about LRT.

Defending the Publicness of Transport Infrastructure Planning in Melbourne

Honig's (2017) concept of the publicness of infrastructure opens new avenues for exploring the powerful role of suburban transport infrastructure as "transitional objects" that help consitute democratic life (p. 11). In the context of transport infrastructure planning, this can ordain new conditions that will shape people's mobility, and the perception of those new mobilities can ignite citizens into a form of political participation to contest those futures. For Hamilton and many other medium-sized cities seeking to set a new trajectory of movement in an increasingly interconnected region, an announced infrastructure project can cast a technology like LRT as a transitional object symbolizing a commitment to a new form of mobility and incentivizing a kind of new regionalism (Addie & Keil, 2015).

Looking to Melbourne, the planning of the West Gate Tunnel toll road was conducted to deliberately undermine the publicness of infrastructure planning. Following the contested and now cancelled East West Link project, there was a considerable effort made by the newly elected Labor government to reduce the visibility of this project and dampen its publicness, which was achieved by subjecting the evaluation of this project to the newly created market-led proposal scheme. This scheme enables private sector actors to propose unsolicited proposals that are then assessed by state governments without direct engagement with suburban inhabitants about the efficacy of such proposals or the availability of an assessment framework to guide the process (Woodcock et al., 2017). Instead, the project proponents can present the project as a "solution" to a transport problem without debate or adherence to an existing transport- and land-use plan. In the case of Melbourne, there is no current regional transport plan to help guide such assessments either.

In response to the level of opacity associated with this process, a series of grassroots campaigns in the middle and inner suburbs emerged to challenge the efficacy of the West Gate Tunnel proposal. These campaigns opposed the project based on the negative amenity impacts it would create, and significant effort was also directed towards challenging the exclusionary processes upon which decisions were being based. A report written by six planning academics (including myself) and signed by twenty-eight built environment scholars from across the State of Victoria was published in the days before contract signing in December 2017 (Woodcock et al., 2017). The aim here was to build a public consciousness of the deficiencies of the project's planning and governance by giving these issues media attention and helping to position this project as politically contestable in the lead up to the 2018 Victorian state election. Building on an existing debate waged across a number of different media outlets in the weeks leading up to the signing of the contracts with the private sector proponent, there were consistent calls for an integrated transport plan to be devised before any new infrastructure investments were announced (Legacy & Woodcock, 2016). A plan would provide the guiding vision and framework to assess unsolicited projects like the West Gate Tunnel against wider social, environmental, and land-use aspirations for the city and region and provide the region with a clear vision for the future.

On the other hand, for those communities in Melbourne that had been asking for an infrastructural response to removing trucks off local residential streets near the Port of Melbourne, the West Gate Tunnel project was met with cautious optimism. As one community organizer I spoke to explained, the result was some groups choosing to "advocate for the best possible outcome for the project, rather than outright protesting it" (Community Organizer 4). This caution was anchored to twenty years of lobbying and ten years of promises in the form of a Truck Action Plan in 2008, and the promise of ramps to reduce local truck traffic on suburban streets as part of the Labor Party's 2014 election platform (Victorian Labor Party, 2014). Neither of these promises had come to fruition, but there was hope that the promise of the West Gate Tunnel could finally address their local concerns. For the resistance campaigns located at the points along the project corridor where access to the toll roads would be provided, the concern was that these trucks would become displaced due to truck operators avoiding tolls (Community Organizer 2). There was also a concern that the project would lock in car dependency without alternative public transport options put in place (McDougall, 2018). The promises coupled to this project became the focus of a multi-dimensional grassroots campaign. Initially, groups formed that were focused on the impacts of the project. Eventually, however, these local efforts would begin to stitch themselves together, drawing from citizen-led campaigns across a range of suburban localities impacted by ongoing motorway construction to form a coordinated effort to contest this project. The effort was facilitated by an existing environmental advocacy group, which attempted to bring these grassroots groups together to achieve a common focus. At the time of writing, these campaigns continue.

Conclusion

The connections between urban infrastructure and citizen participation in city-regions are in constant flux. They can sometimes attract citizen-led politicization, particularly from those affected by proposals or those believing that a different course ought to be invested in. Given the high costs of suburban infrastructure and the risks involved in its construction, there might be interest in managing the political risks that mount. This can be facilitated by achieving wide agreement that a project is necessary so as to minimize political risk. Nearly thirty years of privatization of public assets and public-sector retrenchment has given rise to a system of infrastructure planning in neoliberal planning contexts that is splintered (Graham & Marvin, 2001), opaque and exclusionary, and can oftentimes lack strategic and integrated planning (Dodson, 2009). This splintering has been introduced through complex public and partnership arrangements to finance, build, operate, and maintain infrastructure, while new market-led proposal schemes allow private sector participation to extend into strategic planning. The extent to which this removes infrastructure from direct democratic control by veiling decision-making processes under commercial-in-confidence laws remains an area of ongoing interest.

Yet, as Honig (2017) writes, suburban transport infrastructure planning is inherently a public process. In the planning stages, the promises transport infrastructure embodies contains a transformational and political quality that renders it visible and public. The spatial politics, regionalization, and the jurisdictional challenges – not to mention mediating the promises embodied in an infrastructure project about the future in these complex places – demands that citizen participation is observed in a way that is sensitive to the formation of political suburban subjects emerging in reaction to these projects and processes.

Even so, the publicness of transport infrastructure planning processes like those experienced in Melbourne, where planning decisions and infrastructure assessment are shielded from full public engagement and scrutiny, becomes compromised but never fully destroyed. Drawing on the work of Jacques Rancière, Larkin (2018) describes politics as that which "takes place when those who occupy fixed positions outside a certain order decide to intervene within that order. It is the apportioning that determines who can participate in a system" (p. 187). Urban inhabitants respond to this veiling by politicizing not only the project that may have been the focus of their initial indignation but also by focussing on the processes of infrastructure planning. Certainly, in the context of Melbourne, concerned residents, academics, and grassroots groups looked to insert themselves into the politics of transport infrastructure planning by contesting the process and offering an alternative planning pathway in the creation of an integrated regional transport plan. This helped support a sustained backlash against the State of Victoria, which had yet to commit to a single integrated transport plan upon which the future could be mined.

The publicness of infrastructure planning is also illuminated through the promises infrastructure offers about the future of the urban–suburban relationship. Transport infrastructure is oftentimes city- and region-shaping, and this transformational quality intensifies these promises as the social, environmental, economic, and geographic stakes become significantly higher and multi-scalar in complexity. Infrastructure promises become the symbol of a trajectory the region has set itself on, and so the pressure for the actors in infrastructure planning to "succeed" in landscaping a positive perception of transformation becomes important and deeply political. In the case of Hamilton, some of the questions about the transformational quality of LRT infrastructure, which needed to be situated into a wider regional conversation about the changing ways citizens will be expected to move and how the region is connected, were debated at length by pro- and anti-LRT camps. Infrastructure can generate effects that are sometimes difficult to predict (Filion & Keil, 2017). Yet, the ways this type of infrastructure becomes political is mediated by suburban inhabitants and how they perceive the promises of future infrastructure (see Anand et al., 2018). What the politics of suburban infrastructure planning seemed to achieve in the City of Hamilton was a public debate about the transformational aspect of LRT for a city like Hamilton situated in a region so near to Toronto and the need to plan with and for this transformation. In the case of both city-regions, Melbourne and the GTHA, it was the political participation of citizens and locally elected officials that elevated into public consciousness not only the importance of suburban infrastructure and its planning but also the need to take an integrated and regional outlook to understand the politics that suburban infrastructure promises wield.

REFERENCES

Addie, J.P.D. (2013). Metropolitics in motion: The dynamics of transportation and state reterritorialization in the Chicago and Toronto city-regions. *Urban Geography, 34*(2), 188–217. https://doi.org/10.1080/02723638.2013.778651.

Addie, J.P.D. (2016). Theorising suburban infrastructure: A framework for critical and comparative analysis. *Transactions of the Institute of British Geographers, 41*(3), 273–85. https://doi.org/10.1111/tran.1212.

Addie, J.P.D., & Keil, R. (2015). Real existing regionalism: The region between talk, territory and technology. *International Journal of Urban and Regional Research, 39*(2), 407–17. https://doi.org/10.1111/1468-2427.12179.

Alcoba, N. (2010, December 1). "The war on the car is over"… and so is Transit City: Rob Ford. *National Post*. https://nationalpost.com/posted-toronto/the-war-on-the-car-is-over-and-so-is-transit-city-rob-ford.

Allmendinger, P., & Haughton, G. (2012). Post-political spatial planning in England: A crisis of consensus? *Transactions of the Institute of British Geographers, 37*(1), 89–103. https://doi.org/10.1111/j.1475-5661.2011.00468.x.

Anand, N., Gupta, A., & Appel, H. (Eds.). (2018). *The promise of infrastructure.* Duke University Press.

Appel, H., Anand, N., & Gupta, A. (2018). Introduction: Temporality, politics, and the promise of infrastructure. In N. Anand, A. Gupta, & H. Appel (Eds.), *The promise of infrastructure* (pp. 1–38). Duke University Press.

Baeten, G. (2000). The tragedy of the highway: Empowerment, disempowerment and the politics of sustainability discourses and practices. *European Planning Studies, 8*(1), 69–86. https://doi.org/10.1080/096543100110938.

Bay Observer Staff. (2018, August 15). LRT showdown unavoidable: Ford reaffirms pledge to free Council of transit funding. *Bay Observer.* http://bayobserver.ca/lrt-showdown-unavoidable/ (site no longer exists).

Bissell, D. (2018). *Transit life: How commuting is transforming our cities.* MIT Press.

Boyer, D. (2018). Infrastructure, potential energy, revolution. In N. Anand, A. Gupta, & H. Appel (Eds.), *The promise of infrastructure* (pp. 223–43). Duke University Press.

Cantarelli, C.C., & Flyvbjerg, B. (2013). Mega-projects' cost performance and lock-in: Problems and solutions. In H. Priemus & D. van Wee (Eds.), *International handbook on mega-projects* (pp. 333–60). Edward Elgar.

Craggs, S. (2013, January 2). 5 years later, was the Red Hill Valley Expressway worth it? Community still divided over benefits of the controversial highway. *CBC News.* https://www.cbc.ca/news/canada/hamilton/headlines/5-years-later-was-the-red-hill-valley-parkway-worth-it-1.1275560.

Craggs, S. (2017, April 26). Hamilton LRT clears a major council hurdle, will move ahead. *CBC News.* https://www.cbc.ca/news/canada/hamilton/lrt-environmental-assessment-1.4087610.

Craggs, S. (2018, November 28). Doug Ford says if Hamilton's mayor wants LRT, he'll get LRT. *CBC News.* https://www.cbc.ca/news/canada/hamilton/ford-lrt-1.4924067.

Dodson, J. (2009). The 'infrastructure turn' in Australian metropolitan spatial planning. *International Planning Studies, 14*(2), 109–23. https://doi.org/10.1080/13563470903021100.

Filion, P., & Keil, R. (2017). Contested infrastructures: Tension, inequity and innovation in the global suburb. *Urban Policy and Research, 35*(1), 7–19. https://doi.org/10.1080/08111146.2016.1187122.

Flyvbjerg, B. (2008). Curbing optimism bias and strategic misrepresentation in planning: Reference class forecasting in practice. *European Planning Studies, 16*(1), 3–21. https://doi.org/10.1080/09654310701747936.

Graham, S., & Marvin, S. (2001). *Splintering urbanism: Networked infrastructures, technological mobilities and the urban condition*. Routledge.

Gupta, A. (2018). The future in ruins: Thoughts on the temporality of infrastructure. In N. Anand, A. Gupta, & H. Appel (Eds.), *The promise of infrastructure* (pp. 62–79). Duke University Press.

Harvey, P. (2018). Infrastructures in and out of time: The promise of roads in contemporary Peru. In N. Anand, A. Gupta, & H. Appel (Eds.), *The promise of infrastructure* (pp. 80–101). Duke University Press.

Henderson, J. (2013). *Street fight: The politics of mobility in San Francisco*. University of Massachusetts Press.

Honig, B. (2017). *Public things: Democracy in disrepair*. Oxford University Press.

Larkin, B. (2013). The politics and poetics of infrastructure. *Annual Review of Anthropology, 42*, 327–43. https://doi.org/10.1146/annurev-anthro-092412-155522.

Larkin, B. (2018). Promising forms: The political aesthetics of infrastructure. In N. Anand, A. Gupta, & H. Appel (Eds.), *The promise of infrastructure* (pp. 175–202). Duke University Press.

Legacy, C. (2016). Transforming transport planning in the postpolitical era. *Urban Studies, 53*(14), 3108–24. https://doi.org/10.1177%2F0042098015602649.

Legacy, C. (2017a). Infrastructure planning: In a state of panic? *Urban Policy and Research, 35*(1), 61–73. https://doi.org/10.1080/08111146.2016.1235033.

Legacy, C. (2017b). Is there a crisis of participatory planning? *Planning Theory, 16*(4), 425–42. https://doi.org/10.1177/1473095216667433.

Legacy, C., Curtis, C., & Scheurer, J. (2017). Planning transport infrastructure: Examining the politics of transport planning in Melbourne, Sydney and Perth. *Urban Policy and Research, 35*(1), 44–60. https://doi.org/10.1080/08111146.2016.1272448.

Legacy, C., & Woodcok, I. (2016, October 4). Victoria needs a big-picture transport plan that isn't about winners v losers. *The Conversation*. https://theconversation.com/victoria-needs-a-big-picture-transport-plan-that-isnt-about-winners-v-losers-65567.

Mäntysalo, R., Saglie, I.L., & Cars, G. (2011). Between input legitimacy and output efficiency: Defensive routines and agonistic reflectivity in Nordic land-use planning. *European Planning Studies, 19*(12), 2109–26. https://doi.org/10.1080/09654313.2011.632906.

McDougall, W. (2018, January 6). Our ridiculous frenzy of road construction will swallow up resources for two decades. *Sydney Morning Herald*. https://www.smh.com.au/opinion/our-ridiculous-frenzy-of-road-construction-will-swallow-up-resources-for-two-decades-20180105-h0dwd0.html.

McGreal, R. (2018, November 2). Anchor institutions: "Full speed ahead on Hamilton LRT." *Raise the Hammer*. http://raisethehammer.org/article/3594/anchor-institutions:_full_speed_ahead_on_hamilton_LRT.

Metrolinx. (n.d.). *Hamilton LRT project.* Retrieved 3 September 2018, from http://www.metrolinx.com/en/projectsandprograms/projectpages /Hamilton.aspx.

Moore, O. (2014, October 3). John Tory's transit plan: 53 kilometres, 22 stations and 15-minute service. *Globe and Mail.* https://www .theglobeandmail.com/news/toronto/john-torys-transit-plan-a-new -smarttrack-and-tough-questions/article20923882/.

Moore, O. (2017, April 18). Off the rails: What a billion-dollar LRT means for Hamilton. *Globe and Mail.* https://www.theglobeandmail.com/news /toronto/off-the-rails-what-a-billion-dollar-lrt-line-means-forhamilton /article34731605/.

O'Neill, P.M. (2010). Infrastructure financing and operation in the contemporary city. *Geographical Research, 48*(1), 3–12. https://doi.org /10.1111/j.1745-5871.2009.00606.x.

State Government of Victoria. (n.d.). *West Gate Tunnel project: Project overview.* Victoria's Big Build. Retrieved 3 September 2018, from https://bigbuild .vic.gov.au/projects/west-gate-tunnel-project/about/explore-the-project /overview.

Stone, J. (2014). Continuity and change in urban transport policy: Politics, institutions and actors in Melbourne and Vancouver since 1970. *Planning Practice and Research, 29*(4), 388–404. https://doi.org/10.1080/02697459.2013 .820041.

Swyngedouw, E. (2018). *Promises of the political: Insurgent cities in a post-political environment.* MIT Press.

Victorian Labor Party. (2014). *Project 10,000: Trains. Roads. Jobs.* Office of the Leader of the Opposition. https://www.sanjeev.sabhlokcity.com/Misc /Victorian-Labors-Project-10000.pdf.

Vigar, G. (2013). *The politics of mobility: Transport planning, the environment and public policy.* Routledge.

von Schnitzler, A. (2018). Infrastructure, apartheid technopolitics, and temporalities of "transition." In N. Anand, A. Gupta, & H. Appel (Eds.), *The promise of infrastructure* (pp. 133–54). Duke University Press.

Walks, A. (2015). Stopping the "war on the car": Neoliberalism, Fordism, and the politics of automobility in Toronto. *Mobilities, 10*(3), 402–22. https:// doi.org/10.1080/17450101.2014.880563.

Werner, K. (2017, May 1). Stoney Creek's councillors united on opposing proposed LRT project. *Stoney Creek News.* https://www.toronto.com/news -story/7270122-stoney-creek-s-councillors-united-in-opposing-proposed -lrt-project/.

Woodcock, I., Sturup, S., Stone, J., Pittman, N., Legacy, C., & Dodson, J. (2017). *West Gate Tunnel: Another case of tunnel vision?* The University of Melbourne and RMIT University. https://apo.org.au/sites/default/files/resource-files /2017-12/apo-nid122421.pdf.

Image 9. Shanghai, China (May 2015)

9 Suburban Infrastructures: Benevolent Public Domain and Instruments of Control and Power

PIERRE FILION

Introduction

In this commentary on one aspect of infrastructures first raised in the previous chapter by Crystal Legacy, I address the benevolent, or potentially benevolent, role that Bonnie Honig (2017) attributes to the public domain, which can be understood to include public-sector infrastructures. I agree with this perspective, provided it is the object of certain caveats and considered alongside another antithetical outlook on infrastructures centring on their capacity to control behaviour and, thereby, entrench power relations. My commentary thus proposes a Janus-faced understanding of infrastructures. The first understanding depicts infrastructures as belonging to a universally accessible public domain. Such infrastructures open new opportunities and can thus be perceived as enhancing freedom and quality of life. In this depiction, by fully playing their role of enablers of activities, infrastructures enhance possibilities available to individuals. The second understanding emphasizes the capacity of infrastructures to impose behaviour patterns and therefore play a role of social control within society. Such control compels users of infrastructures to behave according to norms that either are imposed by administrators or conform to commonly accepted conduct standards in infrastructure-related circumstances. The control dimension of infrastructures also pertains to their external impacts, such as when they advantage certain sectors or social groups at the expense of others. Infrastructures can thus be perceived as modes of social control, contributing to common societal norms and to their adherence, while in the process extending the power of dominant interests. From a Foucauldian perspective, they are instruments of micro-power structuring daily life (Foucault, 1991, 1998).

The attention of this chapter is on Canadian suburban infrastructures, with an emphasis on transportation in Toronto-region suburbs. Because Canadian suburbs share many features with their US counterparts, the commentary has continental-wide relevance and may also pertain to the Australian situation addressed previously in chapter 8. It investigates the public domain and control dimensions of suburban infrastructures over the evolution of the post–Second World War suburb, the automobile-dependent form of development that has rapidly come to dominate North American metropolitan patterns. This approach makes it possible to grasp the influence on suburban infrastructure decision-making of different political climates, which are more or less conducive to participatory democracy, the relation between the state and the private economy, the performance of the economic system, and the distribution of wealth. Three stages in the evolution of the post-war suburb are considered here. These are the Fordist suburb, the early neoliberal suburb, and the mature neoliberal suburb. In this commentary, the two first stages serve as background to the third stage, which describes present situations affecting suburbs and explores possible future outcomes. The commentary sets suburban infrastructures within their evolving societal context.

Infrastructures as Public Domain and Social Control and Power

This section explores two dimensions of infrastructures – their public domain and their social control and power roles – within the context of the suburb. It depicts variations in the balance between these two dimensions according to evolving societal contexts. A simplified version of the argument would suggest the existence of an association between advancing neoliberalism and a depletion of the public domain role of infrastructures accompanied by an intensification of their social control function. But as the discussion to follow reveals, the reality is more complicated, as infrastructures are rarely pure manifestations of a universally accessible public domain or of control instruments whose predictable outcomes can be reliably harnessed.

A Benevolent Public Domain

The infrastructure-as-public-domain perspective presents infrastructures as accessible to all and capable of improving quality of life and economic performance. In this benevolent portrayal, infrastructures are a universally accessible state-provided public good. Ideally, these infrastructures are themselves the product of a democratic public domain defined by participatory decision-making processes. Such a depiction of infrastructures, and of the democratic processes undergirding them, is most compatible with periods of social consensus, either of an authentic nature or as a result of the muzzling of discordant perspectives (Newman, 2016); in other words, during periods when conflicting views on infrastructures are suppressed. Consensus prevailed during the 1950s and 1960s modernist era when belief in progress was widespread, but this was also a period that repressed counter-ideologies, such as early manifestations of environmentalism. Generalized enthusiasm for large-scale involvement of the public sector in the provision of infrastructures in the post–Second World War period can be related to the need to catch up after the Great Depression and the war years, which slowed infrastructure development. In the early 1950s, there was a wide gap between rising expectations from the public and the condition of infrastructures. At the time, the public sector also reacted to the poor performance of infrastructures in the hands of private companies by taking them over and reinvesting in formerly privately owned infrastructures. This was notably the case for public transit services.

References to the public-domain understanding of infrastructures become a rallying cry against expensive user fees, the outcome of different forms of privatization and fare setting policies, which restrict access to infrastructures. Examples include toll expressways or lanes and high-fare public-transit services such as those that connect downtown areas to airports (Haughton & McManus, 2012). These infrastructures contribute to a splintering of society between the rich who can benefit from these infrastructures and others who must contend with lesser order infrastructures or no infrastructure at all (Bakker, 2011; Graham & Marvin, 2001). They illustrate a marketization of infrastructures in direct opposition to a universal public-domain understanding of what infrastructures should be. Another consequence of market influence on infrastructure development is oversupply of infrastructures of a type for which, and in places where, people can pay expensive fees, contrasted by deficient provision in other situations (Clarke Annez, 2006). Instead of playing the role of society-wide meeting grounds, infrastructures become instruments of social fragmentation. The defence of state involvement in, and fair access to, public domain–type infrastructures is to be expected when confronted with the adverse consequences of infrastructure marketization. Although this happens infrequently in the North American suburban context, public-domain delivery and operation of infrastructures can also be carried out by cooperative and community-based types of organizations.

While there is no denying the fact that there are many affordability and accessibility advantages to state-provided infrastructures relative to their market-driven provision, one should not exaggerate the public-domain characteristics of public-sector infrastructures. Universally accessible public-domain infrastructures have always been more myth than reality. Low-fare, good-quality public-transit networks are only accessible to people whose origins and destinations are found within areas serviced by such networks, a rare occurrence in North American suburbs. Access to such infrastructures is stratified by location. In a similar vein, being able to use a car is a prerequisite to fully benefit from the accessibility advantages of expressways and arterials. And toll-free expressways are often congested, which means that in the absence of tolls, drivers pay for the privilege of using these expressways with their time. The use of allegedly public-domain infrastructures is nearly always restricted, even when fares are relatively low.

Just as the actual public-domain features of infrastructures fail to live up to idealized models, so does infrastructure decision-making. Examples of democratic participatory and consensual decision-making pertaining to infrastructures are scarce. More usual are decision-making processes dominated by experts or opposing constituencies seeking rewards stemming from infrastructure development or agitating against the negative externalities of infrastructures. What is more, as discussed in chapter 8 of this volume, the privatization of infrastructures pares down democratic decision-making concerning infrastructure planning and development to a minimum, excluding the public from dealmaking between governments and corporations engaged in infrastructure development and operation. Public participation is kept to a minimum, as is the dissemination of information on private infrastructure projects.

As a reaction to the current fashion of infrastructure privatization, fanned, as it is, by neoliberal ideology, we can expect renewed interest for the public-domain understanding of infrastructures. In the present climate, the public-domain depiction offers an alternative to current political and financial thinking on infrastructures, even if, as seen, the depiction this alternative provides has never been fully grounded in reality.

The Social Control and Power Dimension of Infrastructures

I will now consider the antithetical perspective on infrastructures. From this perspective, infrastructures are no longer portrayed as instruments whose purpose, driven by democratic principles, is primarily to improve living conditions for all or at least a majority. In the public-domain view, problems confronting infrastructures do not relate so much to the nature of infrastructures, as to obstacles preventing them from fulfilling their societal advancement mandate, causing an uneven distribution of their benefits. The emphasis turns here to the control and power dimension of infrastructures – to their capacity to shape behaviour and societal outcomes in a manner that mirrors unequal power and economic relations within society. In this interpretation, control is not due to external influences on infrastructures but is a feature inherent in infrastructures.

The discussion here is inspired by Foucault's reflections on power, which he depicted as diffused throughout different aspects of society and embedded in locales not habitually associated with the exercise of power. Foucault insisted on the importance of micro-power manifesting itself in daily situations and thereby setting behavioural norms (Foucault, 1991, 1998). The most visible effect of this capillary distribution of power is the regulation of bodies – their movement, their location, their appearance. For Foucault, such an exercise of power is not necessarily a source of negative consequences – it is, after all, essential to the functioning of any society – but it constricts behaviour by promoting certain patterns of behaviour to the exclusion of others (Gaventa, 2003).

The control and power dimension of infrastructures can also be interpreted in light of more conventional perspectives on power by looking at which interests are advantaged or disadvantaged by infrastructures and their repercussions. It is, indeed, inherent in the nature of infrastructures to distribute their positive and negative externalities unevenly. Certain parts of cities are advantaged by benefits stemming from their presence, while others suffer from the absence of such benefits. And some areas must endure the negative externalities of infrastructures, while others are at a comfortable distance from these effects. The manipulation of the distribution of these benefits and disadvantages by powerful groups causes infrastructures to perpetuate and accentuate social inequalities.

The control dimension of infrastructures takes place at two levels. First, there is the role infrastructures play in assuring the overall functioning of society, and therefore in perpetuating its present organization, including systems that assure collective survival and those that maintain social stratification. The second control dimension relates to behavioural norms confronting people using infrastructures – for example, norms determining what is expected from pedestrians, motorists, and public transit passengers. These norms prevent the emergence of alternative behaviour patterns, just like infrastructures can obstruct societal change on a broader scale. While complying with expected patterns of behaviour when using infrastructures, individuals do not engage in other forms of activities. In Foucauldian terms, these norms discipline the body and are part of a continuum of behaviour-control mechanisms running throughout society. Codes of behaviour operating within infrastructures occupy the interstices of institutional realms, each dominated by recognizable norm-setting authority relations.

Among manifestations of the control and power dimension of infrastructures are attempts to harness the influence of infrastructures on the perception of society and the political orientation of their users. One example is the fierce opposition of the Americans for Prosperity lobby financed by the Koch brothers (who have large holdings in the coal and oil industry and who campaign for conservative and libertarian causes) to new public transit initiatives in the US. The lobby's anti-transit argument revolves around the high cost of these initiatives and their reliance on subsidies. The collectivist nature of public transit is also incompatible with Americans for Prosperity's libertarian ideology, mirrored in its strong support of the individualism and freedom of choice afforded by the car and services such as Uber (Tabuchi, 2018). One might also surmise that this advocacy group's opposition to public-transit initiatives is grounded in an awareness that reliance on collective consumption, such as public transit, shapes political views in a way that is contrary to the group's libertarian ideology.

Uncertain Infrastructure Effects

While power, economic, and quality-of-life advantages can undeniably flow from successful efforts at influencing infrastructure decisions, it is important at the same time to acknowledge the uncertainty of long-term and far-reaching repercussions of infrastructures. Infrastructures are indeed fraught with unexpected consequences. Over time, people may use them in unanticipated ways, and their reverberations on different aspects of society can be sources of unforeseen effects. In terms of time and space, the more distant the reverberations of infrastructures, the more unpredictable they are. Moreover, infrastructures create powerful flywheel effects, meaning that existing systems cast long legacies. It is not accidental that after seventy years of massive road and highway investment, the foremost urban transportation innovation on the horizon, autonomous vehicles, are designed so they can operate on existing road and highway networks. Path dependencies inherent in infrastructures force policymakers to operate within the context of existing infrastructure systems. The combined effects of the unpredictability of the consequences of infrastructures, and of the path dependencies they generate, point to limitations to the extent to which they can be used as instruments to advantage durably the interests of influential social groups.

Suburban Infrastructures

I now turn my focus to the Toronto-region suburbs, which represent a variety of the Canadian and, more generally, North American suburbs. Canadian suburbs share the same general morphological structure as their US counterparts, shaped in both cases by an adaptation of land use to nearly generalized reliance on the automobile. But some traits distinguish Canadian suburbs from US suburbs; for example, suburbs in Canada are generally planned by larger municipal or regional entities and register higher (albeit still very low compared to central cities) public-transit patronage. The present section explores conditions affecting infrastructure development. It also discusses the manifestation of the public domain and social control aspects of infrastructures in three stages of suburban development over the last seventy years: (1) the Fordist suburb, (2) the early neoliberal suburb, and (3) the mature neoliberal suburb. For a similar narrative that chronicles the evolution of suburban areas through Fordism and neoliberalism, see Peck et al. (2014), who focus their attention on Vancouver, and who portray the spread of the suburban lifestyle across the metropolitan region, including the downtown. Peck (2011) also addresses more directly the connection between neoliberalism and the North American suburban phenomenon.

The Fordist Suburb

In its present understanding, the Fordist suburb period is bookended by the late 1940s and the late 1970s; that is, in the first case by the beginning of an all-out public-sector commitment to car-oriented infrastructure and land-use planning in North American suburban development, and in the second case by the rise of the influence of neoliberalism on policymaking. This phase of suburban development was characterized by the invention of a new urban form, fully adapted to generalized automobile reliance – hence, its low density and functionally specialized environment contrasting with the higher density and mixed use of the more public transit–reliant central city. Especially at the beginning of the period, Fordist suburbs represented a minority of their respective metropolitan regions, which means that living there was a matter of choice based on financial capacity and lifestyle preference. In these early years, the Fordist suburb corresponded largely to the post-war middle-class and family-oriented suburban stereotype. Yet, in true Fordist form, public-sector compensatory measures, such as public housing and limited dividend rental units, gave suburban access to some low-income residents.

Fordism was defined by a combination of mass production and measures stimulating demand, thus creating a virtuous cycle between expanding manufactured goods output and expanding markets for their consumption. As the spatial fix of Fordism, the suburb was the primary locale for mass production and mass consumption. It benefited from considerable infrastructure investment promoting reliance on cars and low-density living with its stimulating effect on the accumulation of durable goods. The general prosperity of the Fordist period translated into flush public-sector budgets and plentiful funding for suburban infrastructures. A number of financial circuits supported the development of the Fordist suburb. There were, for example, mortgages, many of which were guaranteed by the Central Mortgage and Housing Corporation (CMHC), a federal government agency. The building of local infrastructures was largely funded by development charges raised from developers. Reliance on development charges meant that local infrastructures could be provided without burdening the municipal tax system. Other municipal infrastructures were financed by municipal tax revenues and grants from senior governments. Finally, major suburban infrastructures with a metropolitan scope such as expressways, subways, commuter trains, and trunk sewer systems were generally paid for by senior governments.

Until later in the Fordist period, open disagreement over infrastructures was limited. Infrastructures were mostly perceived as a universally accessible public domain, hence the wide endorsement of more and improved infrastructures. Such a perception was tied to the apparent unanimity around the suburban middle-class lifestyle. It seemed as if all suburbanites enjoyed, or aspired to, this lifestyle. It stood to reason in these circumstances that infrastructures supporting the suburban middle-class lifestyle would be positively viewed. Of course, the apparent unanimity over this lifestyle did not account for those who did not have access to the car, the condition necessary to be able to fully use suburban infrastructures and partake in the suburban lifestyle. At the time, housewives constituted a large proportion of the car-less suburbanites. The seeming generalized adherence to the suburban lifestyle came at a heavy cost: the repression of non-conforming behaviours and the collective and individual neuroses to which this repression gave rise. These neuroses were an object of predilection for the fiction of the time (e.g., Keats, 1956; Yates, 1961). In reality, suburban infrastructures were a formidable tool of social control, forcing suburbanites into a behavioural matrix organized around the car and ruling out other patterns of conduct and suburban development. Resulting suburban households' consumption patterns coincided with the demand-side requirements of Fordism.

The Early Neoliberal Suburb

In the 1970s, suburbs acquired a majority status within metropolitan regions in demographic, economic, and spatial terms. As a result, they ceased to be a place people chose to pursue an innovative lifestyle departing from then majority urban living norms. Henceforth, suburbs became the de facto residential choice for a majority of metropolitan-region residents. Two consequences ensued from the majority status of suburbs. First, it meant that the lifestyle associated with dependence on the automobile and low-density and functionally specialized land use became generalized across metropolitan regions. Second, suburbs became differentiated in terms of income, age, and ethnic/racial categories, which translated into a diversification of values and preferences. These transformations were accentuated by the income polarization provoked by trade globalization and the reduction of taxation and state redistribution, mirroring governments' growing adherence to neoliberalism (Harvey, 2005; Jessop, 2002). If the previous Fordist period was defined by the emergence of a new suburban form and culture, the early neoliberal suburb was a period of suburban transition.

The period associated with the early neoliberal suburb (from the late 1970s to the 2010s) was marked by successive waves of public expenditure cutbacks. The curtailment of government spending impacted suburban infrastructure development and the overall trajectory of suburbs (Wellman & Spiller, 2012). Local infrastructures, funded by development charges, were an exception as their implementation could proceed as it had in the past, which allowed suburban growth to unfold unimpeded by the deteriorating financial capacity of governments. The situation was, however, different for infrastructures funded by senior governments. Repeatedly over this period, governments implemented budget cuts in reaction to falling revenues due to recessions and reduced tax rates, and reflecting efforts to contain or eliminate deficits. Repercussions on suburban infrastructure development were severe. Expanding suburbs faced a serious lag in infrastructures meant to interconnect them with the metropolitan region, which translated into worsening highway congestion. At the same time, while a diversifying population required both automobile- and public transit-oriented options, the creation and development of suburban public-transit systems were restricted by financial constraints. To be sure, there have been localized suburban public-transit improvements, but they were rarely of a scale sufficient to allow suburbanites to access efficiently the range of destinations they need to reach on a daily basis.

The combination of government budgetary constraints and the private-sector bias of the neoliberal ideology encouraged the privatization of infrastructures (Parkin, 1994). Public administrations relied on public–private partnerships (Siemiatycki, 2013), examples of which include the Viva Transit bus rapid transit system in York Region, north of Toronto, and the Crosstown underground light rail transit (LRT) line in the City of Toronto. There was also the failed public–private partnership (PPP) UP Express (a train linking the Toronto Pearson International Airport to downtown Toronto), which led to a government takeover of the building and operation of the line. In most instances, the implications of this financing and management formula on users were limited, as PPP infrastructures operate within the parameters of existing infrastructure systems with similar fares and subsidy regimes as in the public-sector controlled remainder of infrastructure systems. The situation was different in the case of outright privatizations. In these instances, tolls and fares are expected to cover all costs plus generate a profit to owners. There have not been many examples of such outright privatization in suburban Toronto, largely because of the unpopularity of the Highway 407 experience. The highway was built and originally operated as a PPP. However, in 1998, in order to reduce the provincial deficit on the eve of an election, the Progressive Conservative government sold Highway 407 to a private consortium without restrictions on tolls, so the province could maximize the selling price (Cohn, 2015). It became one of the highways with the most expensive tolls in North America (Hauch, 2012).

The discussion now addresses the public-domain and social-control dimensions of infrastructures in the context of the early neoliberal suburb. The privatization of infrastructures that actually happened, and the possibility of further such privatizations, raised the spectre of splintered accessibility to suburban infrastructures, thus impeding their status as a universally accessible public domain. For example, because of Highway 407's high tolls, low-income car drivers are denied the use of this thoroughfare for regular commuting purposes. The public-domain nature of infrastructures also suffered from the deterioration and incapacity of such infrastructures to keep up with suburban development due to insufficient public-sector investment.

The early neoliberal suburb also felt the social control role of infrastructures. Because of insufficient resources to provide infrastructure options in a way that would accommodate the needs and preferences of an increasingly diversified suburban population, suburbanites faced limited choices. Infrastructure systems imposed a car-oriented lifestyle – the legacy of the Fordist suburb – on a population that could have given rise to different suburban lifestyles had more transportation options been available.

The Mature Neoliberal Suburb

I end with a reflection on contemporary and future suburban infrastructures. I first take stock of the present condition of the neoliberal suburb. Contemporary suburbs are shaped by the intensification over time of the effects of trends originating in the early neoliberal suburb: the growing importance of the suburban realm within metropolitan regions and income polarization at the societal scale. The combined effects of these two factors have produced suburbs that have become more diverse than the central city. Whereas in the previous phase suburbs were engaged in a neoliberal transition, in the present phase they have been shaped by decades of neoliberalism.

The mature neoliberal suburb has not, however, lived up to the expectations of the early political champions of neoliberalism (Kotz, 2015). There was no large neoliberal wave, which would have led to a massive privatization of public-sector infrastructures and services. This is largely due to disillusionment with privatization portrayed, in the infancy of neoliberalism, as a silver bullet capable of catering to suburban infrastructure needs. Unlike PPPs, which despite being widely used have mostly left the public indifferent, all-out privatization has triggered protest against high access costs and the discriminatory effects of such pricing. There is, however, one feature of neoliberalism that has left a deepening mark on the contemporary suburb. As over the previous period, governments are proving unable to solve the transportation infrastructure shortcomings of suburbs. There are different causes to this problem, including the fact that there is no easy solution to the traffic congestion generated by heavy dependence on the car as more highway capacity generates more circulation. But above all, in a neoliberal context it is difficult for governments to direct sufficient resources to address worsening congestion, be it by expanding the expressway network or, more likely, by providing quality public-transit alternatives that can compete with the car.

While local scale infrastructures have been isolated from the uncertainties of government funding by their financial reliance on development charges, the situation changes as suburbs age. When it comes time to renew local infrastructures, suburbs can no longer count on development charges funding. Municipalities must then draw from their general taxation revenues and rely on grants from senior governments when these are available. The financial situation of matured suburbs becomes perilous when the need for infrastructure renewal coincides with deindustrialization, often a consequence of neoliberalism and economic globalization, and of a reduced tax base. Globalization affects infrastructure development and maintenance in another way as well: The outsourcing of production deflates the cost of goods that are imported. Accordingly, the overall consumer price index (CPI) is lowered by these goods. On the other hand, the inflation generated by locally produced goods and services exceeds the CPI. As at a given tax rate government revenues tend to follow the CPI, it becomes increasingly expensive relative to overall public-sector budgets to fund infrastructures. The situation is, of course, exacerbated by the tax reductions that mark the neoliberal era.

The diversity of the mature neoliberal suburb is reflected in attempts to provide different modes of transportation and types of land-use development. Some suburbs have linked densification with new public-transit systems and have improved walking and cycling conditions. Still, automobile orientation remains overwhelmingly dominant. Public transit-, walking- and cycling-conducive developments are islands in a sea of car dependence. Suburban diversity can lead to political confrontations, even taking the form of culture wars, over transportation options. For example, in Brampton, a 600,000 resident suburb to the north-west of Toronto, long-term residents successfully opposed a LRT project supported by immigrant groups residing in this city. The southern part of the city opposed the LRT on the grounds of cost and the presumed damage it would cause to the heritage character of the traditional downtown of this mostly recently developed suburban municipality. Meanwhile, attitudes towards the LRT project were far more favourable in the northern portion of the municipality, where it would have addressed the transportation needs of a large immigrant population. The project was defeated in 2015 by a 6–5 council vote (Criscione, 2015).

To some extent, suburban diversity and its political expression have led to the creation of different transportation infrastructure public domains. But among these domains, the one devoted to the automobile remains overwhelmingly dominant, to the extent that, as in the previous phases of the post-war suburb, it inhibits other transportation public domains. In addition, there are impediments to accessing all transportation infrastructures (including car-oriented infrastructures), and thereby to their achievement of public-domain status. But these impediments are more a consequence of public-sector funding restrictions than of privatization. Inadequate funding is indeed responsible for insufficient attempts to develop infrastructures supporting other modes than the car in response to the diversification of the suburb. Diversification of preferences is thus hampered by a frayed transportation infrastructure public domain. What is more, despite the overwhelming suburban infrastructure emphasis on the automobile, the public domain related to the car remains deficient due to its inability to deal with growing traffic congestion. The future of this situation is uncertain, as new autonomous vehicle technologies will likely increase the use made of these infrastructures, while at the same time augmenting their capacity thanks to faster responding vehicular controls.

The control dimension of the mature neoliberal suburb transportation infrastructures orients behaviour towards the automobile-centric lifestyle, as it did in previous suburban phases. There are some transportation infrastructure alternatives, but they limit users to more spatially confined lifestyles, offering a limited range of activities. The control aspect of transportation technologies favours individualistic over collective options, the automobile over public transportation. Such a tendency is accentuated by the interface between communication and transportation technologies such as Uber. Future development involving a more important role for autonomous vehicles may further accentuate the promotion of individualistic options by transportation infrastructures, reverberating on the perception of society and adherence to political orientations. At this time, the future of suburban transportation seems to point in the direction of the neoliberal and libertarian vision advanced by advocacy groups like Americans for Prosperity. Yet, current and emerging technologies could be used in different ways. For example, taxi-like autonomous vehicle services could be operated by community-based organizations, which would offer better working conditions to drivers while preventing a leakage of profits. Such operations could also be organized to pick up a number of passengers on the same journey, thus reducing congestion and adding a collective dimension to these more economic services. In a similar vein, these autonomous vehicle taxi services could be coordinated with subway, commuter rail, LRT, and bus rapid transit (BRT) systems to address the last kilometre problem of these systems, thereby becoming integrated with public transit networks.

REFERENCES

Bakker, K. (2011). Splintered urbanisms: Water, urban infrastructure and the modern social imaginary. In M. Gandy (Ed.), *Urban constellations* (pp. 62–4). Jovis.

Clarke Annez, P. (2006). *Urban infrastructure finance from private operators: What have we learned from recent experience?* The World Bank.

Cohn, R. (2015, March 30). PC blunder over Highway 407 looms over Liberals on Hydro: Cohn. *Toronto Star.* https://www.thestar.com/news/queenspark /2015/03/30/pc-blunder-over-highway-407-looms-over-liberals-on -hydro-cohn.html.

Criscione, P. (2015, October 27). Brampton council votes down light rail transit plan. *Brampton Guardian.* https://www.bramptonguardian.com/news -story/6058427-brampton-council-votes-down-light-rail-transit-plan/.

Foucault, M. (1991). *Discipline and punish: The birth of a prison.* Penguin.

Foucault, M. (1998). *The history of sexuality, volume 1: The will to knowledge.* Penguin.

Gaventa, J. (2003). *Power after Lukes: A review of the literature.* Institute of Development Studies.

Graham, S., & Marvin, S. (2001). *Splintering urbanism: Networked infrastructures, technological mobilities and the urban condition.* Routledge.

Harvey, D. (2005). *A brief history of neoliberalism.* Oxford University Press.

Hauch, V. (2012, January 31). Think Highway 407 tolls are bad? Trying driving in Orange County, California. *Toronto Star.* https://www.thestar .com/news/gta/2012/01/31/think_highway_407_tolls_are_bad_trying _driving_in_orange_county_california.html.

Haughton, G., & McManus, P. (2012). Neoliberal experiments with urban infrastructures: The Cross City Tunnel, Sydney. *International Journal of Urban and Regional Research, 36*(1), 90–105. https://doi.org/10.1111 /j.1468-2427.2011.01019.x.

Honig, B. (2017). *Public things: Democracy and disrepair.* Oxford University Press.

Jessop, B. (2002). Liberalism, neoliberalism, and urban governance: A state-theoretical perspective. *Antipode, 34*(3), 452–72. https://doi.org/10.1111 /1467-8330.00250.

Keats, J. (1956). *The crack in the picture window.* Houghton Mifflin.

Kotz, D. (2015). *The rise and fall of neoliberal capitalism.* Harvard University Press.

Newman, P. (2016). Sustainable urbanization: Four stages of infrastructure planning and progress. *Journal of Sustainable Urbanization, Planning and Progress, 1*(1), 3–10. https://doi.org/10.18063/JSUPP.2016.01.005.

Parkin, J. (1994). A power model of urban infrastructure: From governments to markets. *Economic and Political Weekly, 37*(25): 2467–70.

Peck, J. (2011). Neoliberal suburbanism: Frontier space. *Urban Geography, 32*(6), 889–919. https://doi.org/10.2747/0272-3638.32.6.884.

Peck, J., Siemiatycki, E., & Wyly, E. (2014). Vancouver's suburban involution. *City, 18*(4–5), 386–415. https://doi.org/10.1080/13604813.2014.939464.

Siemiatycki, M. (2013). The global production of transportation public–private partnerships. *International Journal of Urban and Regional Research, 37*(4), 1254–72. https://doi.org/10.1111/j.1468-2427.2012.01126.x.

Tabuchi, H. (2018, June 19). How the Koch brothers are killing public transit projects around the country. *New York Times.* https://www.nytimes.com/2018/06/19/climate/koch-brothers-public-transit.html.

Wellman, K., & Spiller, M. (2012). *Urban infrastructure finance and management.* Wiley.

Yates, R. (1961). *Revolutionary road.* Little, Brown.

Image 10. Region of Waterloo, Ontario, Canada (Summer 2015)

10 Intertwined Modalities of Suburban Governance in China

FULONG WU

Introduction

This chapter examines China's suburban governance and contextualizes its model in the process of suburban development. Following the analytical framework of governance modalities proposed by Ekers et al. (2012), I interrogate the role of the state, capital accumulation, and possible "private governance" in China's suburban governance. I argue that these modalities are intertwined in suburban development and governance. The study highlights the need to distinguish development organization and management structures, that is, governing suburban development versus the governance of the locality. The former is the governance of development projects and activities, while the latter is the governance of service and welfare provision. The examination of development organization can help to reveal the characteristics of suburban governance. For example, Shanghai used the chance of developing infrastructure for transit-oriented development (TOD) to apply various financial instruments in property development near the TOD (Shen & Wu, 2019). The case used in this research is mainly based on Lingang new town in Shanghai, although the findings are generally applicable to other Chinese cities. When necessary, the particularity of the case is mentioned.

The conceptual framework of governance modalities provides a powerful analytical tool to dissect complex governance into manageable research components. However, it is important to understand that these modalities do not present themselves universally and uniformly across different development contexts. These modalities could be intertwined and configured into an overall development approach in different ways. They could be applied to different scales, for example the mega project of the suburban new town versus residential neighbourhoods. Li et al. (2020) applied the assemblage thinking to the modality of suburban governance in China and examined interactions among various actors during the development of projects. They investigated three types of places – gated community enclaves, new towns, and rural villages and revealed different combinations of actors. As revealed by the Chinese case, while overall suburban development may be initiated by the state, the actual implementation can tactically resort to various market instruments, for example a development vehicle to mobilize capital. In this sense, planning centrality can be coherently maintained together with the adoption of market approaches in China's suburban development (Wu, 2018). Rather than thinking in terms of either neoliberal suburbanism (Peck, 2011) or the authoritarian Chinese state (Harvey, 2005), the understanding of intertwined modalities may help to provide a more nuanced understanding of how contrasting components may be combined through innovation in governance, which is a historical process in response to the ambitions and problems of development. Different political and cultural contexts may lead to different governance innovations and these modalities can be intertwined in various ways.

New Town: A Distinctive Suburban Form in China

China has experienced rampant urban expansion since the development of a land market and the greater autonomy of local government in economic development (T. Zhang, 2000). There are different motivations for relocating to the suburbs or buying suburban properties (Pow, 2009; Shen & Wu, 2013; Wu, 2010; L. Zhang, 2010), including a better quality of life (L. Zhang, 2010), privacy and a middle-class lifestyle (Pow, 2009), packaged service provision (Wu, 2010), and greater affordability and property investment (Shen & Wu, 2013). Therefore, Chinese suburbanization is not just a movement of population to the suburbs but also a process of land conversion from agricultural to urban uses (Lin, 2014). In "land suburbanization," wealthier residents buy properties in the suburbs without physically relocating to live there. Investment in second homes is for asset appreciation rather than lifestyle choice. By contrast, inner urban residents have been relocated to large affordable housing projects built by the municipal government in the suburbs. Their move to the suburbs is not a sign of upward social mobility. In addition, Chinese suburbs are also a place for the influx of migrants from small towns and other regions. The suburban area sees not only residents relocated from the central city but also rural migrants from the countryside in other regions. The suburban area has higher percentages of migrant population than urban built-up areas. As a result, the suburb is a heterogeneous social space (Wu & Shen, 2015), demonstrating diverse landscapes as well as governance modalities.

In terms of development approaches, Chinese suburbs are mixed with formal and informal developments (Wu & Li, 2017). This complexity resulted from the legacy of rural and urban dualism. While urban land is formally state-owned, rural land management has been lax (Wu et al., 2013). The suburban is an interface and transitional zone where formal land uses expand into collectively owned rural area. Self-building of farmers' houses is the norm. To take advantage of increasing housing demand from rural migrants, farmers in peri-urban areas build informal rental housing. For formal development, suburban housing is developed by the real estate market. The traditional form of state-driven suburban industrial development was more self-contained. In industrial satellite towns, workers were state employees of state-owned enterprises, and public services were provided as part of their occupational welfare. However, housing development in new towns is often better designed with larger floor space and better facilities. The developer needs a high-quality design to attract home buyers. Formal development has led to the "designed suburb."

Recent studies on global suburbanism reveal not only different governance types but also diverse built forms (Hamel & Keil, 2015). The stereotype of suburbia is not today's reality, and suburban forms show great diversity (Keil & Addie, 2015). Spatially, the Chinese residential suburb is distinctively different from the image of single family, low-density suburbia, although some estates may replicate Western-design styles. The residential area is often divided into large land plots, built with high-rise apartment buildings, because developers try to reduce the cost of road construction and speed up housing sales. The suburbs thus contain master-planned villa compounds, more "ordinary" commodity housing estates, affordable housing neighbourhoods, and transformed rural villages for rental housing.

To curb earlier rampant sprawl, the Chinese government has tightened land management. The amount of land conversion is restricted annually by a quota allocated from the central government to the municipal government (Tian et al., 2015), which encourages the local government to adopt more compact forms of suburban development such as TOD (Shen & Wu, 2016, 2019). Therefore, besides seeing a new town as an ideal community as in American new urbanism (Duany & Plater-Zyberk, 1993), Chinese new towns are driven by state land policy to use land more efficiently. In addition, as will be seen in this chapter, the new town represents a distinctive Chinese suburban governance model, which provides the most efficient approach to large-scale land development and social management.

New towns represent a more organized way of suburban development than urban sprawl. This distinctive suburban form reflects some new characteristics of suburban governance, which necessarily calls for strategic planning and greater government intervention (Li et al., 2020). To support the development of a deep-water port in Shanghai, a new town, initially called "New Harbour City" (*Haigang Xingcheng*), was built near the port of Yangshan at the border of Shanghai and Zhejiang provinces. This is the origin of today's Lingang new town. Through an international design contest, the design of a German firm, Gerkan, Marg and Partners (GMP), was chosen because of its bold application of the garden city concept, which consists of circular commercial, residential, and industrial zones surrounding the manmade "water-drop lake" named Dishuihu. However, because of its large land plots, wide roads, and scale, the new town is largely dependent upon automobiles for travel. The development of Lingang signals a new stage of suburban governance beyond sporadic market-driven residential development in the inner suburbs.

The State

Suburbanization has long been associated with a localism narrative (Harris, 2004), which describes how the state has played a minimal role in suburban development. However, from the perspective of governance modalities, the state still plays important roles in suburban governance (Ekers et al., 2012), first through housing mortgage and tax policies and second through infrastructure development. In Europe, the welfare state initiated large-scale public housing construction at the periphery of the cities, forming socially isolated public housing estates such as the *banlieues* in France. Therefore, the state plays a role in suburban development to a varying degree in different places.

Similar roles of the state in suburban development have been noticed (Wu, 2015), and indeed may be even more prominent, due to China's strong state capacities. First, the national state formulates a fiscal policy to separate central and local taxes, which gives a significant incentive to the local government which has gained economic autonomy, especially in land development. The central government also makes financial resources available to stimulate investment in fixed assets through overall monetary supply. The 4 trillion yuan stimulus package is such an example. Second, there are state-led infrastructure development projects. Urban expansion and sprawl in the 1990s were directly attributed to the local state's desire to use land development to stimulate

Table 10.1. Population in the Lingang area

Towns	Total population	With household registration (hukou)
1. Nanhui New Town	130,603	28,771
Shengang shequ	34,799	11,602
Luchaogang shequ	35,096	17,169
University students	60,708	
2. Nicheng town	77,134	58,918
3. Shuyuan town	75,378	52,194
4. Wangxiang town	27,754	24,808

Source: Shanghai Pudong Economic Development Report (2017).

economic growth and maximize local revenue (T. Zhang, 2000). Third, more recently, to cope with the affordability crisis, the central state has required local governments to develop social security housing, which is mainly located in peripheral areas due to cheaper land costs. To facilitate urban renewal, municipal governments have also built large-scale relocation housing in the suburbs, indirectly increasing housing stock in suburban areas. Fourth, the development of suburban areas often results from strategic considerations for urban restructuring, which is achieved through "strategic urban plans" and administrative annexation (Wu & Zhang, 2007). China's urban planning is characterized by its strong desire for growth promotion (Wu, 2015). Suburban development often constitutes such a strategic action. For example, the development of Lingang new town is to support the deep-water port at Yangshan and further realize Shanghai's aim to become a global shipping centre.

To understand the precise structure of suburban governance, it is necessary to investigate the administrative structure of the municipality. The municipality of Shanghai consists of sixteen districts (see figure 10.1 which shows the administrative hierarchy related to Lingang). Administratively, Lingang consists of the jurisdiction area of four towns. Table 10.1 shows the population in the area of Lingang. But these four towns do not fall under the management of Lingang new town. They are individually governed by Pudong District for administrative and social management. The overall *development* coordination in the area is achieved through the Lingang Management Committee.

The Chinese model of management committee (*guan wei hui*) is quite unique. The management committee can be set up across various local governments. Town governments are mainly responsible for social management, while the management committee aims mainly for economic development. The actual development work is carried out by various state-owned enterprises (development corporations) at the primary level of land development. The management committee, as a streamlined governing body, provides an overall governance framework. The development corporation does not have planning power. It is a developer. However, as a state-owned enterprise and development vehicle of the municipal government, it receives full support from the management committee. In Lingang, because it is a municipal strategic project, its management committee is a "subordinating agency of the municipal government." However, it is now managed by the Pudong District government for the sake of better coordination.

To cope with the fragmentation of governance, Shanghai tried to create "functional areas," which combined the development zone and town government, in 2010. But the experiment failed because this literally adds an extra layer between the district government (e.g., Pudong New District) and the town government as the basic level of government in the municipal region. The functional area remains in the hands of a streamlined government agency (i.e., the management committee), which is designed for efficiency of development. Planning and development activities can be organized at this level (e.g., the Lingang area), but social management is still performed through the usual administrative hierarchy mentioned earlier. At the development zone level, there is no *full* government except the development agency, in the case of Lingang the Shanghai Lingang Area Development and Construction Management Committee (上海市临港地区开发建设管理委员会, hereafter, Lingang Management Committee). The absence of full government functionality at the level of the development zone is perhaps reminiscent of the notion of "urban entrepreneurialism" (Harvey, 1989; Jessop & Sum, 2000; Peck, 2011).

In terms of actual development, the Lingang area has been mainly developed by two development corporations: Lingang group belongs to the municipal government, while Harbour City (Gangcheng) belongs to the former Nanhui District government (which is now annexed into Pudong District). Because of this historic arrangement, the area was governed by two different governing bodies, Lingang Industrial Park Management Committee directly under the municipality, and Nanhui New Town Management Committee, which was responsible for the construction of the city proper of the new town. To better coordinate the development, these two committees were merged into the present Lingang Management Committee in 2012. The departments under the former Lingang Industrial Park Management Committee were municipal level agencies. After the administrative merger, they were transferred along with the management committee to the Pudong District government. This is a "municipally owned but district managed" (*shi shu qu guan*) model. The purpose is to incentivize the Pudong government to support Lingang development which was initiated, invested in, and owned by the municipal government. Lingang group as a state-owned enterprise (SOE) still belongs to the state asset management committee of Shanghai municipality, which is the major developer of the area. However, for a mega urban project like Lingang, development often requires more than one development corporation.

The establishment of multiple development corporations is also due to the consideration of giving incentives to different governments which can share the costs and benefits. For example, Harbour City was set up initially to give an incentive to the Nanhui District government to participate in the development, while the Lingang group of the municipal government took on the task developing the heavy manufacturing industrial zone. Further, to coordinate with town governments, four plots of one square kilometre land were assigned to each of four town governments to create their own development corporations (see table 10.2 for their investment). Faced with slow development in the Lingang area, Shanghai municipal government required four major development corporations in Pudong District to set up their Lingang corporations. Nearby Lingang, a town belonging to Fengxian District also set up a development corporation (Lingang Fengxian Corporation). As a result, there are multiple development corporations in the area. These development corporations are actual developers in the area, and in the Chinese term they are known as "development actors" (*kaifa zhuti*). In essence, these development actors bear a legal status as "enterprises" and are able to obtain financing from banks. They are similar to the special purpose vehicle (SPV) for development in the West. Table 10.2 lists the investment in fixed assets by these "development actors."

Figure 10.1 shows the administrative hierarchy of Lingang. In parallel with this government structure is a development agency – the Lingang Management Committee, which supervises and coordinates the development. The management committee is located in the jurisdiction area of four towns but does not supervise these towns' governments. The role of the committee is mainly for development, coordinating multiple state-owned development corporations, which belong to different levels of the government, as well as many other "normal" private sector developers in the area. Spatially, the suburban new town as a development zone supervised by the management committee overlaps with the jurisdictions of the town government (figure 10.2). As for the management committee, it governs development according to its designated power as a "state agency." Town governments simply would

Table 10.2. The main "development actors" (corporations) in the Lingang area of Shanghai and their investment in fixed assets in 2015

Development actors	Investment in fixed assets (million yuan)	Notes
Lingang group (equipment and logistics)	4,335.63	Belong to the Lingang group
Minglian Lingang	546.90	From the Minghang District
Lingang Fengxian/Situan town	1,800.51	Joint venture
Lingang Nicheng/Nicheng town	1,485.32	Joint venture
Lingang Shuyuan/Shuyuan town	452.65	Joint venture
Lingang Wangxiang/ Wangxiang town	1,257.59	Joint venture
Nanhui new town/formally Luchaogang town	1,375.66	Joint venture
Harbour City	5,162.35	
Jinqiao Lingang	1,728.12	From Pudong District
Haiyang Gaoxin (Ocean new technology)	488.46	
Lingang City Investment	1,400.30	
Zhangjiang Lingang	1,960.55	From Pudong District
Lujiazui Lingang	2,123.90	From Pudong District
Waigaoqiao Lingang	0	From Pudong District
Others	500.00	
Total	22,853.94	

Source: Shanghai Statistics Bureau (2016).

not be able to perform this level of strategic task as they are low in the administrative hierarchy and historically they have had weak management capacities as rural governments under the county. Development has to be coordinated by a more capable and powerful agency such as the municipally established management committee. Moreover, the actual development has to be carried out by more "experienced" state-owned enterprises belonging to different governments or in the private sector. Within this vast territory, various "development actors" perform their development tasks on the land plots they possess. They govern the development according to their ownership (see Shen et al., 2020).

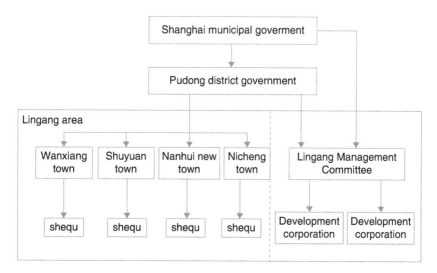

Figure 10.1. The multi-scalar governance of the state

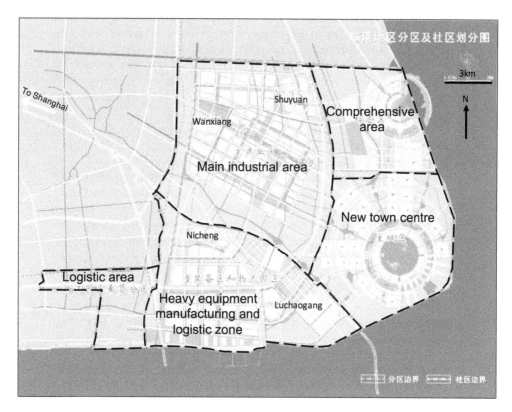

Figure 10.2. The Lingang area – a suburban new town as a development zone

In sum, in addition to macroeconomic and urban policies that have indirect influence on suburban governance, the role of the state in Chinese suburban governance is visible, due to the creation of a development agency ("management committee") and various state development arms or development vehicles. The direct involvement of the state is a salient feature of Chinese suburban governance, which is strategically planned and practically oriented for growth (Wu, 2015). In comparison with certain development corporations in London, such as the new Old Oak and Park Royal Development Corporation (OPDC), which has limited staff, limited registered capital, severe constraints on direct finance from the bank, and limited collateral land assets (see Robinson et al., 2021), the Chinese model is directly involved in development rather than merely coordination. Although the development corporation in the UK has planning power, thus allowing it to intervene and hope to obtain "planning gain," Chinese development corporations do not need to have planning power. Their operation is based more on their capacity for direct finance and strong support from the management committee, which does have the planning power. This structure means that OPDC has to negotiate with external investors through "neoliberal," entrepreneurial, or post-political manoeuvres, whereas the Chinese suburban development model allows the state agency to support state-owned enterprises to perform development duties while leaving the "traditional" government structure to carry the necessary social management.

Capital Accumulation

The hypothesis raised by Harvey (1978) about capital accumulation as the defining force for capitalist urban development and three interlinked capital circuits through which capital is shifted into the new built environment to solve the over-accumulation crisis provides a useful analytical tool. The pursuit of the exchange value of landed properties is the driving force for the formation of a growth machine (Logan & Molotch, 1987). The role of capital accumulation in suburban governance is well documented in the Western literature (Gotham, 2009). In particular, the development and enclosure of suburban housing is examined with reference to the "financialisation of housing" (Aalbers, 2008). The role of the suburb as a space of capital accumulation is also revealed in China (Shen & Wu, 2016). The meaning is twofold.

First, the development of manufacturing sectors in the suburban economies provides an opportunity for the central area to relocate manufacturing industries to the suburbs, consequently upgrading its economic structure and restructuring into the tertiary sector. Chinese suburban development is not equivalent to residential suburbanization. The suburb does not see a shift from industrial development to real estate. Chinese suburbs still show significant industrial growth. Suburban land development is combined with and facilitates industrial development. For example, Songjiang new town develops electronic manufacturing and creative industries. Jiading new town specializes in automobile production. Taihu new town in the city of Wuxi accommodates creative industries such as a film studio base and a software park. Lingang new town set up an advanced manufacturing industries zone. The plan for Lingang was to seize the opportunity of the relocation of heavy equipment manufacturing in the world. Since 2015, Shanghai has been designated by the central government as a global science and innovation centre. Lingang has taken this opportunity to develop "intelligent manufacturing" so as to realize strategic national objectives. Songjiang new town sees both industrial and residential development, including a "university town," an export processing zone, and an upper-market gated community in the style of an English market town (Thames Town). The suburb is thus also a new space for expanding primary capital circulation. This is slightly different from Harvey's notion of the secondary capital circuit in the (suburban) built environment. But the function is similar: sustaining capital accumulation at the metropolitan scale by accommodating relocated industries from the central area.

Second, the development of suburban space represents a "spatial fix" to capital accumulation, that is, suburban landed (property) development as a way to absorb the capital created in China's global workshop. Together, Chinese suburbs see both the development of land-intensive advanced manufacturing industries and new gated residential areas where property development generates enormous profits. The release of suburban land helps to absorb capital and thus sustain the investment in fixed assets, which is a major driving force in addition to export for China's economic growth. Land development In China plays a significant role in balancing local public finance. Known as "land-based finance" (Lin, 2014), land profit is an important source of local revenue. Land revenue constitutes 40 to 60 per cent of local revenue (Wu, 2012, p. 254). The sale of suburban land becomes an important source of local revenue to make up the budget deficit. The financialization of land thus presents an important opportunity to sustain investment in fixed assets that expands the sphere of capital accumulation.

The financialization of land plays a pivotal role in suburban development. This involves state sanction for the land to be a tradable asset with exchange value. Earlier in land reform and the development of the Pudong new district, an important governance innovation was to forcefully attach the attribute of "commodity" to the land. Backed up by the state, land development corporations then put the land into the banks, many of which are state owned, as collateral to obtain development funds. When the land indeed becomes a commodity sold in the land auction market, this exchange value is realized. The land "deed" is created during the process of the land sale. Because suburban land is largely undeveloped open space, agricultural land, or land claimed from the sea in the case of Lingang, the cost of relocation of the original residents is relatively low. It is therefore an ideal place for state development corporations to gain control over its ownership and create the land as a commercial asset. Suburban space thus constitutes a critical link in China's capital accumulations. Suburban land development reveals two important modalities of governance: the state's prominent role in land development, and capital accumulation in the new suburban space.

The changing policy of financing suburban land development is an example of intertwined modalities. The financialization of suburban land is always subject to more strategic control. Aggressive land financialization through creating a pseudo deed has now been banned, owing to the potential financial risk of local government debt. That is, the development corporations might fail to convert the pseudo deed into a genuine deed with market value, leaving the bank with a solvency problem. Moreover, the local government is no longer allowed to use its credit to back up enterprise loans. The new policy requires a clear distinction between government debt and enterprise debt. The development of suburban land thus provides an invaluable angle to observe the critical role of suburban governance in overall capital accumulation in the nation.

Authoritarian Private Governance

Gated residential areas are the ubiquitous landscape of newly developed Chinese suburbs. They reflect the demand for better design and high-quality living environments. Relocation to the suburbs may be driven by different reasons: government relocation programmes (Wu, 2004), desire for privacy and quality of life (Pow, 2009; L. Zhang, 2010), and greater affordability (Shen & Wu, 2013). New communities have been created in places where there was little governance structure. Before the construction of new gated residences, these former rural areas were governed by relatively weak and under-resourced organizations such as "production brigades" for farmers, villagers' committees, and township government (Wu, 2011). Suburban residential development thus leads to the creation of new governance. The characteristic of governance structure, however, is more or less associated with the way in which new gated residential areas are developed.

The most efficient and fastest way of creating residential developments is to assign large land plots to developers in the secondary land market after the SOE has completed the primary development (involving major infrastructure development). High-rise apartment buildings in large plots with wide roads are almost the standard template of suburban Chinese design, which is also due to the consideration of saving the cost of road infrastructure development. While the municipal government is still responsible for providing facilities such as schools and hospitals, these developers tend to build gates and walls along the assigned boundaries of their land plots. They also provide different qualities of amenities such as tennis courts, swimming pools, club houses, and services and maintenance such as landscaping and security. The design of gates is often used to symbolize the quality of their products, which serves for "branding" purpose (Wu, 2010).

To manage the estates, developers need to appoint property management companies, especially when the project is completed and developers leave. The property management companies collect maintenance fees from residents. Hence, there is a need for residents to "manage" these companies. The government sees this relationship as market behaviour, that is, the residents purchase property management services from professionals and specialized companies. The services provided, however, are still limited to housing rather than social needs. Residents choose different estates because of the differentiated prices and qualities of properties and maintenance. But the variety is not sufficiently large to be seen as a "governance market" (Cséfalvay & Webster, 2012), that is, a choice between different types of governance (gated or non-gated, with or without homeowners' associations, strong or weak self-governance, maintenance cost and "property taxes").

To help residents deal with the purchase of property services, a homeowners' association is set up. While there was some speculation about how the establishment of homeowners' associations might lead to a new mechanism of bottom-up governance which might eventually "democratize" China in the 1990s, the government soon realized the problematic implication and restricted homeowners' associations to the role of property management affairs rather than a self-organized political force. Therefore, the legal status of the homeowners' association is absent, that is, it is not a legal corporation and hence cannot have a bank account. A homeowners' association is not an approved association (*shetuan zuzhi*) or non-governmental organization (NGO). Hence, the homeowners' association occupies an awkward position, playing a role in assisting the residents to deal collectively with property related matters. On the other hand, the lowest level of governance component is the residents' committee, which, according to the Organization Law of Residents' Committee is a self-educated, self-monitored, self-organized, self-serviced "mass organization." In reality, the residents' committee is a de facto government (Wu, 2002). Above the residents' committee is the "street office," which is a government agency that reports to the district government. Large gated residential areas may contain several residents' committees. By nature, the residents' committee is a device for the government to manage society rather than acting as a self-organization mechanism (Read, 2012). Traditionally, in old urban areas, the residents' committee assists the state with welfare provision. In relatively wealthier suburban areas, the residents' committee has an easy job, as middle-class residents do not require much social assistance. Other functions such as the neighbourhood watch have been taken over by property management companies. Suburban gated residences do witness a shift of some services from local government – from residents' committee to a market form of service provision. Many residents thus have little contact with the residents' committee or do not participate in the election of the residents' committee. In fact, to strengthen social management, the government has gradually "professionalized" the residents' committee with social workers. The residents' committee thus can be seen as a government agency of social workers.

To sum up, suburban gated residential areas in China are the built form of large development projects. The superblocks are natural units for differentiated residential qualities, styles, and services. They are products of the developers who carried out the development and often bear the imprint of the design feature of their developers. The projects of different developers rarely merge into the "local community" because participatory politics are absent.

In response to the modality of "authoritarian private governance," we do not witness the rise of corporate-style governance over residential areas, in which there is the use of the contract, conveyance, and restriction (CCR) agreement as a mechanism of governance (Low, 2003), and the homeowners' association, replacing the government, becomes the controller as a corporate form of governance (McKenzie, 1994). Because of the absence of participatory democracy, suburban governance is not different from other areas, despite the gated form of residences produced by superblock development and more advanced property management.

Intertwined Modalities of Governance

The three modalities identified by Ekers et al. (2012) are not mutually exclusive. They can be combined in different ways at different scales of governance (Li et al., 2020). To understand the precise configuration of suburban governance, we have to examine the structure of governance at the specific scale and locality. Lingang is an area that is a new town development. But in the Chinese term of governance, the "new town" is not a place but rather a mode of development. It is an assemblage (see Allen & Cochrane, 2010; Halbert & Rouanet, 2014) of various "development actors," including the municipal government-owned corporation (Shen et al., 2020). There is no single government in full charge at this level. In this sense, it is governance – coordination participated in by multiple agencies.

The area is developed mainly by two development corporations: Lingang and Harbour City which are owned by the municipal government and the district government, respectively. Their coordination has to be achieved through the management committee. In addition to these development corporations, there are many developers and investors including developers in the private sector (for example, Haichang from northern China developed the Ocean World theme park). But the two main development corporations do not govern the place. Lingang, as an enterprise, is mainly for heavy manufacturing industries and includes a logistics zone and harbour management. Harbour City is also an enterprise for real estate development in the town centre.

Within this territory, there are four town governments that were formerly responsible for the rural residents. One such "rural" town (Luchaogang town) was absorbed into Nanhui new town (the residential area of Lingang), becoming an urban residential district. They are responsible for administration and social management but are not well resourced and are quite peripheral in governance. Because of the existence of powerful state-owned development corporations and a quasi-government agency (the management committee), none of these four town governments, which have been weak since their origin as rural governments in the former Nanhui County, could govern the whole Lingang area. The question is, Who governs Lingang? With reference to the specific *suburban* governance, no single body can claim to have full control or is willing to take full responsibility.

At the neighbourhood or residential community level (*shequ*), most areas are gated residential areas. As mentioned earlier, self-governance capacity has been weak, while some property services have been subcontracted to the market. Residents are not sufficiently mobilized to influence the governance of Lingang. Major infrastructure and public services are still provided by the municipal government, which wishes to attract new residents to form a new city. At this level, state influence is indirect, through encouraging homeownership and services charged by property management. Homeowners' associations have been set up, while the residents' committee still performs the basic government functions for the street office or town government. The variety of these gated residential areas is not sufficiently large in terms of governance feature. Therefore, there is no choice of governance mode (for example, open versus gated), and the contractual relation is the basic principle of estate property management, although social management is beyond market provision.

However, along with more development, the development area of Lingang is becoming more complex, generating more demand for social services and management. The streamlined governance structure is good for economic development but difficult for dealing with complex social management. Like many other "Economic and Technological Development Zones" (ETDZs) in China, the Lingang Management Committee faces the challenge to take responsibility for comprehensive urban management. Currently, social affairs and management are separated from economic development. The management committee provides an overall management framework, but there is no "Lingang community" and no unified identity and interest of a suburban community based on residency.

Inserting a corporation into the suburban area seems to be an effective way to open up the suburban space for economic growth. But the approach does not work for social development. Over the years, there have been various governance innovations to provide better governance. For example, initially the management committee (in the industrial park) was directly under the municipal government, creating a coordination problem with local residential development (in the new town). This structure has been adjusted through merging two management committees and transferring management to the district government. Further, the municipal government initiated a policy of "double exemption" (shuangte), which means that decision-making is exempted from municipal government approval and tax is exempted from submission to the municipal government. The Lingang Management Committee can retain all the taxes from municipal development corporations for its own use. However, from the discussion above, it can be seen that there is a lack of social innovation. The local "community" has not gained autonomy for decision-making in their local affairs. Derived from this approach of governance is the fragmentation between industrial and residential uses. As in many development zones, Lingang strives to integrate industrial function and urban life in the new town (*chan cheng ronghe*). The current problem of spatial and governance fragmentation is a reflection of the particular way of intertwined governance modalities, which are more state-led, more for capital accumulation as a municipal-level strategy, and show underdeveloped social innovation and local communities.

Conclusion

The Chinese suburb represents more than a single modality of governance. Rather than thinking in either neoliberal or developmental terms, suburban governance opens up the possibility of understanding intertwined modalities that reflect continuing state centrality in urban governance while market instruments are deployed. The feature of "state entrepreneurialism" in China (Wu, 2018), however, has various manifestations in the suburbs, due to the geographical specificity of the suburb.

Formerly, as a rural area inhabited by famers outside the powerful state system (Wu, 2011), the suburbs were a weak society. Their governance mechanism has been entirely removed or transformed along with rapid urbanization. Because of underdeveloped infrastructure and the peripheral locations, these places lack possibilities for large-scale, self-built activities. The initiative has to come from the powerful state rather than a local property owner class or development industry (Logan & Molotch, 1987). Although located in the peripheral area, the suburb plays a pivotal role in sustaining capital accumulation. The opening of suburbs is driven by the state to expand a new space for economic growth – including new industries in the suburbs and new financial services in the central city, and to accommodate the inflow of rural migrants for industrial development. The development is thus more *strategic* to upgrade the central city for new tertiary industries and for suburbs to become industrial, than simply for development profit. The Chinese suburbs thus are characterized by a landscape of large development zones and industrial parks, similar to the landscape of "post-suburbia" (Wu & Phelps, 2011) where post-Fordist economies see many heterogeneous and flexible spaces in the suburbs (Fishman, 1987).

The development of new towns through TOD is a more effective and environmentally acceptable form of compact development. The new town is thus created thanks to the state's attempt to control urban sprawl (Shen & Wu, 2016, 2019). Development is a combination of state and market mechanisms, which is actually carried out by state-owned development corporations in the primary market of land development. Large land plots are further assigned to industrial and real estate development projects, which involve more visible market forces. The land market in the suburbs is tactically used to pave the way for infrastructure development, which in turn supports industrial development. To attract home buyers, developers boast of their products as an ideal living environment and present them as a new community. But the development of gated residential areas has its practical reason as a cost-effective way of developing large land parcels. The primary state developer only develops rather sparse road networks and leaves the secondary (market) developers to develop subdivisions. These subdivisions are in the form of enclosed residential estates for more visible identities as the products of their developers. Some services are provided by property management companies initially subcontracted by the developers and subsequently transferred to newly founded homeowners' associations. However, major services and facilities are still provided by the municipal and district governments. Although the purchase of property management services bears some similarity to "private governance" (McKenzie, 1994), self-governance through homeowners' associations is still weak. These neighbourhoods are still fragmented and do not form "local communities" as a political force. Market provision and management through self-governance and contracts have not become the dominant form of governance in Chinese suburbs, despite the ubiquitous landscape of gating. Such an approach of mega-urban projects led by the state across multi-scales produced mixed impacts on local residents. Some hugely benefited from these projects through generous compensation, while others are left out and marginalized because they are less directly "relevant" to these projects (Wang & Wu, 2019).

In sum, the three modalities of the state, capital accumulation, and private governance are intertwined: suburban Chinese governance is still state-led and state-sanctioned, but at the same time market instruments are deployed through the real estate market, development corporations, and landscaped and gated residential areas. The governance of the suburban world thus represents a variegated configuration of different governance modalities, which cannot be simply characterized as state-led or market dominated.

ACKNOWLEDGMENTS

The chapter was initiated prepared as a keynote for the conference of "After Suburbia: Extended Urbanization and Life on the Planet's

Periphery" held in October 2017 at York University. I would like to acknowledge grant support from the Major Collaborative Research Initiative "Global Suburban Governance, Land and Infrastructure in the Twenty-First Century" from the Social Sciences and Humanities Research Council of Canada, of which Roger Keil is the PI. In addition, I acknowledge the support of the UK ESRC research project "Governing the Future City" (ES/N006070/1), of which Jenny Robinson is the PI, and my ESRC project, "The Financialisation of Urban Development and Associated Financial Risks in China" (ES/P003435/1). I thank Jenny Robinson, Phil Harrison, and Allan Cochrane for their inspiration and collaboration on Governing the Future City project; Zhigang Li and Jie Shen for their collaboration on the MCRI suburban project: and Yuemin Ning, Xiang Luo, and Zheng Wang for their collaboration on Lingang.

REFERENCES

Aalbers, M.B. (2008). The financialization of home and the mortgage market crisis. *Competition and Change*, 12(2), 148–66. https://doi.org/10.1179 /102452908X289802.

Allen, J., & Cochrane, A. (2010). Assemblages of state power: Topological shifts in the organization of government and politics. *Antipode*, 42(5), 1071–89. https://doi.org/10.1111/j.1467-8330.2010.00794.x.

Cséfalvay, Z., & Webster, C. (2012). Gates or no gates? A cross-European enquiry into the driving forces behind gated communities. *Regional Studies*, 46(3), 293–308. https://doi.org/10.1080/00343404.2010.505917.

Duany, A., & Plater-Zyberk, E. (1993). The neighbourhood, the district, and the corridor. In P. Katz & S.J. Vincent (Eds.), *The new urbanism: Toward an architecture of community* (pp. 17–20). McGraw-Hill.

Ekers, M., Hamel P., & Keil., R. (2012). Governing suburbia: Modalities and mechanisms of suburban governance. *Regional Studies* 46(3), 405–22. https://doi.org/10.1080/00343404.2012.658036.

Fishman, R. (1987). *Bourgeois utopias: The rise and fall of suburbia*. Basic Books.

Gotham, K.F. (2009). Creating liquidity out of spatial fixity: The secondary circuit of capital and the subprime mortgage crisis. *International Journal of Urban and Regional Research*, 33(2), 355–71. https://doi.org/10.1111 /j.1468-2427.2009.00874.x.

Halbert, L., & Rouanet, H. (2014). Filtering risk away: Global finance capital, transcalar territorial networks and the (Un)making of city-regions: An analysis of business property development in Bangalore, India. *Regional Studies*, 48(3), 471–84. https://doi.org/10.1080/00343404.201 3.779658.

Hamel, P., & Keil, R. (2015). *Suburban governance: A global view*. University of Toronto Press.

Harris, R. (2004). *Creeping conformity: How Canada became suburban, 1900–1960*. University of Toronto Press.

Harvey, D. (1978). The urban process under capitalism. *International Journal of Urban and Regional Research, 2*(1–3), 101–31. https://doi.org.10.1111/j.1468 -2427.1978.tb00738.x.

Harvey, D. (1989). From managerialism to entrepreneurialism: The transformation in urban governance in late capitalism. *Geografiska Annaler, 71B*(1), 3–17. https://doi.org/10.1080/04353684.1989.11879583.

Harvey, D. (2005). *A brief history of neoliberalism*. Oxford University Press.

Jessop, B., & Sum, N.L. (2000). An entrepreneurial city in action: Hong Kong's emerging strategies in and for (inter)urban competition. *Urban Studies, 37*(12), 2287–313. https://doi.org/10.1080/00420980020002814.

Keil, R., & Addie, J.P.D. (2015). "It's not going to be duburban, it's going to be sll urban": Assembling post-suburbia in the Toronto and Chicago regions. *International Journal of Urban and Regional Research, 39*(5), 892–911. https:// doi.org/10.1111/1468-2427.12303.

Li, Z., Chen, Y., & Wu, R. (2020). The assemblage and making of suburbs in post-reform China: The case of Guangzhou. *Urban Geography, 41*(7), 990– 1009. https://doi.org/10.1080/02723638.2019.1598732.

Lin, G.C.S. (2014). China's landed urbanization: Neoliberalizing politics, land commodification, and municipal finance in the growth of metropolises. *Environment and Planning A, 46*(8), 1814–35. https://doi.org/10.1068/a130016p.

Logan, J.R., & Molotch, H.L. (1987). *Urban fortunes: The political economy of place*. University of California Press.

Low, S. (2003). *Behind the gates: Life, security, and the pursuit of happiness in fortress America*. Routledge.

McKenzie, E. (1994). *Privatopia: Homeowner associations and the rise of residential private government*. Yale University Press.

Peck, J. (2011). Neoliberal suburbanism: Frontier space. *Urban Geography, 32*(6), 884–919. https://doi.org/10.2747/0272-3638.32.6.884.

Pow, C-P. (2009). *Gated communities in China: Class, privilege and the moral politics of the good life*. Routledge.

Read, B.L. (2012). *Roots of the state: Neighborhood organization and social networks in Beijing and Taipei*. Stanford University Press.

Robinson, J., Harrison, P., Shen, J., & Wu, F. (2021). Financing urban development, three business models: Johannesburg, Shanghai and London. *Progress in Planning, 154*, Article 100513. https://doi.org/10.1016 /j.progress.2020.100513.

Shanghai Pudong District Government. (2017). *Shanghai Pudong economic development report*. Shanghai Pudong District Government.

Shanghai Statistics Bureau. (2016). *Shanghai Pudong District statistics handbook.* Shanghai Statistics Bureau.

Shen, J., Luo, X., & Wu, F. (2020). Assembling mega-urban projects through state-guided governance innovation: The development of Lingang in Shanghai. *Regional Studies, 54*(12), 1644–54. https://doi.org/10.1080/00343404.2020.1762853.

Shen, J., & Wu, F. (2013). Moving to the suburbs: Demand-side driving forces of suburban growth in China. *Environment and Planning A, 45*(8), 1823–44. https://doi.org/10.1068/a45565.

Shen, J., & Wu, F. (2016). The suburb as a space of capital accumulation: The development of new towns in Shanghai, China. *Antipode, 49*(3), 761–80. https://doi.org/10.1111/anti.12302.

Shen, J., & Wu, F. (2019). Paving the way to growth: Transit-oriented development as a financing instrument for Shanghai's post-suburbanization. *Urban Geography, 41*(7), 1010–32. https://doi.org/10.1080/02723638.2019.1630209.

Tian, L., Guo, X., & Yin, W. (2015). From urban sprawl to land consolidation in suburban Shanghai under the backdrop of increasing versus decreasing balance policy: A perspective of property rights transfer. *Urban Studies, 54*(4), 878–96. https://doi.org/10.1177/0042098015615098.

Wang, Z., & Wu, F. (2019). In-situ marginalisation: Social impact of Chinese mega-projects. *Antipode 51*(5), 1640–63. https://doi.org/10.1111/anti.12560.

Wu, F. (2002). China's changing urban governance in the transition towards a more market-oriented economy. *Urban Studies, 39*(7), 1071–93. https://doi.org10.1080/00420980220135491.

Wu, F. (2004). Intraurban residential relocation in Shanghai: Modes and stratification. *Environment and Planning A, 36*(1), 7–25. https://doi.org/10.1068/a35177.

Wu, F. (2010). Gated and packaged suburbia: Packaging and branding Chinese suburban residential development. *Cities, 27*(5), 385–96. https://doi.org/10.1016/j.cities.2010.06.003.

Wu, F. (2011). Retreat from a totalitarian society: China's urbanism in the making. In G. Bridge & S. Watson (Eds.), *The new Blackwell companion to the city* (pp. 701–12). Blackwell.

Wu, F. (2012). Urbanization. A. So & W. Tay (Eds.), *Handbook of contemporary China* (pp. 237–62). World Scientific Publishing.

Wu, F. (2015). *Planning for growth: Urban and regional planning in China.* Routledge.

Wu, F. (2018). Planning centrality, market instruments: Governing Chinese urban transformation under state entrepreneurialism. *Urban Studies, 55*(7), 1383–99. https://doi.org/10.1177/0042098017721828.

Wu, F., & Li, Z. (2017). The paradox of informality and formality: China's suburban land development and planning. In R. Harris & U. Lehrer (Eds.), *Suburban land question* (pp. 145–66). University of Toronto Press.

Wu, F., & Phelps, N.A. (2011). (Post-)suburban development and state entrepreneurialism in Beijing's outer suburbs. *Environment and Planning A, 43*(2), 410–30. https://doi.org/10.1068/a43125.

Wu, F., & Shen, J. (2015). Suburban development and governance in China. In P. Hamel & R. Keil (Eds.), *Suburban governance: A global view* (pp. 303–24). University of Toronto Press.

Wu, F., & Zhang, J.X. (2007). Planning the competitive city-region: The emergence of strategic development plan in China. *Urban Affairs Review, 42*(5), 714–40. https://doi.org/10.1177/1078087406298119.

Wu, F., Zhang, F.Z., & Webster, C. (2013). Informality and the development and demolition of urban villages in the Chinese peri-urban area. *Urban Studies, 50*(10), 1919–34. https://doi.org.10.1177/0042098012466600.

Zhang, L. (2010). *In search of paradise: Middle-class living in a Chinese metropolis.* Cornell University Press.

Zhang, T. (2000). Land market forces and government's role in sprawl. *Cities, 17*(2), 123–35. https://doi.org/10.1016/S0264-2751(00)00007-X.

Image 11. Johannesburg, South Africa (October 2016)

11 Governing Cities in a Post-Suburban Era: New Challenges for Planning?

PIERRE HAMEL

Since the Second World War, cities in the US have undergone a series of transformations that have radically changed the form, content, and extension of urban landscapes. The same occurred in Canada, albeit a few years later. Defined as an "urban nation" as early as the middle of the nineteenth century, with central cities achieving the functionality and cultural meaning of city-regions, Canada has since been radically altered through a "metropolitan revolution." Previously, the metropolis was considered "a place with readily discernible edges, its lifestyle sharply distinguished from that of rural 'rubes' and 'hicks,' many of whom had obtained the benefits of electricity only a decade before" (Teaford, 2006, p. 1). With the sweeping changes occurring after the war – often expressed in technological and cultural terms – the industrial metropolis was irrevocably losing ground in favour of new forms of urban expansion intensified by a booming housing industry, giving the opportunity to an increasing number of households to acquire a home in the suburbs (Teaford, 2011).

In "post-urban America" following the "metropolitan revolution" through – among other things – intense processes of dispersed suburbanization, a new ethos emerged in correspondence with questioning the historic meaning of the urban core as the locus of centrality. This has been explored at length by historians who have adequately grasped the transformations of social behaviours brought in on a daily basis in geographic and cultural terms:

> For most Americans, the real center of their lives is neither an urban nor a rural nor even a suburban area, as these entities have traditionally been conceived, but rather the technoburb, the boundaries of which are defined by the locations they can conveniently reach in their cars. The true center of this new city is not some downtown business district but in each

residential unit. From that central starting point, the members of the household create their own city from the multitude of destinations that are within suitable driving distance. (Fishman, 1987, p. 185)

Even if this account of the current city-regions landscape conveys a technological reductionism, it nonetheless emphasizes some of the issues at stake in recent forms of metropolitanization. Following the previous patterns of suburban developments in relation to industrialization processes and their structural mobility consequences, these new forms can be associated either with the paradigm of the "postmetropolis" – space production being influenced both by post-Fordism and postmodernity (Soja, 2000) – and/or the paradigm of a post-suburban era, when outer suburbs were expanding faster than cities and inner suburbs (Lucy & Phillips, 1997; Phelps & Wu, 2011). In this context, urban planning as defined at the turn of the twentieth century has to face several challenges. This is what I would like to address within an exploratory perspective aimed to better understand the new challenges planning practices cannot avoid.

The post-suburban realities within which planners are immersed bring in new concerns and invite urban researchers to re-examine the principles and values bequeathed by the planning tradition. This observation is hardly an original one. With the emergence of modernity – provoking an historical fracture with totality and introducing an existential devastation for individuals along with a new sense of freedom (Martuccelli, 2017) – everyone must adapt to a constantly changing situation, having to break more or less radically with path dependency.

Post-suburbanization processes are certainly part of modernizing trends. And these trends, when considering human settlements, are nowadays defined through the development of the periphery. Consequently, it is not only the meaning of historical city core that has been transformed, but also urbanization and with it the urban as such. This involves putting aside the "centralist bias in urban theory" (Keil, 2018, p. 42) and revisiting our understanding of cities from the multiplicity of influences and forces at play. The form and content of peripheral growth is certainly at stake, and with this come the necessity to revisit past representations of cities, as well as our understanding of their configuration.

That said, it is not my intention in this chapter to assess how urban theory has tried to apprehend cities during the last decades through competing perspectives, namely postcolonial perspective, assemblage theory, and planetary urbanism (Storper & Scott, 2016). My objective is far more restricted. In connection to the current processes of urbanization and the way they are adapting to and/or transforming cities, I raise the question of the role of planning and planners in suburban governance in the post-suburban era. Are planners still in an advantageous position to contribute to the future of cities as has been the case for the city of the industrial era? To address this question, it is necessary to both recall the principal tenets of planning and to consider the post-suburban processes with which contemporary cities are dealing. For that matter, urban planners have been sidelined as their mediating role grew in importance, having to share it, nonetheless, with an expanding number of people. This can be associated with the presence of governance networks and the deliberative turn taken by planning. But it is also the right of city dwellers to be involved in participatory processes that is illustrative of this.

The chapter is subdivided into three sections. First, I need to go back to the planning model as defined in connection to modern industrial cities. Second, I will shift focus to the structural and contextual changes brought in by the post-suburban era in order to highlight the several adjustments urban planning had to carry out. Finally, I will ask if those adjustments are still compatible with the mission planning was expected to fulfill when it came up as "the twentieth-century response to the nineteenth-century industrial city" (Campbell & Fainstein, 1996, p. 5).

Planning Practice and Planning Theory

If, during the industrial era, urban planning played a significant role in the design and construction of cities based on a rational and functionalist model, in the context of "post-suburban realities," such a model is being revisited, reflecting the new uncertainties with which planning practices must deal. At the end of the nineteenth and beginning of the twentieth century, the planning profession emerged as a specialized field of expertise, with planners relying mainly on scientific knowledge to advise decision-makers. Convinced that they could alleviate the contradictions of capitalism, modernist planners believed it was possible "to diminish the excesses of industrial capitalism while mediating the intramural frictions among capitalists that had resulted in a city inefficiently organized for production and reproduction" (Beauregard, 1989, p. 383). But before paying attention to the more recent challenges that planners now face, it is useful to recall the main components of planning practices and theory. This will give us better insight into the role that mediation planners have been invited to fulfill since the 1980s. In addition, this will help us to better understand how procedural and normative dimensions of planning cannot easily be conceived separately.

Defined as a professional practice, as is common in the Anglo-American tradition, urban planning was above all orientated towards serving the public interest, even though this intention could not be pursued without compromises. The need for compromise results from the fact that public and private interests can diverge, to say the least. But it is also that pluralism prevails within liberal democracies, introducing the convergence of interests as a real issue. Among other things, this follows from an imbalance of power incorporated in the decision-making system. But even more importantly, this power imbalance is experienced when liberal democracy meets capitalism. Hence, two competing values are involved, equality and liberty, even though the encounter between democratic equality and capitalist inequalities has contributed to the emergence of the welfare state (Dubet, 2000).

Planning processes are fraught with uncertainty, and they copy, so to speak, the way democratic practices or democratic performances are transforming the political scene (Lefort, 1981). However, it should not be forgotten that if planning, especially physical planning, can be defined with reference to the categories of rationality and decision influenced by a Weberian vision of modernity, it always implies a "simplified 'model' of the real situation" (Faludi, 1996, p. 71). For that matter, the initial intentions of planning are difficult to achieve with confidence. At the same time, simplification of the "real situation" as Faludi envisions – it is always impossible to include all the components at the basis of it – is certainly unavoidable, even though the rational planning model at the base of planning theory was considered by some thinkers an efficient tool to overcome "the evils of unreason that had overtaken inter-war Europe" (Friedmann, 1996, p. 15).

Historically, the main components of planning were entrenched in the project of modernity as conceived by the thinkers of the Enlightenment (Pinker, 2018). Advancing the ideals of reason, science, humanism, and progress, the Enlightenment succeeded in promoting a major social transition from traditional communities to modern societies, and planning became one of the operators of this process. In the face of needing to accommodate utilitarianism with the principles of humanism, planning was confronted with several uncertainties. On the road to modernity, including the design of modern institutions, planning could benefit from several intellectual contributions based on humanistic values, but also on the natural and social sciences, even though this knowledge was not oriented initially to contribute to planning theory.

As explained by John Friedmann (1987), planning theory is a specialized field of knowledge relying on four different traditions of thinking able to link knowledge and action: social reform, social mobilization, policy analysis, and social learning. By calling upon these traditions, it has been possible to define the basic comprehensive model of planning and decision-making as being one founded on rationality – even though various meanings of the notion prevail (Mannheim, 1940, p. 51) – to support the professional behaviour of planners.

The rational comprehensive model defined in the policy analysis tradition contributed to the foundation of planning and became the dominant approach, even though it is "naïve" to represent it in reference to a sort of "golden era ... during the early postwar years" (Campbell & Fainstein, 1996, p. 9). At the same time, it would be presumptuous to assume that it has been totally abandoned following criticisms raised by practitioners, citizens, and social activists going back to the 1960s. Nonetheless, planning practices took into account communicative approaches and various modalities of "activist planning" (Davidoff, 1965; Sager, 2018).

Still, planners and theoreticians have had a hard time trying to solve the numerous difficulties with which practitioners were confronted. Exploring several avenues – from advocacy to strategic planning – planning theory and planning practices going back to the 1960s went on with diversified experiments, dealing with the radical uncertainty inherent to modernity and the modern experience (Martuccelli, 2017). Is it possible to assume the limits of scientific rationalism while at the same time reaffirming the foundational dimensions of modernity promoted by the Enlightenment? To what extent one can combine the modern tenets of rational planning with postmodern concerns? In underlying concerns about identity, subjectivity, and gender, the postmodern critique added to the difficulties with which modern planning was already dealing. Nonetheless, one cannot forget many of the harsh critiques made regarding the feasibility of planning in its rational form (Friedmann, 1987). Its limited scope falls under the difficulty of predicting the future, but also under conditions imposed by a "bounded rationality" (Simon, 1957), where contextual or environmental limitations unavoidably constrict effective behaviours.

Regarding the dominant rational comprehensive model, one can also recall the difficulty for planning to reduce uncertainty. According to the epistemological principles of rationalism, action should evolve within the limits of predictability inferred by knowledge. Ideally, action is the result of rational principles supporting thinking. Thus, to achieve unfailing action, it is tempting to rely on perfect knowledge. Paradoxically, however, with perfect knowledge it would be impossible to act. Most of the time, if someone were aware of all the consequences of their actions, they would be reluctant to go forward (Hirschman, 1970). In fact, it is our ignorance of the future that gives us the confidence to act (Friedmann, 1979). In that respect, it is vital to break with the dependency of action on knowledge as implied in the rational model. Consequently, it is less important to predict the future than to face it. What is needed, then, is to situate the relationship between knowledge and action within a dynamic process. And in retrospect, this brings about another understanding of planning; one where it can play a mediating role between knowledge and action (Friedmann, 1979).

Such a view regarding knowledge pertaining to the rationalist approach is not frivolous. On the one hand, it is the result of shortcomings of technocratic planning in the 1950s and 1960s. On the other hand, it brings to the fore another understanding of planning, where planning, as just mentioned, can play a mediating role between knowledge and action. Therefore, this perspective is compatible with the opening to new forms of social practices and collective action where new democratic concerns are introduced and democratic purposes targeted.

If the critique addressed to the rational model was introduced initially from outside of the professionals' circle, it was not long before it was also adopted by planners themselves. Due to several factors, starting with the emergence of ethical concerns brought to the fore by the American Planning Association (APA) at the end of the 1980s (Lucy, 1996), planners began to reflect on their own practices. This reflection occurred alongside a reassessment of planning and the introduction of new ideas, among them the promotion of a pragmatic approach using citizen participation as a strategic variable in defining urban policies and/or urban projects. If this started already in the 1960s, it was strongly institutionalized in the 1980s (Hamel, 1986).

At the outset, the introduction of participatory mechanisms questioned the legitimacy of planners themselves and their professional expertise. Nonetheless, it revealed an occasion to counter criticisms and explore new models of action. The expectations were numerous: bring new knowledge and innovative ideas to planners (Brody et al., 2003); reduce participatory uncertainty and enlarge consensus around the substantive content of projects and/or their implementation (Christensen, 1985); make planning compatible with the requirements of democratic deliberation (Forester, 1999); and contribute to the improvement of the policy process (Bherer & Breux, 2012). In fact, the criticisms involved in the new orientation taken by planning and planning theory were nourished by concerns coming from the insights of feminist, postmodern, and postcolonial thinking. As Sandercock (1998) underlines, in the past, modernist planning had often been characterized by its "anti-democratic, race and gender-blind, and culturally homogenizing practices" (p. 4). In addition, environmental considerations and issues were oftentimes ignored or underestimated. Thus, a major paradigm shift occurred in the wake of such critiques, suggesting new understandings of what should be the task of planners or planning practices. Accordingly, we are today faced with a different understanding of planning and its challenges, with environmental issues increasingly at the forefront.

The few elements of criticism I have just recalled, albeit in a schematic way, bear witness to the various changes with which planners have had to deal in the last few decades. In its modern version, planning is supposed to cope with the challenges brought about by an industrialized society. The increasing number of social relationships, starting with the individualization of social interactions, thus require more coordination than the prevailing situation, which means that individuals must accept the need to subordinate their own initiatives, when interacting with others, to a planning authority. This is at least what was suggested by Mannheim (1940), who spoke of the "growing necessity for planning" (p. 157) while comparing this obligation to the facility of coordination through "individual adaptation" when small numbers of interactions daily were the rule.

In retrospect, at the turn of the twentieth century – when modern planning emerged – planning was adhering to well-defined requirements to cope with industrialization and urbanization processes. In fact, the expectation of planning was based on consensus building around the built environment, infrastructures, and the best strategies to support economic development. In many ways, planning was then converging with a technocratic model of decision-making. Being aware of the contradictions inherent in capitalist economic development, technocrats nonetheless accepted many constraints introduced by modernization. However, in some ways, they adhered to a reformist orientation defending the regulating role of the state towards the economy – but while also trying to improve the living conditions of the "lower classes" (Fainstein & Fainstein, 1996, p. 274).

Working hand in hand, planners and technocrats shared a similar understanding of social contradictions filled by class conflicts – a situation that has been subsequently transformed, considering that social inequalities are no longer exclusively defined in classist terms, but are seen through more diversified lenses. As recalled by Dubet (2000), if during the thirty glorious years ("les trente glorieuses") classical social inequalities – related to class belonging – tended to decrease, at the same time other inequalities related to gender, culture, and identity took over and multiplied. Consequently, the challenges to be met for planning now tend to be defined in new terms. This is what we should now turn to, considering particularly the characteristics of the post-suburban era.

Structural and Contextual Changes in the Post-Suburban Era

This new era can be apprehended in relation to a turning point of modernity where the meaning of change itself is undergoing a significant transformation. In many ways, in the contemporary context, a deepening modernity as apprehended in sociological terms does not mean that the structural features of the modern condition are radically different in comparison to what has been experienced by citizens in modern industrial societies. Going back to the emergence of modern civilization, a sense of continuity prevails, even though in the contemporary era a new stage of individuation has been reached (Martuccelli, 2017). With the current situation starting in the 1990s, the experience of the modern condition has been going through a series of "inflexions" that can be found in all spheres of social life. This is also true with social relations to space.

In comparison to previous periods of urbanization, the post-suburban age revisits the relations between city, suburbs, and outer suburbs. Relying on extended urbanization produced from the outside in, post-suburban areas and post-suburban developments imply "an emancipation from the core city" (Musil, 2007, p. 150). The post-suburban age also involves a redefinition of centrality. New relationships are emerging between the centre and its peripheries. When the "growing necessity for planning" was expressed by the elites facing the initial contradictions of capitalism, it was possible to think in systemic and functionalist terms based on a reasonably predictable future. With the multiplication of centres and "in-between cities" (Sieverts, 2011) at the scale of the city-region, the experience of sub/urbanity is conducted under a regime of growing uncertainty. As the traditional city-core functions are drained by multiple suburban peripheries, city centres are at the same time redesigned through an increased global connectivity. However, such restructurings are punctuated by permanent retrofitting adaptations, giving free rein to never ended transactions in line with an open horizon.

This has been examined by numerous studies conducted under the "new regionalist" approach, emphasizing the multiple components and processes involved in rethinking the urban at the scale of the city-region (Keil, 2018; Keil et al., 2017). Territorial diversity, presence of multiple scales for managing city-regions, competing social and political representations, and new models of agency in search of efficient collective action are called upon for questioning the feasibility of metropolitan governance. For several reasons, urban realities are the result of new forms of urbanization. As Brenner and Schmid (2015) underlined, "(t)he basic nature of urban realities – long understood under the singular, encompassing rubric of 'cityness' – has become more differentiated, polymorphic, variegated and multiscalar than in previous cycles of capitalist urbanization" (p. 152). This is exemplified and deepened by Alessandro Balducci, Valeria Fideli, and Camilia Perrone, who are looking at the Italian debate going back to the 1960s when urban regionalization had started to redefine the relations between centre and periphery, as was the case in the Milan urban region (see chapter 4 in this volume). For them, processes of urban regionalization go with a revision of the "linear relationship between territory and authority," raising the issue of boundaries in their interactions with power.

Under epistemological considerations, sub/urban realities are fraught with multiple dimensions – material and immaterial, technological and ideological, objective and subjective – involved in thinking spatiality. This is what the concept of the urban as elaborated by Henri Lefebvre (1970, 1974), relying on the hypothesis of the complete urbanization of society, brought to the fore. If the urban issue is worldwide, it is nonetheless circumscribed by the economic, social, and political realities of the different countries where it is expressed (Lefebvre, 1970). From then on, the urban is before all apprehended through its possibilities as a site making encounters possible: "The urban as a form is transforming what is being united (concentrated)" (p. 230).

As a global phenomenon, particularly in relation to a post-suburban era, suburbanization has been discussed widely (e.g., Hamel & Keil, 2019; Keil, 2018). Its convergence with planetary urbanization informed by geographical and cultural local diversity helped to revise the assumption upon which planning was based: introducing rational comprehensive decision-making in the liberal production process of cities. Relying on rationality and scientific modes of legitimation (Beauregard, 1989), planning has promoted the main values advanced by the thinkers of the Enlightenment – universality, order, rationality, and standardization – but also concerns for human fulfilment and progress (Pinker, 2018).

Nonetheless, in North America, defined as a model for collective action but also as a profession, urban planning has not been represented in abstract terms but, before all, in reference to specific problems with which city dwellers deal: "Although the planning of the built environment in the United States can be traced to the seventeenth-century European colonists ... the impetus for the institutionalization of planning grew out of social problems related to massive immigration, large-scale manufacturing, and the lack of controls over the built environment" (Beauregard, 1989, p. 381). The history of US urban planning reveals that planners never stop responding and adapting to diversified challenges that punctuate the transformation of capitalism (Abu-Lughod, 1999; Monkkonen, 1988). At the outset, the inherent contradictions at hand – as recalled by Foglesong (1996), building on David Harvey's analysis in *Social Justice and the City* – arise because of the way capitalism at the same time "engenders and constraints demands for state intervention in the sphere of the built environment" (p. 169). But these contradictions are added to a more fundamental one that I mentioned previously concerning the relations between capitalism and democracy. While the economic rules of the market system go hand in hand with social inequalities, the functioning of democracy depends upon the promotion of equality and social justice. And there is no formal solution to this contradiction for adapting cities to both requirements – at least within liberal democracies – except to invent compromises.

Urban studies have paid a great deal of attention to the consequences of such compromises regarding the built environment, questioning the modalities and finalities of planning processes. Over the years, researchers in this field as well as planning theorists have examined the modalities of regulation elaborated by the state to manage the compromises. They have also suggested new perspectives either to better understand the possibility of collective action or to better define the conditions of its implementation. In so doing, planning theories have led to a refinement or a revision of the dominant rational comprehensive planning model, building on the innovative contributions of the social sciences in response to new social, cultural, and ethical concerns, adopting successively or in an overlapping way several perspectives (pragmatic, radical, feminist, post-colonialist, postmodernist, strategic, and communicative). The diversity of approaches is certainly informative about the nature of compromises involved and the necessity to explore new potential courses of action. Nonetheless, most of the work done in the field of planning mentioned above did not take seriously enough the changes brought in by post-suburbanization. This is what I shall turn to now.

Challenges of Planning in the Post-Suburban Era

With increased suburbanization, the requirements for metropolitan governance are on the rise. Consequently, peripheries – reinforcing social, geographical, economic, and cultural diversity – as well as central cities are experiencing new modes of cooperation for adapting urban settlements to unexpected changes, whether these changes are coming from environmental degradation, technological innovations, or social transformations. Metropolitan governance is taking place at the scale of city-regions, where multiple administrative units and/or fragmented parcels of territories are sharing a common destiny. Fragmentations and divisions seem to be the rule everywhere, even though, in a paradoxical way, new forms of collective agency are being expressed: "All metropolitan areas share underlying needs to govern themselves, which stem from the strong interdependencies and externalities generated by urbanization; but all such regions have fragmented political geographies for addressing these problems. This is why some version of the same basic mishmash is found in countries whose political and administrative systems are otherwise quite different" (Storper, 2014, p. 118).

Urban governance has been explored from the 1980s onwards by local states as a new perspective of planning and management for revising the traditional organizational principle of political hierarchy. A consensus is prevailing in that regard within the urban studies literature (Levy-Faur, 2012). However, even though experimentations through governance emphasized the increased role of private and civil-society actors in public affairs, it did not imply that the state was less important. In fact, it is the other way around, with public interventions by the state greatly increasing over the last few decades. Consequently, some contend that spheres of public intervention are more and more wide ranging (e.g., Pinson, 2015).

Hajer and Versteeg (2008) have made a similar claim. According to them, even though governance networks are contributing to define the context within which policymaking is occurring, "it is the old-modernist order which remains the strongest symbolization of political democracy" and that is promoted by the media as the "strongest symbolization of political democracy" (p. 9). Nonetheless, besides the legitimacy of the traditional representative and policymaking system, new forms of cooperation and deliberation are involved in the redefinition of the cultural and political landscape of politics. Thus, it is not surprising to see almost everywhere the emergence of a "complicated form of politics hidden in the shadow of the well-established existing political institutions" (Hajer & Wagenaard, 2003, p. 98). The relations between state institutions and civil society actors within city-regions are, through governance processes, increasingly involved in new forms of decision-making. And those relations are the result of several adjustments taking place on the political scene, simultaneously beyond and in connection to the weight of hierarchical organizations and political institutions.

For a better understanding of the issues at play with governance in relation to planning, one must come back to the recent discussion around deliberative democracy and democratic deliberation (see Chambers, 2009; Mansbridge, 2007). Some time ago, it was observed that Western democracies have taken a deliberative turn (Blondiaux & Sintomer, 2002). Sometime later, it was noted that deliberative democratic theory is now experiencing a similar institutional turn (Chambers, 2009). And with this turn, the question of democratic legitimacy is being pushed aside in favour of more pragmatic concerns with "the nuts and bolts" of democratic institutions, considering at first the functioning and role of "mini-publics." According to Chambers (2009), "(m)ini-publics are exercises in deliberative democracy in which citizens come together to discuss and decide on public policy. They are mini because they are small scale, manageable, and indeed designed settings. They are publics because there is usually some claim that deliberation mirrors, represents, or speaks for some larger public. Deliberative opinion polls, citizen juries, consensus conferences, and citizens' assemblies are just a few examples of such forums" (p. 129–30). These mini-publics, however, do not respond to all the challenges raised by the "democratic invention" (Lefort, 1981). Even if they could contribute to overcoming the gap between the normative theory of deliberative democracy and practical experiments aimed at improving democratizing processes (Mutz, 2008), they cannot meet all the conditions required by democratic accountability for promoting mass democracy. As Chambers (2009) underlined, mini-publics are not easily transferred to mass democracy.

In the 1980s, planning theories were certainly aware of the deliberative turn of democracies, as well as of the institutional turn of deliberative theory (Forester, 1989). But researchers were focusing mostly on communicative dimensions and narratives at play, leaving aside the structural transformations of the urban fabric underway. Within the post-suburban era, dialogue between planners, citizens, and local public administration is thus entering a new stage of relations that was previously difficult to apprehend. These relations are combining representative democracy and dialogical spaces, raising the issue of urban expertise on new grounds. All of this is occurring in a regional context where local communities are involved in new forms of compromises and "rearrangements" in relation to "intra-regional territorial and local-regional conflicts and conversation" (Boudreau et al.,

2017, p. 380). This aspect is brought in differently in comparison to previous periods of urbanization (Phelps & Wu, 2011).

Experimentations with "hybrid" or "open" forums deploying mechanisms of public consultations used largely by planners over the last decades for making governance more inclusive and obliquely increasing its legitimacy have not been entirely efficient. The most convincing explanation for this is given by Farias (2016), who argues that these experiments with public consultation were ineffective because the forums were primarily framed through "the open, direct and participative language of technical democracy to hide the fact that decisions were being made on the basis of cold economic calculations and arrogant expert epistemologies" (p. 561). But even more decisively, it is because these forums are intended to provide "consensual mechanisms aimed at producing decisions about the planning and design of city reconstruction" (p. 561). In other words, these forums are not designed to explore the uncertainty in the planning process but are for promoting "consensual management of existing conflicts" (p. 561).

Researchers have underlined for some time that the performance and results of hybrid forums and consultation processes are biased because of the unequal access marginal communities and lay citizens have to resources (Fung & Wright, 2001; Hamel, 2008; Talisse, 2019). Nonetheless, Farias (2016) interjects that open forums are connected both to representative democracy and to heterogeneous actors, including actors who do not share the same confidence in these forums, or who do not recognize themselves in the values and interests promoted by their organizers about an ideal speech situation, either due to historical or social reasons. Finally, what is at stake beyond the implementation of technical democracy are the democratic challenges of cities defined in terms of "multiplicities of assemblages." Despite the critique addressed to the notion of "urban assemblages" (Brenner et al., 2014), this notion helps to better understand the multiplicities of situations and processes attached to urban restructuring, allowing us to open the box of "socio-technical uncertainty" involved in the making of sub/urban landscapes.

If democratic issues have been central to the definition of planning practices and have been addressed critically by planners over the years, with the rise of sub/urbanization processes of governance, global challenges are introducing new concerns informing post-suburban realities. This comes not only from the inequalities related to the traditional North–South divides, but also from the fact that social mobility at the scale of the globe has transformed the former international view of planning. As Roy (2011) recalled, "if the mid-twentieth century was characterized by national planning cultures, then today's policy world is marked by crossing of borders" (p. 410). Through the formal and informal presence of resources, people, and ideas coming from elsewhere, global processes are open to multiple channels of influence in governing city-regions. Increased mobilities and several "technologies of crossing" are deployed by immigrants in creating what Miraftab (2011a) refers to as "transnationalism from below, above, in between, and sideways" (p. 376).

Planning "models-in-circulation" are not only diversified in a cultural sense; they are also part of a "transnational framing of the global processes" (Miraftab, 2011a, p. 376). However, this does not mean that disparities in resources, social divisions, and spatial inequalities are tending to disappear. We are far from ratifying the vision of "one unitary world" with converging values. The experiments by planning practices in this perspective can be associated with "worlding tactics" in front of planetary capitalism, supplying "some vision of the world in formation" (Ong, 2011, p. 11). This perspective emphasizes the intricate networks of interactions between people – especially those who are disadvantaged – localities, and resources: "If the city is a living, shifting network, then worlding practices are those activities that gather in some outside elements and dispatch others back into the world" (Ong, 2011, p. 12). With such an understanding, planning is no longer exclusive to professional planners.

This reading is clearly in rupture with the initial rational model of decision-making upon which the planning profession was based. Using a "universal language," planning theory was then paying little attention to contexts and, in theory, voluntarily "ignoring political interests" (Baum, 1996, p. 367). Nowadays, such a position must be abandoned. Not only because local contexts and pluralism are gaining in legitimacy, being promoted by social actors and social movements and/or endorsed by the postmodern critique, but more importantly because requests are being made of civil-society actors everywhere to be involved in decision-making processes at several stages of planning. In addition, the situation created by the rise of neoliberalism converging with the weakening of the political sphere has had important negative consequences regarding state legitimacy. However, if the state is stepping aside in the face of the rising power of financial capitalism – keeping regulation at a minimum – the quest for city-regional cooperation in urban matters is at the same time clearly on the rise, introducing new concerns for planning (Keil et al., 2017).

Conclusion

Within this chapter, I have gone through the planning perspectives – albeit schematically – as elaborated by planning theories and planning practices for solving the problems of modern industrial cities. Coping with the rise of these cities, planning has referred to modern rational decision-making for improving the territorial organization of urban settlements. In so doing, the challenge has also been to regulate the contradictions of modern capitalism (Foglesong, 1996; Harvey, 1973).

In the post-suburban era, social problems are showing up in a new light in comparison with those faced in the industrial context. From a professional standpoint, planners are no longer able to project an image of exclusivity regarding city form, definition, or orientation. This had already been admitted in the 1980s through the communicative and deliberative turn in which planning practices were engaged (Forester, 1999). However, despite a "benevolent" representation of planning, or even with the "profession's self-ascribed narrative of political innocence" (Miraftab, 2011b, p. 860), the reality was much different: "This innocent image of planning contradicts the wretched record of planners as facilitators of state agendas for social control and planning decisions that systematically displace disadvantaged populations through zoning and urban renewal projects to create exclusionary cities" (p. 860). Communicative and/or participatory inputs inserted in new approaches to planning were not enough to reverse past trends.

The critiques addressed to planning practices as they have been deployed since the 1960s brought to the fore the necessity of creating a forum for informal actors and citizens who were concerned by the form and content of urbanism. Incidentally, as new claims of democratization increased with social mobilization, planners accepted to include residents in planning processes, for example through public hearings. However, this recognition rarely saw any significant redistribution of resources. This is particularly prominent in the case of suburban governance because the consequences for urban mobility, urban sustainability, and social justice are growing in importance in relation to the "global variety of suburban constellations" (Keil, 2018, p. 63).

Returning now to the question I raised in the introduction – Are planners still in an advantageous position to contribute to the future of cities as has been the case for the city of the industrial era? – the answer is certainly negative. Not only have planners recognized that they need to share their responsibility with social actors – even though their willingness to do so was often insufficient and biased – but it is also their role as mediators and/or facilitators that has impaired their image.

The communicative and deliberative turn in planning took place simultaneously with changes in the meaning of government and the rise of new political challenges as expressed through the "problématique" of governance. Converging to a certain extent with deliberative democracy, deliberative governance provides an occasion to go beyond an abstract definition of democracy, looking for its application with possible benefits for ordinary citizens. Thus, planning practices, despite several shortcomings, may still contribute to deepening the understanding of democracy through various empirical fields. But this is under the condition that we no longer accept or see planning practices as the exclusive prerogative of experts promoting an elitist normative model for global sub/urbanization governance and planning.

ACKNOWLEDGMENTS

I would like to thank Roger Keil and Fulong Wu for their detailed comments on an earlier draft of this chapter.

REFERENCES

Abu-Lughod, J. (1999). *New York, Chicago, Los Angeles: America's global cities.* University of Minnesota Press.

Baum, H.S. (1996). Practicing planning theory in a political world. In S.J. Mandelbaum, L. Mazza, & R.W. Burchell (Eds.), *Explorations in planning theory* (pp. 365–82). Rutgers.

Beauregard, R.A. (1989). Between modernity and postmodernity: The ambiguous position of US planning. *Environment and Planning D: Society and Space, 7*(4), 381–95. https://doi.org/10.1068/d070381.

Bherer, L., & Breux, S. (2012). The diversity of participation tools: Complementing or competing with one another? *Canadian Journal of Political Science, 45*(2), 379–403. https://doi.org/10.1017/S0008423912000376.

Blondiaux, L., & Sintomer, Y. (2002). L'impératif délibératif [The deliberative imperative]. *Politix: Revue des sciences sociales du politique, 57,* 17–35. https://www.persee.fr/doc/polix_0295-2319_2002_num_15_57_1205.

Boudreau, J.-A., Hamel, P., Keil, R., & Kipfer, S. (2017). North Atlantic urban and regional governance. In R. Keil, P. Hamel, J.-A. Boudreau, & S. Kipfer (Eds.), *Governing cities through regions: Canadian and European perspectives* (pp. 377–84). Wilfrid Laurier University Press.

Brenner, N., Madden, D.J., & Wachsmuth, D. (2014). Assemblage urbanism and the challenges of critical urban theory. *City, 15*(2), 225–40. https://doi.org/10.1080/13604813.2011.568717.

Brenner, N., & Schmid, C. (2015). Towards a new epistemology of the urban? *City, 19*(2–3), 151–82. https://doi.org/10.1080/13604813.2015.1014712.

Brody S.D., Godschalk D.R., & Burby, R.J. (2003). Mandating citizen participation in plan making: Six strategic planning choices. *Journal of the American Planning Association 69*(3), 245–64. https://doi.org/10.1080 /01944360308978018.

Campbell, S., & Fainstein, S.S. (1996). Introduction: The structure and debates of planning theory. In S. Campbell & S.S. Fainstein (Eds.), *Readings in planning theory* (pp. 1–14). Blackwell.

Chambers, S. (2009). Rhetoric and the public sphere: Has deliberative democracy abandoned mass democracy? *Political Theory, 37*(3), 323–50. https://doi.org/10.1177/0090591709332336.

Christensen, K.S. (1985). Coping with uncertainty in planning. *Journal of the American Planning Association, 51*(1), 63–73. https://doi.org/10.1080 /01944368508976801.

Davidoff, P. (1965). Advocacy and pluralism in planning. *Journal of the American Institute of Planners, 31*(4), 331–8. https://doi.org/10.1080 /01944366508978187.

Dubet, F. (2000). *Les inégalités multipliées*. Éditions de l'Aube.

Fainstein, S.S., & Fainstein, N. (1996). City planning and political values: An updated view. In S. Campbell & S.S. Fainstein (Eds.), *Readings in planning theory* (pp. 265–87). Blackwell.

Faludi, A. (1996). Rationality, critical rationalism, and planning doctrine. In S.J. Mandelbaum, L. Mazza, & R.W. Burchell (Eds.), *Explorations in planning theory* (pp. 65–82). Rutgers.

Farias, I. (2016). Devising hybrid forums. *City, 20*(4), 549–62. https://doi.org /10.1080/13604813.2016.1193998.

Fishman, R. (1987). *Bourgeois utopias: The rise and fall of suburbia*. Basic Books.

Foglesong, R.E. (1996). Planning the capitalist city. In S. Campbell & S.S. Fainstein (Eds.), *Readings in planning theory* (pp. 169–75). Blackwell.

Forester, J. (1989). *Planning in the face of power*. University of California Press.

Forester, J. (1999). *The deliberative practioner: Encouraging participatory planning processes*. MIT Press.

Friedmann, J. (1979). *The good society*. MIT Press.

Friedmann, J. (1987). *Planning in the public domain: From knowledge to action*. Princeton University Press.

Friedmann, J. (1996). Two centuries of planning theory: An overview. In S.J. Mandelbaum, L. Mazza, & R.W. Burchell (Eds.), *Explorations in planning theory* (pp. 10–29). Rutgers.

Fung, A., & Wright, E.O. (2001). Deepening democracy: Innovations in empowered participatory governance. *Politics & Society, 29*(1), 5–41. https://doi.org/10.1177/0032329201029001002.

Hajer, M., & Versteeg, W. (2008, August 28–31). *The limits to deliberative governance* [Paper presentation]. American Political Science Association Annual Meeting, Boston, MA, United States.

Hajer, M., & Wagenaard, H. (2003). A frame in the fields: Policymaking and the reinvention of politics. In M. Hajer & H. Wagenaard (Eds.), *Deliberative policy analysis: Understanding governance in the network society* (pp. 88–110). Cambridge University Press.

Hamel, P. (1986). Les pratiques planificatrices dans le contexte actuel: Comment interpréter l'appel à la participation? [Planning in today's societies: What does participation mean?]. *Revue internationale d'action communautaire, 15*(55), 66–76. https://doi.org/10.7202/1034436ar.

Hamel, P. (2008). *Ville et débat public: Agir en démocratie* [City and public debate: Acting in democracy]. Les Presses de l'Université Laval.

Hamel, P., & Keil, R. (2019). Toward a comparative global suburbanism. In B. Hanlon & T.V. Vicino (Eds.), *The Routledge companion to the suburbs* (pp. 51–61). Routledge.

Harvey, D. (1973). *Social justice and the city*. Johns Hopkins University Press.

Hirschman, A.O. (1970). *Exit, voice, and loyalty: Responses to decline in firms, organizations, and states*. Harvard University Press.

Keil, R. (2018). *Suburban planet*. Polity Press.

Keil, R., Hamel, P., Boudreau, J.-A., & Kipfer, S. (Eds). (2017). *Governing cities through regions: Canadian and European perspectives*. Wilfrid Laurier University Press.

Lefebvre, H. (1970). *La révolution urbaine* [The urban revolution]. Gallimard.

Lefebvre, H. (1974). *La production de l'espace* [The production of space]. Éditions Anthropos.

Lefort, C. (1981). *L'invention démocratique, les limites de la domination totalitaire* [Democratic invention, the limits of totalitarian domination]. Fayard.

Levy-Faur, D. (2012). From "big government" to "big governance"? In D. Levy-Faur (Ed.), *The Oxford handbook of governance* (pp. 3–18). Oxford University Press.

Lucy, W.H. (1996). APA's ethical principles include simplistic planning theories. In S. Campbell & S.S. Fainstein (Eds.), *Readings in planning theory* (pp. 479–84). Blackwell.

Lucy, W.H., & Phillips, D.L. (1997). The post-suburban era comes to Richmond: City decline, suburban transition and exurban growth. *Landscape and Urban Planning, 36*(4), 259–75. https://doi.org/10.1016/S0169-2046(96)00358-1.

Mannheim, K. (1940). *Man and society in an age of reconstruction: Studies in modern social structures*. Harcourt, Brace & World.

Mansbridge, J. (2007). "Deliberative democracy" or "democratic deliberation"? In S.W. Rosenberg (Ed.), *Deliberation, participation and democracy: Can the people govern?* (pp. 251–71). Palgrave Macmillan.

Martuccelli, D. (2017). *La condition sociale moderne: L'avenir d'une inquiétude* [The modern social condition: The future of a concern]. Gallimard.

Miraftab, F. (2011a). Symposium introduction – Immigration and transnationalities of planning. *Journal of Planning Education and Research, 31*(4), 375–8. https://doi.org/10.1177/0739456X11425001.

Miraftab, F. (2011b). Beyond formal politics of planning (debates and developments). *International Journal of Urban and Regional Research, 35*(4), 860–2.

Monkkonen, E.H. (1988). *America becomes urban: The development of US cities and towns, 1780–1980*. University of California Press.

Musil, R. (2007). Globalized post-suburbia. *Belgeo, 1*, 147–62. https://doi.org/10.4000/belgeo.11718.

Mutz, D. (2008). Is deliberative democracy a falsifiable theory? *The Annual Review of Political Science, 11*, 521–38. https://doi.org/10.1146/annurev.polisci.11.081306.070308.

Ong, A. (2011). Introduction: Worlding cities, or the art of being global. In A. Roy & A. Ong (Eds.), *Worlding cities: Asian experiments and the art of being global* (pp. 1–26). Wiley-Blackwell.

Phelps, N.A., & Wu, F. (Eds). (2011). *International perspectives on suburbanization: A post-suburban world?* Palgrave Macmillan.

Pinker, S. (2018). *Enlightenment now: The case for reason, science, humanism and progress*. Penguin Books.

Pinson, G. (2015). Gouvernement et sociologie de l'action organisée: Action publique, coordination et théorie de l'État [Governance and sociology of organized action: Public action, coordination and theory of the state]. *L'année sociologique, 65*(2), 483–519. https://doi.org/10.3917/anso.152.0483.

Roy, A. (2011). Commentary: Placing planning in the world – Transnationalism as practice and critique. *Journal of Planning Education and Research, 31*(4), 406–15. https://doi.org/10.1177/0739456X11405060.

Sager, T. (2018). Planning by intentional communities: An understudied form of activist planning. *Planning Theory, 17*(4), 449–71. https://doi.org/10.1177/1473095217723381.

Sandercock, L. (1998). *Towards cosmopolis*. John Wiley & Sons.

Sieverts, T. (2011). The in-between city as an image of society: From the impossible order towards a possible disorder in the urban landscape. In D. Young, P.B. Wood, & R. Keil (Eds.), *In between infrastructure: Urban connectivity in an age of vulnerability* (pp. 239–50). Praxis Press.

Simon, H. (1957). *Administrative behavior: A study of decision-making processes in administrative organization* (2nd ed.). Macmillan.

Soja, E. (2000). *Postmetropolis: Critical studies of cities and regions*. Blackwell.

Storper, M. (2014). Governing the large metropolis. *Territory, Politics, Governance, 2*(2), 115–34. https://doi.org/10.1080/21622671.2014.919874.

Storper, M., & Scott, A.J. (2016). Current debates in urban theory: A critical assessment. *Urban Studies, 53*(6), 1114–36. https://doi.org/10.1177/0042098016634002.

Talisse, R. (2019). New trouble for deliberative democracy. *Les ateliers de l'éthique, 12*(1), 107–23. https://doi.org/10.7202/1042280ar.

Teaford, J. (2006). *The metropolitan revolution: The rise of post-urban America.* Columbia University Press.

Teaford, J. (2011). Suburbia and post-suburbia: A brief history. In N.A. Phelps & F. Wu (Eds.), *International perspectives on suburbanization: A post-suburban world?* (pp. 15–34). Palgrave Macmillan.

After Suburbia | The Path Ahead

Image 12. Four students jumping high

12 An Atlas of Suburbanisms

MARKUS MOOS

The images that you see as chapter dividers in the second half of this volume are outputs from the Global Suburbanisms research project lead by Roger Keil. I joined the project in 2010 as team lead of one of the foundational areas of the project focusing on socio-spatial analysis of Canadian suburbanisms. The team included faculty members (Pierre Filion, Richard Harris, Ute Lehrer, Pablo Mendez, Alan Walks, Elvin Wyly) and research assistants (Anna Kramer, Liam McGuire, Michael Seasons, Robert Walter-Joseph) from several Canadian universities.

I recall one of our initial team meetings at the University of Waterloo's School of Planning, where we had fruitful discussions about the various ways to define the suburbs. Alan Walks proposed (and later published on) suburbanisms as a way of life, building on Henry Lefebvre's concept of urbanism as a way of life (see Walks, 2013). The focus of this approach is on the reciprocal relationship between place and particular ways of living. For our quantitative project, this meant we would focus on analysing the geographies of specific ways of living that had been, at least historically, associated with suburbs.

Anna Kramer (now faculty at McGill University) innovatively proposed the use of Venn diagrams and dot density maps to operationalize our variables, which aimed to measure, of course very coarsely, the diverse and intersecting suburban ways of living. At least initially, we settled on housing type, commuting patterns, and tenure to operationalize what are historically archetypical North American suburban ways of living. The intent here was never to obtain an exact boundary of what constitutes the suburbs or precisely capture all dimensions of suburban ways of living. Instead, we asked the following question: If we were to map some of the characteristics that are often (stereotypically) associated with suburbs, what kinds of geographies would we get? We first published this work in 2012 as an online "Atlas of Suburbanisms," which received more attention than we could have possibly imagined (see uwaterloo.ca/atlas-of-suburbanisms). Media outlets from across the US and Canada were interested in the ideas and maps contained in the Atlas, and several university libraries added the resource to their online collections.

The result of this quantitative project was a more nuanced picture of metropolitan areas' urban social geographies than could have been revealed using methods based on a more traditional urban/suburban dichotomy. Although suburban ways of living as defined by single-detached housing, automobile commuting, and home ownership generally remain decentralized, we also found suburban ways of living in urban places and urban ways of living in suburban places. Although we still ended up categorizing neighbourhoods as more or less suburban, I think we succeeded in developing an alternative (complementary) method of analysis that reveals greater complexity in the metropolitan landscape than could have been obtained by focusing on traditional, often binary, place-based definitions of suburbs alone.

We reworked and refined the analysis of these maps over several years, eventually culminating in an edited book that maps out the geographies of North American suburbanisms in all major US and Canadian metropolitan areas (see Moos and Walter-Joseph, 2017). Brought together by an overarching fictional narrative of a team of urban planning consultants, *Still Detached and Subdivided? Suburban Ways of Living in 21st-Century North America* details the actual social geographies of a large chunk of the North American metropolitan system through the conceptual lens of suburbanisms.

We launched the book in the fall of 2017 in Toronto's urban core, helping to bring the suburbs, literally and metaphorically, to the centre of urban research. The launch even included a musical contribution that is still available online to view (see Generationed City, 2017). The images included in the present volume were first modified from *Still Detached and Subdivided* with the help of research assistants Nicole Yang and Kourosh Mahvash to be displayed as large panels at the book launch, and later at the final Global Suburbanisms conference, "After Suburbia," held at York University in October 2017. The images include select mappings of suburbanisms, as well as renderings of potential suburban-made solutions to some of the social and environmental issues currently associated with suburban living.

Our project, and eventually the book and its launch, benefited from the involvement of a great number of faculty members and students working in various capacities – too many to name here individually. I wish to wholeheartedly thank everyone involved.

REFERENCES

Generationed City. (2017, October 23). *Still detached and subdivided? Book Launch* [Video]. YouTube. https://youtu.be/bss-UDpNgcw.

Moos, M., & Walter-Joseph, R. (Eds.). (2017). *Still detached and subdivided? Suburban ways of living in 21st-century North America*. Jovis.

Walks, A. (2013). Suburbanism as a way of life, slight return. *Urban Studies, 50*(8), 1471–88. https://doi.org/10.1177/0042098012462610.

Image 13. Exploring methodologies

13 Decolonizing Suburban Research

ROB SHIELDS

North American suburbs have been treated as sites of a collective amnesia concerning previous patterns of occupation and occupants. As "greenfield sites," they often either lack history, or local history is shallow, rarely extending back before the agricultural tenants of the last century. A number of critics have also pointed out that issues of indigeneity, migrancy, and ethnicity have received less attention in urban and suburban research than they should have (Gururani & Dasgupta, 2018; Keil, 2013; Roy, 2011). This chapter considers the challenges of researching past suburban occupation, the evidence for which has often been erased along with removal of the area's flora and fauna and, commonly, even the topsoil (see also Cariou & McArthur, 2009). Recent literature on the "decolonization" of urban and suburban research that draws on postcolonial critiques and settler colonial studies to open a new vantage point on suburbia suggests that North American suburbs are not greenfield sites after all; rather, they are intersectional sites of a "colonial matrix" or "logic" that combines capitalism, colonialism, nationalism, and modernity. This background provides the basis for new methods that are being developed to study suburbs. The tensions of these forces emerge in the more conspicuous disfunctionality of suburbia – its consumerism, commuting times, energy expense, and even household divisions of labour inside those tract houses.

The viewpoints of Indigenous peoples in Canada and the United States have been consistently excluded from official histories of place and place-making. Often removed from areas near settlements during colonial times, Indigenous populations were segregated into reserves or ghettoized in inner city neighbourhoods. Those who had previously called areas developed into suburbs home were excluded or disregarded. Although historical Indigenous terms and place names were often used to name picturesque suburban streets and parks, Indigenous history itself was either not celebrated or treated as an archive of a past culture of extinct traditions (Casagranda, 2013). North American peri-urban areas still bear the "interspatiality" (McIlvenny et al., 2009), or marks and appropriations, from an erased Indigenous spatialization (Shields, 1991). Settlers or the descendants of settlers, myself included, have been increasingly struck by the general Indigenous absence in these areas that only seems to become more tangible the more pre-existing occupation or history is repressed.

Background: Postcolonial and Decolonializing Approaches

I begin with a review of approaches that are influencing decolonial research on North American suburbs. Decolonial urban and suburban research builds not only on previous scholarship on the built environment, social spatializations, and urbanization but also on postcolonial criticism of nationalism and racism that emerged in the humanities, critical race studies, and Native studies in the second half of the twentieth century. After the colonies of former European empires gained their independence, the vantage point of postcolonial writers and critics reversed from the point of view of the metropolitan centres of these former European empires to a view from the peripheries (Gilmartin & Berg, 2007).[1] As a term, *postcolonial critique* particularly refers to critical analyses of the legacy of British imperialism and relations with Commonwealth countries. The legacy of these different sources and waves of critical thought is a somewhat confusing set of terms for the uninitiated. These need to be understood as reflecting their different disciplines, and the different places and times in which these critiques arose.

Decolonial and decolonizing research agendas have focused firstly on the treatment of Indigenous societies in North American cultural and educational systems in which there was an imbalance of knowledge and perspectives between Eurocentric hierarchies of prestige and authority versus colonial, peripheral, or subaltern perspectives and priorities. The challenge of decolonial research on suburbs is epistemic. Decolonial approaches not only dispute historical facts but also critique dominant approaches, disciplines, and cultural institutions that have hidden the ethics of research and the representation of knowledge by ignoring moral dimensions of right and wrong, all while claiming objectivity. Despite their limited vantage point, Eurocentric researchers and institutions have claimed a universality and value neutrality for their arguments. To contest this, decolonial theory has drawn on historical geography and has used geographical relations, travel, and diaspora in ways that have allowed researchers to begin to reappraise the role of academic disciplines in imperialism and in forming spatial prejudices (Gilroy, 1993; Said, 1978). Decolonial writers advocate for the development of new analytical perspectives that expose and replace the "colonial matrix" of not only nationalism, capitalism, and colonialism but also modernity that is found in the methods of late twentieth-century academic research (Mignolo, 2011, p. xxvii). Decolonial theory argues that this colonial logic has persisted after political decolonization through economic and cultural globalization and has perpetuated a hierarchy of racial, gender, and epistemic values that privilege dominant white, Euro-American "Western" elites (Lugones, 2003; Maldonado-Torres, 2007; Quijano, 2007).

Decolonial theory identifies alternatives as well as existing trajectories that have continued to exist within and at the margins of dominant forms of modernity, or that delink from the colonial matrix. Alternatives to the four aspects of nationalism, modernity, capitalism, and colonialism are argued to lie with those marginalized by this matrix, including subaltern, Indigenous, and migrant populations. However, many projects under the banner of "emancipation" are questioned as pathways that only lead back to the same totalizing logic as modernity. Instead, decolonial critics argue that we should avoid the abstraction and universality of the existing media and academic claims. Knowledge should be more nuanced, recognize local specificity, and foster a plurality of voices rather than a simple hierarchy led by Euro-American experts only. In this way, decolonial research and critique will offer a better basis for understanding the two faces of modernity, which includes development and, at the same time, recognizes exploitation and thus acts as a call for action to resolve injustices.

In the last two decades, Indigenous research has taken up the decolonization agenda (see Smith, 2012). It is counterposed to research and cultural production that blindly serves the needs of the capital or state for statistical surveillance and for social management and ideological control of culture. The dominant model and its experts have tended to "label" and criminalize economically and culturally marginalized Others. Colonial knowledge projects presuming the universal superiority of dominant Euro-American culture have attempted to extract and exploit marginalized knowledges, from Indigenous agricultural techniques to medicinal remedies to cultural motifs. Indigenous cultures have been articulate in rejecting research that is "done" to them rather than with or for them (Maori elder cited in Smith, 2012, p. xi). Rather than continuing Victorian-era missions to conquer, normalize, and assimilate populations into roles established for them by a dominant elite, "decolonization" became a mantle for research approaches that try not only to expand knowledge but also to enlighten, to liberate, and to work in the interests of their most vulnerable research participants. This amounts to a critique that is more than methodological. Decolonial critiques challenge and shame existing value systems and expose the vulnerability of Western moral systems to this critique (Snyman, 2015).

In addition to redistributive justice, which has long been a focus of political economists, a decolonial approach adds cultural recognition as an essential component of urban and suburban research. It supports demands for recognition by Indigenous and diasporic communities such as migrant labourers and temporary foreign workers and immigrants. Decolonization research has sensitized policymakers and innovators to ongoing relations of exploitation (Snyman, 2015) and has become more inclusive in the values, legal entitlements, and social arrangements that are created at the scale of the city-region. It also validates municipalities as lead actors in responses to social problems such as homelessness, points to the importance of overlooked or erased aspects of places, and demands that researchers reflect on and state the position from which they come as researchers.

Settler Colonial Society

In North America, the role of settler colonization in clearing and assembling land is evident. Histories of places written by European settlers focused on what O'Brien (2010) calls "firsting and lasting" – that is, naming the first settlers and categorizing Indigenous occupants as the last of their cultures, thus naturalizing the takeover of areas by the new arrivals. Drawing on empirical research in Australia and North America, settler colonial studies critiques Eurocentric political economy and history (Wolfe, 1999). It demonstrates the role of occupying land as a precondition to accumulation by dispossession and as a spatial fix for capitalism (Harvey, 2005). Colonialism is at the centre of Western economic history and essential to understand the success of capitalism. Settler colonialism eliminates residents or Indigenous populations who are replaced by settlers and/or slaves and other migrants (Wolfe, 1999, 2006, 2011). As Tuck and Yang (2012) explain, "Settler colonialism implicates everyone." It denies "the existence of Indigenous peoples and the legitimacy of claims to land ... the long-lasting impacts of slavery ... [and] requires arrivants to participate as settlers" (as cited in Tuck & McKenzie, 2015, pp. 69–70).

A growing literature over the last two decades on urban studies and indigeneity has developed well beyond work on the social and health problems of Indigenous city dwellers (Auger, 1999; Browne et al., 2009; Peters, 2004). The new contributions of settler colonial studies to the disciplines of history, sociology, geography, and planning are still not in the mainstream (Patrick, 2015, p. 534), but a growing chorus of scholarship now considers cities and suburbs as Indigenous land. This empirical research is changing our perspectives on cities such as Chicago (Bang et al., 2014; LaGrand, 2002), Detroit (Mays, 2015), Sydney (Gulson & Parkes, 2009), and San Diego (Pulido, 2000).

Thrush's (2016, 2017) analyses of Seattle and London show that even imperial centres were taken up and partly shared by Indigenous geographies and cultural networks. His Indigenous history of the Seattle area argues that Indigenous title to the land has been reduced and encroached upon to remake Seattle and Tacoma into cities where Indigenous heritage and people are repressed (Thrush, 2017). By appropriating this invisibility and repression, however, Indigenous writers and cultures more broadly have sought to replace a sense of absence, urban haunting, and settler guilt with "Indigenous mourning, and imagined spectral ancestries with actual genealogies embedded in the land" (Boyd & Thrush, 2011, p. xx; see also Thrush, 2013, 2016). As Coulthard (2014) has shown, these writers, activists and researchers try to activate the remaining interspatiality of previous patterns of occupation and land use:

> Place is a way of knowing, experiencing, and relating with the world ... ways of knowing often guide forms of resistance to power relations that threaten to erase or destroy our senses of place. This ... is precisely the understanding of land and/or place that not only anchors many Indigenous peoples' critique of colonial relations of force and command, but also our visions of what a truly post-colonial relationship of peaceful coexistence might look like. (pp. 79–80)

The insight that knowledges are embedded in place or "co-produced" in place-based practices (Basso, 1996; TallBear, 2013) is taken up by Tuck and McKenzie (2015) as a decolonial and Indigenous method. Indigenous methods are characterized by a dynamic orientation in which ecological forces are treated as agential and causal (e.g., Bang et al., 2014). Louis (2007, p. 133) provides an excellent overview of common points across Indigenous methodological and ethical thinking (see also the literature review by Drawson et al., 2017). To avoid continuing colonial processes of erasure and spatialized oppressions (Lipsitz, 2011 p. 3; Said, 1994), critical place research informed by Indigenous knowledge and methods posits that "places are ... not always justly named ... are not fixed … [nor] understood by objective accounts" (Tuck & McKenzie, 2015, p. 14). Places both have and are practices (Deyhle, 2009).

Decolonizing Suburban Research

The argument for place-based research is directly relevant to suburban researchers. The postcolonial reversal of the point of view of analysists suggests we look at suburbia not from the urban perspective as "less than the city" but from a decolonial perspective as a site partly hidden by dominant understandings. These mask the process of dislocation of what may have existed before a suburb was developed. Settler colonial studies highlights the displacement of people and the erasure of the signs of their presence and history. Suburbia is argued to re-enact colonial settlement by mirroring an anxious "escape" from threatening environments (Veracini, 2012). Combined with the forced "standardizing ideals of whiteness, masculinity, and Britishness" (Frew, 2013, p. 281), the settlement process of suburbia takes on the spatial and social forms of separation. Consequently, the sociocultural "status" of people in Canada, for example, has been forced into the three broad categories of "Canadians, Indians and Immigrants" (Thobani, 2007). This is played out in North American suburbia generally as "settlers" move in after developers have bought out or displaced previous groups and have uprooted the existing ecosystem.

There are numerous examples of decolonial suburban research in the last five years. For example, Keeler (2016) has identified an exclusion from home ownership that contributed to the absence of Indigenous families in post-war North American suburbs, concluding that "[e]ventually, in public memory and in public history, suburbs became white and the opportunity for envisioning suburban Indians was rapidly closed" (p. 7). This absence is also reflected in the scholarship on those suburbs where indigeneity became a blind spot. This lacuna buttressed in turn a blindness in other research fields towards the programs and policies of relocation of Indigenous peoples. This created not only a racialized geography, but the hidden historical layers also create a historical geography. This space–time "topology" is characterized not only by "horizontal" dualisms such as core–periphery, urban–suburban, and settled–natural but also by "vertical" hierarchies of racial preference, structural racism, and settler privilege.

This is part of a broader settler–colonial formation and topology that characterizes in different but similar ways both the United States and Canada. As Coulthard (2014) argues, "through gentrification, Native spaces in the city are now being treated as *urbs nullius* – urban space void of Indigenous sovereign presence" (p. 176). These comments about urban space can be extended to suburbs, which sometimes were developed through the forcible eviction of residents on the urban fringe. This is illustrated below in the case of Minneapolis–St. Paul's post-war expansion, and in the case of Winnipeg's redevelopment of Rooster Town.

Suburbia has also previously been examined as a racial project (Sugrue, 2004; Vallejo, 2012; Wiese, 2005). However, Keeler (2016) provides one of the only studies of "suburban Indians" that "brings Indian people to the centre of suburbanization" (p. 8). Keeler and other critics (e.g., Keil, 2013) note that, in the past, many studies have been located within the stereotypical dualisms of the urban–suburban topology such as core–periphery, Black–white, or old–new. This includes identifying socio-economic factors such as education and employment that contribute to access to housing and home ownership in general, and in post-war suburbia specifically. Suburbs are often thought to be marked by higher rates of home ownership and/or the goal of home ownership, and as places that have been "remade and repopulated throughout the twentieth century" (Keeler, 2016, p. 80) and up until today. In what follows, six case studies will illustrate decolonial and Indigenous studies of North American suburbia.

Case Studies

Indigenous Suburbia: Minneapolis–St. Paul

In North America today, the majority of all Indigenous people live outside of rural, reservation environments and instead reside in metropolitan areas; increasingly, these Indigenous people live in suburbs (Keeler, 2016). For example, according to the 2010 United States census, in Minnesota, one fifth of all individuals who identified as single-race American Indian and 38 per cent of those who identified as American Indian in conjunction with one or more races lived in a suburb of the major city centre, Minneapolis–St. Paul. Despite federal Indian policies and housing policies that sought to first confine Indigenous people to reservations – or, in the case of the Dakota peoples, to exile them to reservations – and then later to relocate them to urban areas, many Indigenous people have remained in or moved between suburbs (Keeler 2016).

The city-region of Minneapolis–St. Paul demonstrates that "the places we think of today as suburbs have much longer and complex histories as Indian places as well, a juxtaposition that should also be acknowledged rather than overlooked and, in many cases, erased"

(Keeler, 2016, p. 19). The dominant understanding of North American suburbs is that they were produced through private speculative development of agricultural land aided by public financial and spatial zoning policies. However, the work of Keeler (2016) and other researchers demonstrates that the expansion of North American cities necessarily involved the displacement and exclusion of many residents with less power to benefit from the development process but often with no less claim on the land.

Indigenous Veterans: Exclusion from American Suburbia

In the US, in the years following the Second World War until 1964, the Veterans Administration's home loan program allowed returning soldiers to purchase new suburban houses with government guaranteed mortgages. However, other policies such as the Termination Bill passed by the US Congress in 1953 (HCR-108) and the Indian Relocation Act of 1956 streamed Indigenous GIs and their families into inner-city temporary rental housing (Keeler, 2016). The objectives of these measures were to empty reserves and assimilate Indigenous people in cities, thereby extinguishing their status rights. Dominant, white Americans moved into suburbs, stripping city-centre neighbourhoods of their middle-class constituencies. The precarity of inner-city rental housing conditions worked against the purported objective of assimilation to produce long-term disadvantage and, in some cases, immiseration (Keeler, 2016, pp. 146–7).

In Canada, there is a gap in research on housing and homelessness for Indigenous veterans. In a systematic literature review of this topic, only one paper was found (see Serrato et al., 2019). Canada's Indian Act (1985) stripped those who had been away from reserves for four years of their Indian status. Most were in fact required to renounce their status upon enlisting with the Canadian Armed Forces. As a result, returning Indigenous veterans of the Second World War no longer qualified for benefits under the Act and faced discrimination under the programs of Veteran's Affairs (Ellis, 2019; Sheffield, 2007). And for those who were able to maintain their Indian status after the war, their pensions remained subject to administration by Indian Agents. This contradiction and double set of bureaucratic reviews and approval processes added challenges for Indigenous claimants (Sheffield, 2007, p. 71).

In addition, the Indian Affairs Branch decided to use the Veterans' Land Act (1970) to subsidize the branch's overstretched welfare budget for on-reserve housing. While making houses available to veterans may have improved their quality of life in the short term, the program was intended to help veterans re-establish themselves in a livelihood that provided long-term stability (Sheffield, 2007, p. 49). As in the US, these programs reduced the ability of Indigenous families to pursue post-war suburban housing, work, and educational opportunities by sequestering them on reserves or allowing them only marginal participation in urban life.

Erased Suburbs: Rooster Town, Winnipeg

In other cases, existing Métis communities were displaced as expanding suburbs brought aspiring families into contact and conflict with remaining area residents through schools (see Peters et al., 2018; see also https://roostertown.lib.umanitoba.ca). Burley (2013) has researched the erasure of "Rooster Town" (see figure 13.1) in 1959, a displacement that occurred on the southwestern fringe of Winnipeg, Manitoba, along a railway right-of-way. In Rooster Town, "suburban anxiety was reinforced by a deeply embedded sense that Aboriginal people did not belong in the city and by a history of municipal efforts, from the city's incorporation, to remove their visible presence" (Burley, 2013, p. 4). Inexpensive land around cities appreciated in value due to speculation. Poor and racialized residents "moved into deteriorating inner-city neighbourhoods" as they were displaced by "upwardly mobile middle- income families … in pursuit of their suburban dreams" (Burley, 2013, p. 4). Unable to find work or services in rural areas, Métis families moved to the urban edges. Denied title to their lands after the defeat of the 1885 North-West Rebellion, they had been pushed into interstitial spaces, such as public road allowances, and excluded from public services including schooling. Métis inhabitants of the Prairies also did not have Indian status under the Indian Act (1985), making it easy for authorities to deem them squatters.

Harris (2004) notes other examples of suburban expansion that overran unplanned communities, whether squatters or Métis on legally subdivided land. Similar cases can likewise be found in Hamilton (Bouchier & Cruikshank, 2003), Kingston (Osborne & Swainson, 1988), Ottawa (Tomiak, 2016), and Vancouver (Mawani, 2003). Even in the 2010s, the area near Rooster Town was the site of land conflict between environmental and Indigenous "land defenders" and developers who used what Wilt (2018) has called SLAPP lawsuits, "strategic lawsuits against public participation," to pursue individuals. Cities and suburbs are sites of settler occupation and active colonial assimilation and cultural pressure. Suburbia has become a space of anxiety and non-belonging for many. One female Indigenous participant in Nejad et al.'s (2019)

Figure 13.1. Grant Park Plaza, Winnipeg, Manitoba, 1967. A Métis settlement on the site of "Rooster Town." Photo courtesy of the City of Winnipeg Archives, with thanks to Gerry Olenko.

study of Winnipeg comments, "I would probably feel less safe walking in suburbia … people are more skeptical or curious about you."

Municipal Colonialism: The Oka Crisis, 1990

One of the best-known recent clashes between Indigenous peoples and developers was the confrontation at Oka, outside Montreal, Quebec, in March 1990. Members of the Haudenosaunee (Mohawk) Confederacy erected barricades to prevent the expansion of a golf course onto sacred land. The Haudenosaunee claim that the area, set aside as a Sulpician mission to the Indigenous peoples, was the subject of a 1717 wampum belt agreement, but this has never been acknowledged by the Crown. Parts of the land were later sold by the Church to settlers and to the Town of Oka, west of Montreal (see figure 13.2; for a detailed history, see Morgan, 2018). A police assault to clear the barricade in 1990 ended in the death of an officer and the deployment of almost 3,000 Canadian soldiers, one of the largest and most expensive military operations of

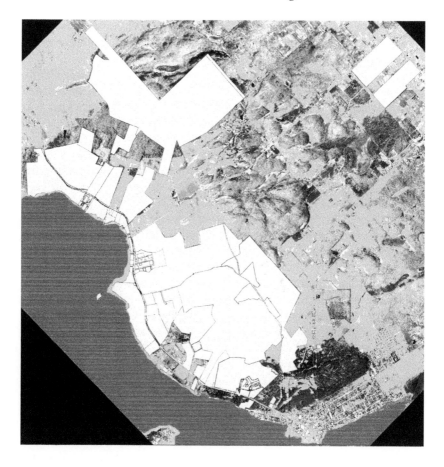

Figure 13.2. Map of Kanesatake reserve, 2007. The reserve was enlarged from a single town-centre site (1925) by adding together the land sections indicated in white. In 1990, the block adjacent to the Pines Golf Course was added. The golf course is the dark area between the reserve lands and the town site of Oka, Quebec (lower left), where the Kanehsatà:kehró:non band also owns a patchwork of downtown properties hidden due to scale. Since 1990, additions (blocks in white) have expanded the reserve and stitched together the largest parcels. Map created by the author based on 2014 digital cadastral data, Surveyor General, Natural Resources Canada.

the last fifty years. One of the main bridges into Montreal was block-aded, and a seventy-eight-day standoff followed (Betasamosake Simpson & Ladner, 2010; see also ESRI Canada Education, 2018).

Stanger-Ross (2008) terms this event an example of "municipal colonialism." The Oka Crisis embroiled the entire state in a peri-urban land development and land use conflict over a golf course. This was a suburban confrontation (Betasamosake Simpson & Ladner, 2010) and a moment of interruption of the dominant settler-colonial matrix. Coulthard (2014) argues that such eruptions of colonial violence must be understood as part of a cyclic pattern focused on maintaining the state's access to land and resources. Coercion alternates with more conciliatory moments (Epstein & Coulthard, 2015), including the efforts of the Truth and Reconciliation Commission of Canada (2015).

Suburbs have been a point of intersection between the land development economy, the reproduction of labour and the dispossession of resources, a lack of cultural recognition, and the inequalities of domestic arrangements within families. Developers and the industries built around the expansion of infrastructure and cities depend on land being made available and underwritten by the state at an attractive price (Shields, 2015). Suburbia is a site of both exclusion and the creation of not just a stereotypical "suburban middle class" but of a domesticated, planned environment and range of modern citizens who are workers in a division of labour that is economic and sexual. Unsurprisingly, gendered tensions are felt in family roles and relations but are rarely considered in the broader context of a place-based approach to the colonial matrix that is usually still theorized abstractly from above, at the scale of the state (Giraldo, 2016). Coulthard (2014) argues that sexism and gendered violence are aspects of an ongoing colonial matrix (see also Epstein & Coulthard, 2015).

The Caledonia Land Dispute, 2006–21

More recently, in 2006, a residential development near Caledonia, Ontario, purchased from the Province of Ontario was occupied. The dispute stems from a large tract of land granted by the Crown to the Six Nations of Grand River in 1784 for loyalty during the American War of Independence (see figure 13.5). Before a grant of legal title was concluded, however, the term of Frederick Haldimand, then governor-in-chief of British North America, ended. The Ontario government subsequently interpreted the document simply as a licence to occupy the land. The leader of the Six Nations, Joseph Brant, insisted that the title to the land was absolute and sold parts off to prove this point.

This disagreement has persisted to the present day (although Indigenous groups have recently forced developers to cancel their plans after defeating them in court in 2021). Government documents show that the land was surrendered in 1848, although the chiefs petitioned within three weeks' time, arguing that they had agreed only to lease the land back to the Crown. When the Province of Ontario sold the land to a developer in 1993, the Six Nations started litigation against Canada and Ontario for an accounting of the land and money involved. The spring 2006 occupation was initially to raise awareness about this legal suit. The Ontario government quickly bought back the contested land to hold in trust until negotiations settled the claim. However, violence, damage, and blockades of the area's main road followed, which disrupted emergency access to the Town of Caledonia. In return, town residents blockaded the same road. They were separated from the Six Nations by police but eventually fights broke out and led to injuries. CAD$1.5 million in damage to a power substation led to a local declaration of a State of Emergency. Blockages of the local highway and railroad continued on and off since 2006, and the conflict became a local feud. New developments are proposed on nearby parcels of land and protests follow as the municipality continues to support land development by approving subdivision and rezoning proposals. While there is little agreement on the facts, nor on their implications (Desjardins, 2017), a subsequent judgment in a CAD$20-million class-action lawsuit by residents and businesses found that the police failed to intervene to properly protect citizens.

In both Oka and Caledonia, there were two distinct systems of land tenure in conflict (Aaron, 2006). One is an Indigenous land right, even without a specific written deed that survives subsequent government interpretations, encroachments, and malfeasance. The other is, in the case of Caledonia, rooted in the release of land under the Ontario Land Titles Act by the Crown. Moreover, the most recent scholarship now speculates that Canada has two territorial sovereignties, a set of Indigenous sovereignties based on uninterrupted occupation and unextinguished claims (Russell, 2017) and the sovereignty of the Canadian state granted by the Crown (Nichols, 2019; Nichols & Hamilton, 2019). Wampum belt treaties testify to the recognition of Indigenous sovereignty by the Crown, but Section 94 of the Canadian Constitution grants blanket sovereignty to the Government of Canada. The more powerful Crown system of title has steadily encroached on Indigenous land for over 200 years. However, this historical conflict was exposed by the Caledonia proposal to develop over 200 houses, a relatively large proposal for the local area. The proposal for a suburban residential subdivision by the land development industry brought the conflict out into the open. Even though the land is adjacent to a town, rather than a major city, it is accurate to understand this as a conflict about suburbia.

The Ontario government was unwilling to act to resolve the Caledonia conflict and reconcile the two systems of tenure and sovereignty. Government negotiators were not given mandates to negotiate, only to listen, and mediators were never sought. The lead negotiator for the Six Nations, Haudenosaunee Chief Allen MacNaughton, commented in 2016 that "Canada hasn't learned much, the province hasn't learned much and neither has the municipality" (as cited in Moro, 2016). Previous models for negotiations have been unworkable because calling for the extinguishment of Indigenous rights also demands an extinguishment of cultural identity (Desjardins, 2017). There is a persistent gap in knowledge of the local history and understanding of others' points of view in these and other conflicts.

In the case of the Caledonia land dispute, one can observe how different orders of the state not only play different roles but even become the locus of colonial appropriation in different ways and at different times. While the Canadian federal government may be conciliatory (e.g., Truth and Reconciliation Commission of Canada, 2015), the Ontario provincial government has been noncommittal and its police even appeared to withhold their full participation, perhaps mindful of past bloody confrontations in which police forces were found guilty and held accountable for the violence of their individual officers. These two orders of government manage blame by attempting to "pass the buck" to each other. The role of the government at the level of the municipality, however, is less well examined. Small municipalities often have much less breadth and depth of expertise and rely on part-time councillors and local capacity without the benefit of consultants. Moreover, the majority of their electorate is non-Indigenous; it is thus unsurprising that Desjardins (2017) found at least one councillor who was categorized as in "strong support of non-native rights" (p. 124).

The further significance of the Caledonia land dispute is that the costs of policing the standoff, litigating the class action and Six Nations suits, not to mention lost tax revenues and "compromised … imagination, vision and capacity for a bright economic future" (Desjardins, 2017, p. 84) in the area have been borne by the public taxpayer via the state (Nadler, 2011). Despite the inaction of the province and the multitude of ambiguities and disagreement over the facts, the developer was compensated to an extent through the repurchase of the land by the province. Local Indigenous groups, area residents, and builders have subsequently spent well over a decade embroiled in continuing conflict.

Settler Colonial Suburbia: Mill Woods, Edmonton

Despite these cases and our increasing awareness of many other smaller conflicts of a similar nature, there is a persistent sense of a lost history of previous occupation of the land that appears as much in names and local stories as it does in the uncanny absence of continuous and coherent histories of place. Mill Woods, a suburb of Edmonton, Alberta, is situated on part of the historical land of the annulled Papaschase Indian Reserve No. 136 (Shields et al., 2019, 2020). The reserve was created as part of Treaty 6 in 1877. The Plains and Woods Cree, the Assiniboine, and other bands agreed with the Crown to surrender land in return for economic aid. By the early 1880s, most of the reserve's starving populace was induced to take "scrip," accepting payment to cede their Treaty rights.

A century later, parts of the annulled reserve were the site of an idealistic project to create a new suburb of affordable housing for workers and later immigrants. Now, almost fifty years after its founding, Mill Woods is home to an ethnically hyper-diverse population of over 80,000, with over 40 per cent of residents identifying as a visible minority (Statistics Canada, 2012), effectively defying stereotypical twentieth-century North American suburb images of white, nuclear families. Local, white descendants of European settlers first populated the suburb, only to be replaced by an increasingly ethnically diverse immigrant population, cementing Mill Woods's character as a community of new settlers. Today, Edmonton has one of the highest proportions of Indigenous people in its population. Over 1,600 Métis may live in Mill Woods (Andersen, 2009), but, similar to the Minneapolis–St. Paul case, few who are registered as Indians under the Indian Act (1985) reside there (Shields et al., 2020). Most First Nations Edmontonians live near the city centre and are in turn stereotyped as dependent on welfare services concentrated there. Zwicker (n.d.) argues that "colonialism literally changes shape over time, moving from a logic of exclusion ('Indian' reserves outside the city limits) to a logic of containment (inner-city poverty is disproportionately Indigenous)." Yet Mill Woods is dense with unexplained Cree names applied to streets, neighbourhoods, and parks. These are interspatial traces of an absence that is still felt and legible.

Although Mill Woods is not a homogenous social or spatial unit and does not have a simple shared history, a focus on a spatial, socioeconomic, and racial/ethnic definition of suburbs has distracted attention from the history of the site itself – and others like it. This has excluded settler colonialism from suburban history. Research has

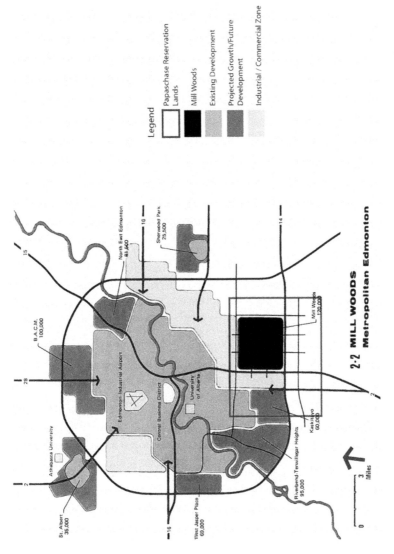

Figure 13.3. Mill Woods development concept, 1971, with overlay of Papaschase reserve lands. Map by Kieran Moran, 2017.

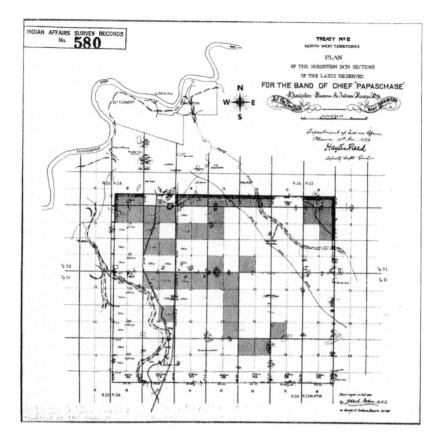

Figure 13.4. Papaschase Reserve No. 136. The current city of Edmonton extends north and south of the river and the historic settlement and trading post (marked H.B. Co's Post). The north–south railway line remains and marks the approximate centre line of the present-day city-region, while the north edge of the proposed reserve marks the approximate east–west centre line. Mill Woods was developed from the 1970s, overlapping the southeast quarter of the proposed reserve. Map courtesy of Library and Archives Canada, 2007.

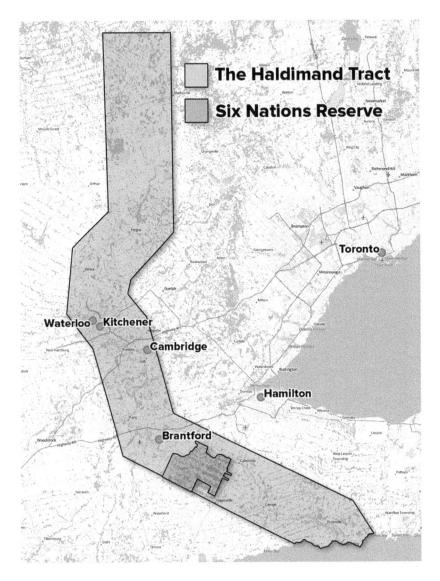

Figure 13.5. Map of Haldimand Tract (1784) and Six Nations reserve, 2021.
Map courtesy of AlternativesJournal.ca.

tended to cast suburbs as quintessentially modern. In the case of Mill Woods, this obscures both the differences of the modern suburb from the reserve and their similarities.

The Papaschase peoples of Treaty 6 participate in an oral history of the place, which has different temporal dimensions from official Canadian history, stretching back much further and grounded in an unchanging sense of place. Alongside new immigrants, they unequally participate in the costs and benefits of development. The claims of the Papaschase and other Indigenous groups to this site are reasserted in recent "Idle No More" protests (CBC News, 2013). Mill Woods is an example of accumulation by dispossession, obscured by a national historical amnesia and the rhetoric constructed by planning professionals and politicians.

Interspatial Methods for Settler Colonial Suburbia

These case studies raise the question of how to research the layered past of both migration and displacement that has been both erased and forgotten, but which may still haunt suburbs in the form of street names, park names, and neighbourhood toponymy (Boyd & Thrush, 2011). Here we can draw on the insights of ethnographic and discourse analysis – that it is essential to attend to what is not expressly said in any text. Inconsistent and puzzling presences and absences found on maps, in archives, and in actual places provide the opportunity for an immanent critique that produces questions about a particular place. The suburban cases reviewed in this chapter show how different orders of government, working at different scales, also have contradictory stances. The examples illustrate the importance of developing approaches that can work with ambiguities and in spite of contradictions where there is no agreement on the historical or present facts.

One research method is to compare present-day site visits with historical maps (see, for example, figures 13.3 and 13.4), photos, and other representations of the area found in archival sources. For example, newspaper descriptions at the time of Treaty 6 adhesion in the 1870s can be compared to planning documents at the time of the creation of Mill Woods in the late 1960s. Maps can also be read for shared sites: Overlaying historical surveys and present-day maps makes the common areas between these materials easily discernable on visual inspection. This also highlights the disappearance of local place names and changes to landscape features. Keeler (2016) followed hundreds of specific Indigenous family names enumerated in the historical records of Minneapolis–St. Paul through to present day to show the persistent presence of Indigenous peoples. This effort to trace historical continuity is a method to contest the spatial and experiential and spatial everyday understanding that sees suburbs as a radical break with the past.

Zwicker (n.d.) refers to tracing both continuity and ruptures as "remapping," but this remapping risks recolonizing knowledge with a single new and authoritative map. Instead, it can be useful to put present-day experience (of absence) and the historical times of suburbs into a topological relation to consider both the geographical and historical changes (Tuck & McKenzie, 2015). If a remapping is successful, "[t]he task is not to recover a static past, but to 'acknowledge the power of Native epistemologies in defining our moves toward spatial decolonization'" (Goeman, 2013, cited in Tuck & McKenzie, 2015, p. 135). A decolonial plurality is preferred to forcing any of these narratives into one oversimplified statement of facts. For example, creating a reductive economic analysis ignores the importance of cultural recognition. Instead, an inclusive approach provides the research basis for political and cultural processes of engagement by communities.

Donald (2004) uses a metaphor of the artistic concept of *pentimento* for this: "The history of Aboriginal people before and after contact with Europeans has been 'painted over' by mainstream interpretations of official history ... however, Aboriginal history and memory have begun to show through the official history of Canada" (p. 23). We might think of this as recognizing and researching both horizontal and vertical geographies. For example, not just core and periphery but also the history of the suburban site itself and its layers of residents and occupants.

Oral histories and the gathering of further visual and archival materials could tell us more inclusive stories of the diversity of suburbs and other places. A decolonial research method might interview and gather Indigenous and migrant community histories or popular narratives and help to rebalance our understanding of the intersecting human and non-human roles in the local ecology. This in turn has the potential to repopulate spaces with *en place* local stories that could potentially combine into new, clear-eyed knowledge of places – a re-spatialization of suburbs. Informal sources, oral histories, and ephemera are valuable to understand the suburban areas of North America that articulate city to country, urban to rural, and one entire set of powers and land uses to another set.

Conclusion

There is much scope for queer, feminist, Indigenous, and settler interrogation of the legacy of colonialism, even in – in fact, especially in – suburbia as a built form and dominant cultural site in North America (see, for example, Lugones, 2012). The significance of indigeneity, migrancy, and ethnicity for North American suburbs as social spaces is poorly captured by the transportation, density, and infrastructural categories conventionally used in discussions of the "suburban." The legacy of the state's treatment of Indigenous people is emblematically reflected in the examples of suburbs as settler enclaves. Colonial strategies of displacement of Indigenous peoples have created the spaces of these suburbs, in which few Indigenous people reside. Indigenous responses in the form of First Nations land title claims, civil unrest, and violence respond to the ongoing "discursive management" of the "indigene" by settler cultures (Goldie, 1989). In recent years, Canadian and American suburbs have experienced land title claims in the courts as well as civil unrest and barricades on their streets, golf courses, and parks, resulting in disruption and even deaths (Doucette, 2016). The entrenchment of land claims and the normalization of these struggles in legal and media culture indicates an ongoing justification by settler societies of the "dispossession, oppression, and effacement" (Goldie, 1989) of Indigenous peoples and a blindness to difference in the history of urban and suburban research and planning.

NOTE

1 The roots of postcolonial criticism stem from further back in the 1930s when critiques of relations with colonized societies appeared, notably in French language publications on the inequalities and cultural impacts of colonialism and struggles to decolonize Caribbean colonies and Algeria. The work of Aimé Césaire and Franz Fanon produced a diagnosis of the conflicted identity forced on colonized people summed up in Fanon's (1986) *Black Skins, White Masks*.

   Even further back, W.E.B. Du Bois (1967), a lost forebearer of urban social geography, developed statistical and interviewing methods to conduct studies of racialized communities in turn-of-the-twentieth-century suburbs in *The Philadelphia Negro*. Via the development of cultural studies in the UK by figures such as Stuart Hall (1981) and Homi Bhabha (1991), and the dissemination of South Asian intellectuals such as Ashis Nandy (1983), issues of race, subaltern status, and colonialism came to figure alongside political economic analyses of contemporary societies.

REFERENCES

Aaron, B. (2006, May 27). Tough to reconcile Caledonia land dispute. *Toronto Star*. https://www.aaron.ca/tough-to-reconcile-caledonia-land-dispute/.

Andersen, C. (2009). *Aboriginal Edmonton: A statistical story – 2009.* Aboriginal Relations Office, City of Edmonton. https://www.edmonton.ca/public -files/assets/document?path=PDF/Stat_Story-Final-Jan26-10.pdf.

Auger, J. (1999). *Walking through fire and surviving: Resiliency among Aboriginal peoples with diabetes* [Master's thesis, University of Alberta]. ERA: Education and Research Archive. https://doi.org/10.7939/R3BG2HJ2W.

Bang, M., Curley, L., Kessel, A., Marin, A., Suzukovich, E., & Strack, G. (2014). Muskrat theories, tobacco in the streets, and living Chicago as Indigenous land. *Environmental Education Research, 20*(1), 37–55. https://doi.org/10.1080 /13504622.2013.865113.

Basso, K.H. (1996). *Wisdom sits in places: Landscape and language among the western Apache.* University of New Mexico Press.

Betasamosake Simpson, L., & Ladner, K. (2010). *This is an honour song: Twenty years after the blockades.* ARP Books.

Bhabha, H. (Ed.). (1991). *Nation and narration.* Routledge.

Bouchier, N., & Cruikshank, K. (2003). The war on the squatters, 1920–1940: Hamilton's boathouse community and the re-creation of recreation on Burlington Bay. *Labour/Le Travail, 51* (Spring), 9–46.

Boyd, C.E., & Thrush, C. (2011). Introduction: Bringing ghosts to ground. In C.E. Boyd & C. Thrush (Eds.), *Phantom pasts, Indigenous presence: Native ghosts in North American culture and history* (pp. vii–xl). University of Nebraska Press.

Browne, A.J., McDonald, H., & Elliott, D. (2009). *Urban First Nations health research discussion paper.* National Aboriginal Health Organization. https:// open.library.ubc.ca/media/stream/pdf/52383/1.0084587/3.

Burley, D.G. (2013). Rooster Town: Winnipeg's lost Métis suburb, 1900–1960. *Urban History Review/Revue d'histoire urbaine, 42*(1), 3–25. https://doi.org /10.7202/1022056ar.

Cariou, W., & McArthur, N. (Directors). (2009). *Overburden* [Film]. Winnipeg Film Group.

Casagranda, M. (2013, May 9–11). From Empire Avenue to Hiawatha Road: (Post)Colonial naming practices in the Toronto Street Index. In O. Felecan (Ed.), *Proceedings of ICONN 2* (pp. 291–302), Baia Mare, Romania. Cluj-Napoca: Editura Mega. https://onomasticafelecan.ro/iconn2 /proceedings/3_06_Casagranda_Mirko_ICONN_2.pdf.

CBC News. (2013, January 16). Idle No More protesters block QEII Highway. *CBC News.* https://www.cbc.ca/news/canada/edmonton /idle-no-more-protesters-block-qeii-highway-1.1368673.

Césaire, A. (1955). *Discours sur le colonialisme.* Présence Africaine.

Coulthard, G. (2014). *Red Skin, white masks: Rejecting the colonial politics of recognition.* University of Minnesota Press.

Desjardins, N. (2017). *Making meaning of the 2006 territorial conflict in Caledonia and reflecting upon the future of the impacted communities* [Master's thesis, University of Ottawa]. uO Research. http://doi.org/10.20381/ruor-20934.

Deyhle, D. (2009). *Reflections in place: Connected lives of Navajo women.* University of Arizona Press.

Donald, D. (2004). Edmonton Pentimento: Re-reading history in the case of the Papaschase Cree. *Journal of the Canadian Association for Curriculum Studies, 2*(1), 21–54. https://jcacs.journals.yorku.ca/index.php/jcacs/article/view/16868.

Doucette, J. (2016). *Pigs, flowers and bricks: A history of Leslieville.* Joanne Doucette.

Drawson, A.S., Toombs, E., & Mushsquash, C.J. (2017). Indigenous research methods: A systematic review. *The International Indigenous Policy Journal, 8*(2), Article 5. https://doi.org/10.18584/iipj.2017.8.2.5.

Du Bois, W.E.B. (1967). *The Philadelphia Negro: A social study.* B. Blom.

Ellis, N.R. (2019). *Indigenous veterans: From memories of injustice to lasting recognition: Report of the standing committee on Veterans Affairs.* House of Commons Canada. https://www.ourcommons.ca/Content/Committee/421/ACVA/Reports/RP10301835/acvarp11/acvarp11-e.pdf.

Epstein, A.B., & Coulthard, G. (2015, January 13). The colonialism of the present: An interview with Glen Couthard. *Jacobin.* https://jacobinmag.com/2015/01/indigenous-left-glen-coulthard-interview/.

ESRI Canada Education. (2018). *ArcGIS: The Kanesatake resistance map* [Map]. ArcGIS. https://www.arcgis.com/home/webmap/viewer.html?webmap=a42212e6d8ef4beeac6238676cf0e694.

Fanon, F. (1986). *Black skin, white masks.* Pluto Press.

Frew, L. (2013). Settler nationalism and the foreign: The representation of the exogene in Ernest Thompson Seton's *Two Little Savages* and Dionne Brand's *What We All Long For. University of Toronto Quarterly, 82*(2), 278–97. https://doi.org/10.3138/UTQ.82.2.278.

Gilmartin, M., & Berg, L.D. (2007). Locating postcolonialism. *Area, 39*(1), 120–4. https://doi.org/10.1111/j.1475-4762.2007.00724.x.

Gilroy, P. (1993). *The Black Atlantic: Double consciousness and modernity.* Harvard University Press.

Giraldo, I. (2016). Coloniality at work: Decolonial critique and the postfeminist regime. *Feminist Theory, 17*(2), 157–73. https://doi.org/10.1177/1464700116652835.

Goeman, M. (2013). *Mark my words: Native women mapping our nations.* University of Minnesota Press.

Goldie, T. (1989). *Fear and temptation: The image of the Indigene in Canadian, Australian, and New Zealand literatures.* McGill-Queen's University Press.

Gulson, K.N., & Parkes, R.J. (2009). In the shadows of the mission: Education policy, urban space, and the colonial present in Sydney. *Race Ethnicity and Education, 12*(3), 267–80. https://doi.org/10.1080/13613320903178246.

Gururani, S., & Dasgupta, R. (2018). Frontier urbanism: Urbanisation beyond cities in South Asia. *Review of Urban Affairs, 53*(12), 41–5.

Hall, S. (1981). Cultural studies: Two paradigms. In T. Bennet (Ed.), *Culture, ideology and social process: A reader* (pp. 57–72). Open University.

Harris, R. (2004). *Creeping conformity: How Canada became suburban, 1900–1960.* University of Toronto Press.

Harvey, D. (2005). *The new imperialism.* Oxford University Press.

Indian Act, RSC 1985, c 1–5. https://laws-lois.justice.gc.ca/eng/acts/i-5/.

Keeler, K. (2016). *Indigenous suburbs: Settler colonialism, housing policy, and American Indians in suburbia* [Doctoral dissertation, University of Minnesota]. University of Minnesota Digital Conservancy. https://hdl .handle.net/11299/199061.

Keil, R. (Ed.). (2013). *Suburban constellations.* Jovis.

LaGrand, J. (2002). *Indian metropolis: Native Americans in Chicago, 1945–1975.* University of Illinois Press.

Library and Archives Canada. (2007). *Indian reserves – Western Canada* [Database]. https://www.collectionscanada.gc.ca/databases/indian -reserves/index-e.html.

Lipsitz, G. (2011). *How racism takes place.* Temple University Press.

Louis, R.P. (2007). Can you hear us now? Voices from the margin: Using Indigenous methodologies in geographic research. *Geographical Research, 45*(2), 130–9. https://doi.org/10.1111/j.1745-5871.2007.00443.x.

Lugones, M. (2003). *Pilgrimages/Peregrinajes: Theorizing coalition against multiple oppressions.* Rowman and Littlefield.

Lugones, M. (2012). Methodological notes toward a decolonial feminism. In A.M. Isasi-Díaz & E. Mendiette (Eds.), *Decolonizing epistemologies: Latina/o theology and philosophy* (pp. 68–86). Fordham University Press.

Maldonado-Torres, N. (2007). On the coloniality of being: Contributions to the development of a concept. *Cultural Studies, 21*(2–3), 240–70. https://doi .org/10.1080/09502380601162548.

Mawani, R. (2003). Imperial legacies (post)colonial identities: Law, space and the making of Stanley Park, 1859–2001. *Law Text Culture, 7,* 99–141.

Mays, K. (2015*). Indigenous Detroit: Indigeneity, modernity and racial and gender formation in a modern American city, 1871–2000* [Doctoral dissertation, University of Illinois at Urbana-Champaign]. IDEALS. http://hdl.handle .net/2142/78653.

McIlvenny, P., Broth, M., & Haddington, P. (2009). Communicating place, space and mobility. *Journal of Pragmatics, 41*(10), 1879–86. http://doi.org /10.1016/j.pragma.2008.09.014.

Mignolo, W. (2011). *The darker side of Western modernity: Global futures, decolonial options.* Duke University Press.

Morgan, J. (2018). *Restorying Indigenous–settler relations in Canada: Taking a decolonial turn toward a settler theology of liberation* [Doctoral dissertation, Saint Paul University]. uO Research. http://doi.org/10.20381/ruor-21858.

Moro, T. (2016, March 4). Natives recall Caledonia tensions 10 years later. *Hamilton Spectator*. https://www.thespec.com/news/hamilton-region /2016/03/04/natives-recall-caledonia-tensions-10-years-later.html.

Nadler, S. (2011, August 11). The true costs of Caledonia. *Globe and Mail*. https://www.theglobeandmail.com/opinion/the-true-costs-of-caledonia /article590038/.

Nandy, A. (1983). *The intimate enemy: Loss and recovery of self under colonialism*. Oxford University Press.

Nejad, S., Walker, R., Macdougall, B., Belanger, Y., & Newhouse, D. (2019). "This is an Indigenous city; why don't we see it?" Indigenous urbanism and spatial production in Winnipeg. *The Canadian Geographer, 63*(3), 413–24. https://doi.org/10.1111/cag.12520.

Nichols, J.B.D. (2019). *Reconciliation without reflection? An investigation of the foundations of Aboriginal law in Canada*. University of Toronto Press.

Nichols, J.B.D., & Hamilton, R. (2019). The tin ear of the court: Ktunaxa Nation and the foundation of the duty to consult. *Alberta Law Review, 56*(3), 729–60. https://doi.org/10.29173/alr2520.

O'Brien, J. (2010). *Firsting and lasting: Writing Indians out of existence in New England*. University of Minnesota Press.

Osborne, B., & Swainson, D. (1988). *Kingston: Building on the past*. Butternut Press.

Patrick, L. (2015). Review: Unlearning the colonial cultures of planning. *Journal of Planning Education and Research, 35*(4), 534–5. https://doi.org /10.1177/0739456X15614586.

Peters, E.J. (2004). *Three myths about Aboriginals in cities*. Canadian Federation for the Humanities and Social Sciences.

Peters, E.J., Stock, M., Barkwell, L., & Werner, A. (2018). *Rooster Town: The history of an urban Métis community, 1901–1961*. University of Manitoba Press.

Pulido, L. (2000). Rethinking environmental racism: White privilege and urban development in Southern California. *Annals of the Association of American Geographers, 90*(1), 12–40. https:// doi.org/10.1111/0004 -5608.00182.

Quijano, A. (2007). Coloniality and modernity/rationality. *Cultural Studies, 21*(2), 168–78. https://doi.org/10.1080/09502380601164353.

Roy, A. (2011). Slumdog cities: Rethinking subaltern urbanism. *International Journal of Urban and Regional Research, 35*(2), 323–38. https:// doi.org /10.1111/j.1468-2427.2011.01051.x.

Russell, P.H. (2017). *Canada's odyssey: A country based on incomplete conquests*. University of Toronto Press.

Said, E. (1978). *Orientalism*. Pantheon.

Said, E. (1994). *Culture and Imperialism*. Vintage

Serrato, J., Hassan, H., & Forchuk, C. (2019). Homeless Indigenous veterans and the current gaps in knowledge: The state of the literature. *Journal of*

Military and Veterans' Health, 27(1), 101–11. https://www.homelesshub.ca
/resource/homeless-indigenous-veterans-and-current-gaps-knowledge
-state-literature%C2%A0.

Sheffield, R.S. (2007). Canadian Aboriginal veterans and the Veterans'
Charter after the Second World War. In P.W. Lackenbauer, R.S. Sheffield, &
C. Mantle (Eds.), *Aboriginal peoples and military participation: Canadian and
international perspectives* (pp. 77–98). Canadian Defence Academy Press.

Shields, R. (1991). Imaginary sites. In D. Augaitis & S. Gilbert (Eds.), *Between
Views* (pp. 22–6). Walter Phillips Gallery, Banff Centre for the Arts.

Shields, R., Gillespie, D., & Moran, K., (2020). Edmonton, Mill Woods,
Amiskwaciy Waskahikan. In J. Nijman (Ed.), *The life of North American suburbs:
Imagined utopias and transitional spaces* (pp. 245–68). University of Toronto
Press.

Shields, R., Moran, K., & Gillespie, D. (2019). Edmonton, *Amiskwaciy
Wâskahikan*, and a Papaschase suburb for settlers. *The Canadian Geographer*,
64(1), 105–19. https://doi.org/10.1111/cag.12562.

Smith, L.T. (2012). *Decolonizing methodologies: Research and indigenous peoples*
(2nd ed.). Zed Books.

Snyman, G. (2015). Responding to the decolonial turn: Epistemic
vulnerability. *Missionalia*, 43(3), 266–91. https://doi.org/10.7832/43-3-77.

Stanger-Ross, J. (2008). Municipal colonialism in Vancouver: City planning
and the conflict over Indian reserves, 1928–1950s. *The Canadian Historical
Review*, 89(4), 541–80. https://doi.org/10.3138/chr.89.4.541.

Statistics Canada. (2012). *Population and dwelling count highlight tables, 2011
census*. https://www12.statcan.gc.ca/census-recensement/2011/dp-pd/hlt
-fst/pd-pl/index-eng.cfm.

Sugrue, T. (2004). *The origins of the urban crisis: Race and inequality in postwar
Detroit*. Princeton University Press.

TallBear, K. (2013). *Native American DNA: Tribal belonging and the false promise
of genetic science*. University of Minnesota Press.

Thobani, S. (2007). *Exalted subjects: Studies in the making of race and nation in
Canada*. University of Toronto Press.

Thrush. C. (2013). "Meere strangers": Indigenous and urban performances in
Algonquian London, 1580–1630. In E.A. Fay & L. von Morzé (Eds.), *Urban
identity and the Atlantic World* (pp. 195–218). Palgrave Macmillan.

Thrush. C. (2016). *Indigenous London: Native travelers at the heart of empire*. Yale
University Press.

Thrush, C. (2017). *Native Seattle: Histories from the crossing-over place*.
University of Washington Press.

Tomiak, J. (2016). Unsettling Ottawa: Settler colonialism, indigenous
resistance, and the politics of scale. *Canadian Journal of Urban Research*,
25(1), 8–21. https://www.jstor.org/stable/26195303.

Truth and Reconciliation Commission of Canada. (2015). *Canada's Residential Schools: Reconciliation* (Vol. 6). McGill-Queen's University Press.

Tuck, E., & McKenzie, M. (2015). *Place in research: Theory, methodology, and methods*. Routledge.

Tuck, E., & Yang, K.W. (2012). Decolonization is not a metaphor. *Decolonization: Indigeneity, Education and Society, 7*(1), 1–40. https://jps .library.utoronto.ca/index.php/des/article/view/18630.

Vallejo, J. (2012). *Barrios to burbs: The making of the Mexican American middle class*. Stanford University Press.

Veracini, L. (2012). Suburbia, settler colonialism and the world turned inside out. *Housing, Theory and Society, 29*(4), 339–57. https://doi.org/10.1080 /14036096.2011.638316.

Veterans' Land Act, RSC 1970, c V-4. https://laws-lois.justice.gc.ca/eng /acts/V-1.5/page-1.html.

Wiese, W. (2005). *Places of their own: African American suburbanization in the twentieth century*. University of Chicago Press.

Wilt, J. (2018, April 30). Chilling public protest: Is a lawsuit against Winnipeg land defenders a new form of legal intimidation? *Briarpatch*. https:// briarpatchmagazine.com/articles/view/chilling-public-protest -rooster-town-slapp.

Wolfe, P. (1999). *Settler colonialism and the transformation of anthropology: The politics and poetics of an ethnographic event*. Cassell.

Wolfe, P. (2006). Settler colonialism and the elimination of the native. *Journal of Genocide Research, 5*(4), 387–409. https://doi.org/10.1080 /14623520601056240.

Wolfe, P. (2011). After the frontier: Separation and absorption in U.S. Indian policy. *Settler Colonial Studies, 7*(1), 13–51. https://doi.org/10.1080 /2201473X.2011.10648800.

Zwicker, H. (n.d.). *Amiskwaciwâskahikan*. The Edmonton Pipeline Project. Retrieved 1 June 2018, from https://archive.artsrn.ualberta.ca /edmontonpipelines/edmontonpipelines.org/edmontonpipelines.org /projects/amiskwaciwaskahikan/index.html.

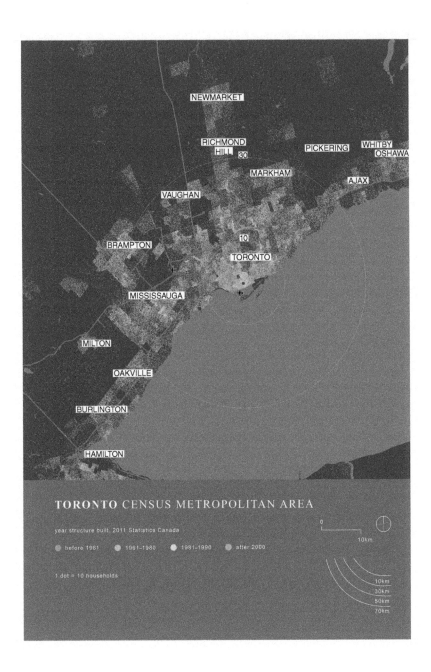

Image 14. Residential dwellings by year built, Toronto

14 Seeing through the Darkness of Future Past: "After Suburbia" from a Historical Perspective

ILJA VAN DAMME AND STIJN OOSTERLYNCK

Introduction

"After suburbia" evokes associations of the all-too-self-conscious urban futurology that is currently plaguing urban studies. Theorists of many disciplinary denominations eagerly fill the pages left blank between the urban now and the increasingly popular literary and cinematic dystopia that somehow seems to tap into a globally shared unease about our urban future (Rickards et al., 2016). By rushing in to define and understand the current process of accelerating planetary sub/urbanization, new academic argot, concepts, and ideas have rapidly crowded out well-known notions of the urban, often to the despair of more classic-minded theorists (Storper & Scott, 2016). Already at the start of the twenty-first century, Taylor and Lang (2004) famously compiled a list of over 100 concepts describing recent urban change, but urban theory has not slowed down since then (Harding & Blockland, 2014; Jayne & Ward, 2017).

However, we do not have to engage in theoretical speculation or wait upon the future of suburbia to know more about what comes after: Our expanding urban historical knowledge can help us to see "through the darkness of future past," as one of the characters of David Lynch's *Twin Peaks* puts it so enigmatically.[1] In other words, looking at the past offers us a privileged perspective *as if* we would be looking at future sub/urban developments, the difference being, of course, that we are much better informed about past results and historical outcomes. We are not so naïve as to claim that a study of the past can help "predict" or "navigate" the future, as some more positivist-inclined theorists would like us to believe (e.g., Guldi & Armitage, 2014). Yet, having data about various outcomes and different historical evolutions and contexts urges us to sharpen our conceptualizations and definitions and be more precise and critical about causalities and long-term forces shaping our current and future sub/urban predicaments. Simply put, the study of the past can contribute to theory-building about the sub/urban and its possible futures.

In this chapter, we start from the observation that history and historical analysis is relatively absent from urban theoretical debates (Sewell, 2005). The discipline of urban history has evolved quite independently from hotly debated issues in urban theory (Ewen, 2016). Yet, isolation and estrangement have worked both ways. Very recently, still, the position of urban history was put on trial for supposedly failing to have made any significant contribution to the field of urban studies (Harris & Smith, 2011). Ready to claim otherwise, Michael Smith (2010), one of the advocates of the usefulness of historical perspectives for contemporary urban theory, dubbed this tendency to ignore the past and the added value of historical research a "recentist" fallacy. It entails, specifically, a mode of doing theory that is focused on the urban present and future at the detriment of studying and comparing to earlier periods. Concepts like "sprawl" or "squatter" are basically framed and theorized in an ahistorical fashion, creating not only an empirical but an *a priori* theoretical neglect of the past to the extent that a postulated "newness" of urban phenomena eclipses older, temporal patterns, trends, and cycles over longer periods of time.

Our aim here is similar; namely, to highlight the potential contribution of both urban history as a discipline and historical analysis as a particular epistemological approach – which can be found in other social science and humanities disciplines as well – to theory-building on the sub/urban. We argue that the theoretical contribution of history and historical analysis revolves crucially around the notion of *time*. By making sense of time, we develop two concrete interventions about how our thinking and theorizing about "after suburbia" could profit from a historically attuned understanding.

First, we propose that researchers compare the suburban phenomenon more frequently across time in addition to comparisons across space, which is more usually done in contemporary urban studies (Madgin & Kenny, 2015; Robinson, 2013). Similar to comparisons across space, comparisons across time contribute to sub/urban theory-making. We challenge the "recentism" in many accounts of suburbia, which we argue leads to an overly restrictive definition and misunderstanding of what suburbia is all about. More specifically, by taking into account the already available wealth of historical data and analysis on the sub-urban phenomenon through history, we argue that a one-sided linear conception of suburbia as a socio-spatial configuration in the process of *necessarily* becoming "urban" in the future is eventually untenable (for a somewhat similar critique, see Harris in this volume). Crucial in recognizing such teleological biases in urban theory is conceptualizing time as being inherently eventful, meaning that structures and processes can change in the *longue durée*.

This brings us to our second intervention. Comparing sub/urban phenomena – whether through space or time – is not undisputed, since it is said to simply add to the empirical "exceptions" and theoretically endless "particularism" of cities around the world (for some general critiques on the "comparative gesture" in urban studies, see Peck, 2015). Moreover, comparisons might sharpen our conceptions and definitions in urban theory but do not necessarily lead to better insights into causes, since contexts and structures can differ widely from period to period and from place to place. Yet, taking our cue from historians Sewell (2005) and Corfield (2013), we contemplate how comparing across time might actually give us a better theoretical understanding of continuity, newness, and change. Time itself can be seen as a structuring agent of socio-spatial processes. Usually discussed under the umbrella term of "path dependency" (Molotch et al., 2000; Oosterlynck, 2012; Sorenson, 2015), previous evolutions are reckoned to have left "traces," which structure causalities and the agency and relative freedom of people to act on these causalities. It is important, however, not to understand this weight of the past as inevitable – as locking people in – but to think of temporality as radically eventful, always subject to discontinuities and contingencies rather than to inevitable iron-clad social laws.

Comparing "After Suburbia" through Time

In this section, we propose to look at suburbia, or the suburban phenomenon, as a trans-historical category of urban comparison and understanding. In order to arrive at such a conceptualization, we need to – as we will argue below – define the suburban phenomenon as the ever-changing and varying contact zone between the dominant macro-categories of "city" and "countryside." Such a definition both requires and allows us to move beyond unproductive sub-disciplinary divisions between urban and rural research. Moreover, it takes into account that historical evolutions are inherently variable in their outcomes and lack a clear linear or teleological arch. Recognizing historical diversity in suburban developments challenges the idea that suburbs will *necessarily* become urban in the future, and brings in aspects of continuity, newness, and change (Harris, 2010).

It can easily be argued and evidenced that the history of the suburban phenomenon is as old as the city itself, yet its complex and multifarious historic record has only been partially written (Vaughan et al., 2009). Urban theory has mainly stressed the Anglo-Saxon evolutionary model of the last 200 years, and this dominant paradigm has had a huge impact on the way we theorize about and historicize our subject (Harris & Vorms, 2017). A dominant mode of recentist framing has simply denied or ignored that our subject of inquiry existed before industrial modernization took off in the nineteenth and twentieth centuries. This indicates that we have to move – in a metaphorical sense, at least – from suburbia's dominant "core" meanings in England and North America to consider the much more peripheral and largely unexplored ways we might think and theorize about suburbia and its ever-after.

Once upon a time – as all straightforward narratives tend to start – the history and afterlife of suburbia was believed to be surprisingly one-dimensional. The historical study of suburbs, suburbanization, and suburbanism was connected to those places, processes, and lifestyles that typified "out-of-town" living in fast growing cities like London, Manchester, New York, and Boston (Short, 2019). Suburbs were planned, residential, bourgeois areas sustained by transportation lines starting in the central city and populated by white, distinction-seeking elites that had grown tired of the obnoxious smells, sights, and sounds of growing modern industry and city life. The process of suburbanization globalized on the rhythm of the nineteenth- and twentieth-century transport revolutions (from carriages to trolleys and commuter trains to cars), eventually to have outlived itself in a democratic, but watered-down, uniform, suburban "wasteland." Finally, with global, post-industrial restructurings of economies in the 1970s and 1980s, suburbia was transformed into so-called technoburbs, edge-cities, exurbs, post-suburbias, and other sprawl-like urban mutations of our present-day (Phelps, 2019).

In hindsight, it is easy to see what is wrong with this historical narrative. Most importantly, it is recentist because it only considers the modern, industrial period of history, and it is teleological because it retrofits and presents a skewed and misleading Anglo-Saxon historical trajectory as a general, linear model. What is more, a surprisingly huge number of flat distinctions and dichotomies were holding this classic, but biased, historical account of suburbia afloat. Basically, twentieth-century suburban living was (and still is!) stereotyped in novels, in movies, on television, and through other popular media as the "triumph" of white, conformist, male-dominated, middle-class ideals, emulating the preferences for greenspace and "anti-urbanism" of the Anglo-Saxon nobility and bourgeoisie in the centuries before (Harris, 2019).

Revisionist "new suburban" research in the last few decades has been especially influential in undermining such established, clichéd distinctions and turning away from Anglo-Saxon, homogenizing treatments of suburbia (see Keil, 2018). However, concern with the empirical diversity of suburbs and processes of suburbanization on a worldwide scale – as well as with the growing internal and spatial differentiation of suburbs and suburban living – has been mainly applied to a present-day, global research agenda. Despite the insightful and much-needed call of McManus and Ethington (2007) to compare suburbs not only through space in the present day but also *through time* as ever-changing, "interactive ecologies" (pp. 327–37), a comparative, trans-historical approach to the suburban phenomenon has yet to emerge. Why is this the case?

The reason such a historical viewpoint has not materialized in sub/ urban research is connected to two issues – one related to more largely shared definitional problems in pinpointing what suburbia is, and the other to sub-disciplinary fixations and paradigmatic biases within urban research itself. In the first case, a certain elusiveness in defining and naming what suburbia actually is has certainly marred the emergence of its trans-historical understanding (on suburban definitions, see Forsyth, 2019). Trying to grasp the suburban phenomenon with a definition that clearly demarcates it tends to restrict and over-specify certain characteristics (relating to location, legal status, functions, density, housing, class, ethnicity, culture, lifestyle), which, from a comparative historical viewpoint, reduces complexity, multi-dimensionality, and freezes what is, basically, always in flux and motion. Confusion about what one is actually talking about becomes even more pronounced from the moment that the safe contours of Anglo-Saxon and recentist framing are left behind. This entails also losing our attachment to the etymological meaning of the word *suburb* itself – referring, in Latin, to what was "under," "close to," "up to," or "under" the city. Instead, as indicated recently by Harris and Vorms (2017), one now has to deal with a multiplicity of affiliated words and non-English "linguistic trajectories" (*faubourg, banlieu, Stadtrand, borgata, favela, jiaqu, buitenwijk*), all shaped in response to different histories and varying cultural contexts.

Despite these definitional difficulties, however, the fact remains that for most urban studies – urban history being no exception in this – definitions are almost always both entry point and impediment to understanding (past) reality. It is well recognized among the wider social sciences how defining phenomena is a conscious act of creating "black-boxed signifiers" (Callon & Latour, 1981), which, nevertheless, always need to be unpacked further by concrete (historical) research and contextualization. The suburban phenomenon could easily be added to such a list of free-floating macro-categories in urban studies – recognized as a "process" and shaped by "structures" – without necessarily adhering to the reductionist definitions and language-sensitive labels we place on its functioning in urban theory and analysis.

But, if we argue this to be the case, why has the suburban phenomenon specifically not given rise to more concrete historical research? This is where sub-disciplinary fixations directly connected to how suburbia is commonly understood play out. Keil and Shields (2013) cleverly propose to bypass definitional issues by thinking about suburbia through fluid material boundaries, temporal geographical fringes, and shifting spatial contact zones between the macro-categories of "city" and "countryside." Such a flexible approach, which sees suburbia in the first place as an open-ended category of geographical comparison, would work for comparisons through time as well. It invites us to pose questions and reflections such as the following: How did the contact zone between city and countryside evolve through time? Which type of suburban functions and material forms were recurrent through history, or were rather related to temporal phases of urban expansion and contraction? And so on.

Unfortunately, the suburban phenomenon before industrial modernization has not attracted much attention. Historians fixated on the macro-category of "city" forgot about the "countryside," and so-called rural historians developed other paradigms and research interests that had little in common with their urban counterparts (Van de Walle, 2019). So, although historians, in principle at least, would agree on the existence of a geographically and historically changing contact zone between "city" and "countryside," it was for reasons of disciplinary attachments between rural and urban historians that this was never really studied extensively. Urban historians studied changes in densely populated settlements more or less enclosed by walls and specific institutional setups ("cities"), while rural historians studied small-scale hamlets and villages and a rural "way of life." Both hardly bothered about the continuum in between. This hybrid spatial zone between city and countryside was only taken seriously from the moment industrial modernization supposedly introduced suburbanization as something "new."

Without a doubt, however, the inside/outside problematic and hybridity between "city" and "countryside" has been an essential dynamic everywhere in history: from medieval guilds excluding industrial production taking place just outside cities in urban Europe to the strict legal distinctions being used in present-day urban China for restricting rural *hukou* households from migrating to cities (see, for instance, Epstein, 2001). Everywhere in urban history, fortified walls, moats, gates, and checkpoints were less of a military strategy than they were one aimed at supervising, taxing, and controlling the inflow and concentration of people and goods from the countryside. As is well known, some sort of urban political control and governance over the surrounding hinterland was crucial for any type of town, if only to secure a steady inflow of food for its hungry citizens and to organize proto-industrial types of entrepreneurship in the countryside (Stabel, 2015). Already in the third millennium BCE – right at the start of the infamous "urban revolution" theorized by V. Gordon Childe – cities had a suburban belt used for residential, economic, social, cultural, and other purposes. Comparing through time how suburban phenomena evolved can form a starting point for thinking about "after suburbia" in its necessary historical context.

"After suburbia" itself intrudes as a "transitional moment" in the relational tension between city and countryside, often unnoticed by both urban and rural historians alike. Yet, it can be researched and compared as both common geographical "in-between" zone and time-laden passageway between the two: an emergent and unstable material space and contingent timeframe placed within an existing, complex network of political, social, economic, and symbolic relationships between town and country. Hence, "after suburbia" as material and transitional "place in time" and "in-between space" should never be isolated from its wider surroundings and already-existing rural and small-town settlement structure – often ignored or thought to be "virgin" or "greenfield" territory – nor should it only, or even in the first place, be understood in relation to the urban category of which it is often understood as becoming a decentralized part. Thus, we have to move beyond the dominant tendency of "methodological cityism" in urban theory (Angelo & Wachsmuth, 2015), which simply defines suburbs as the bleak mirror image or opposition of central cities and ignores their much more complex relationships with wider regions.

It is exactly here that the trans-historical definition of the suburban phenomenon as a shifting, "in-between" contact zone between city and countryside also shifts our perspective of suburbanization. The suburban phenomenon and its aftermath in history is as much entangled with existing villages nucleating and encroaching upon the city along the major thoroughfares and socio-economic axes leading to town as it is with so-called exploding cities extending their tentacles towards the countryside. Therefore, suburbanization as an inherently discontinuous and contingent "process in time" should equally be understood as a transitional phase of settlement expansion and contraction that is, again, not taking place in a socio-economic vacuum but is being shaped in relation to bigger patterns of demographic and economic growth, population movement, and migration affecting a wider metropolitan region during certain historical epochs. "After suburbia" in this context can then mean further densification, material build-up, and diversification of activities and functions – thus, the suburban phenomenon giving way to "urbanization" as it is commonly understood in urban theory (see the "urban revolution" thesis of Brenner, 2014; and for a critique, see Harris in this volume). But "after suburbia" can also imply abandonment – people moving away or distancing themselves from the city, both imagined and real – and "reruralization," meaning a weakening of operations, contact ties, and mental identifications with an urban centre, which effectively shrink or displace the suburban phenomenon as such. Equally, it can also stand for stasis, immobility, or a continuation of a current situation for longer periods of time. These different historical scenarios or possibilities are – in our global age of accelerating sub/urbanization – often overlooked. Nevertheless, they merit scholarly attention since they are common in global history.

If one looks, for instance, only at the sub/urban trends of European cities and towns between 400 and 1500 (as detailed by Clark, 2009), one can notice both periods of growth (between the eleventh and thirteenth centuries) and contraction (between the late fourteenth and fifteenth centuries). Moreover, variations between individual European regions and individual cities and towns tend to bring even more detail to sub/urban histories. Cities and their suburban developments could simply disappear (for instance, after warfare, geological, or demographical shocks) or experience discontinuous histories of change. For example, suburban tracts were often turned into farmland or given back to nature and only sometimes resurface again as sub/urban settlements centuries later (a common European trajectory between the fifth and eighth centuries, after the collapse of the Roman Empire). Similarly, from a cultural-historical perspective, communities living in the "in-between zone" between city and countryside could preserve "rural identities," identify themselves with the city, or hold on to suburban distinctiveness, even after, for instance, the material-morphological outlook, economic functions, and demographic characteristics of the spatial field had changed (Epstein, 2001; Stabel, 2015).

Incorporating a wider and more diverse set of historical trajectories and cultural experiences in scholarly accounts of the suburban seems paramount for future research. While the current state of the field has slowly but steadily moved away from its dominant anglophile and cliché interpretations, a comparative move towards a trans-historical comparison of the suburban phenomenon and its aftermath is still awaiting progress. Integrating a much longer, *longue durée* timeframe and a much wider scope of comparison will open up deeply rooted forms of theory-making and result in a more adequate, trans-historical definition of suburbanization as an always shifting and spatially variegated contact zone between city and countryside.

Approaching the suburban phenomenon by comparing it through time is not, however, without its problems, especially when this approach is used for causal analysis. Comparative causal research on macro-phenomena such as waves of suburbanization requires the phenomena to be as equivalent (i.e., all relevant factors kept constant) and independent (i.e., separated in space and time) as possible. As Sewell (2005) notes, however, "attempts to assure equivalence in historical cases will actually result in decreasing the independence between cases – and vice versa" (p. 96). Historical comparative research may then be useful for generating new ideas about the causal dynamics of particular factors, but the "experimental temporality" that informs it (i.e., setting-up a "historical experiment" where conditions are kept *ceteris paribus*) is less adequate for the causal analysis of interdependent sub/urban phenomena. The methodological strategy of "comparing through time" does not, however, exhaust the contribution of urban history and urban historical analysis, as the latter also has much to contribute to a better understanding of how suburbia as a contact zone is structured *by* time. In the next section, we therefore discuss how time is a structuring agent of socio-spatial continuity as well as change in society, and how the concept of "path dependency" helps capture this in specific cases of suburban development (see also Balducci et al. in this volume).

Structuring "After Suburbia" by Time

In an impressive account of the relationship between history and the social sciences, Sewell (2005) notes that "social theory badly needs a serious infusion of historical habits of mind" (p. 1). More specifically, he feels that "historians' complex and many-sided understanding of the temporalities of social life has scarcely found its way in social theoretical debate" (p. 1). Indeed, understanding the suburban phenomenon in its manifold empirical expressions and evolutions also requires us to trace their specific historical trajectories – that is, the processes of sub/urbanization through which they are formed or the processes of de-surbanization through which they may also dissolve. In this context, social scientists readily think about the notion of path dependency, which refers to "how previous events affect the probability of future events to occur" (Frenken & Boschma, as cited in Martin, 2010, p. 2). Ironically, the notion of path dependency has its roots in economic development studies rather than in history, notably in evolutionary economic accounts of technological change in the context of the demise of Fordism (Storper, 1999). Nevertheless, path dependency offers a good analytical framework to examine how socio-spatial processes of sub/urbanization and de-surbanization are structured by time. However, to avoid shallow, empiricist applications of the notion of path dependency and show that history matters "in a nontrivial sense" (Martin, 2010, p. 2), we need to explore in more detail the theoretical understanding of temporalities from which historians tend to work and from which urban theory can be critiqued and reformulated.

Sewell (2005) argues that, for historians, time is both "fateful" and "contingent." While the future cannot, in principle, be predicted due to unforeseen, contingent events, neither can the past be reversed and "irrevocably alters the situation in which it occurs" (p. 7). Moreover, the effect of any action is "dependent upon its place in the sequence" of actions (p. 7). In this context, Sewell argues for an "eventful temporality," meaning a notion of time wherein events matter, and chronology and context are considered the reference base of all social theorizing. Contemporary social scientists, however, often adopt a teleological temporality, in which history is no more than the "temporal working out of an inherent logic of social development" (p. 83). This tendency can be observed in many urban theories linking urbanization to capitalist development and leads to the historically flawed idea that suburbanization is "one directional urbanization," meaning "countryside" in the process of becoming "city." This, according to Sewell, goes against the way urban historians work through "historical contextualization." Suburbanization cannot be subsumed under some general, "transhistorical progressive laws" about urbanization, capitalist or otherwise (p. 83), but should be approached as an "event." This means that the analytical focus should be on the "actions and reactions that constitute the happening," in this case, instances of suburbanization, and on "the concrete and specifiable conditions that shape or constrains the actions and reaction" (p. 84).

However, adopting what Sewell calls an "eventful temporality" does not have to contradict social scientists' concerns about causal analysis. In fact, Sewell's eventful temporality is not so much a critique of social science in general, but of positivist social science and the teleological temporality on which it operates, which, admittedly, is dominant in many areas of social science research. What we need to aim for in any historical analysis of suburban (and other) phenomena is a "multiple causal narrative" that improves our understanding of the "conjunctural, unfolding interactions" or "events" and defies teleological explanations due to the focus on the multiplicity of causal factors that is at work in any given case (p. 98). This eventful temporality is entirely consistent with the layered ontology of critical realism, which rejects the "flat ontology" of positivist philosophies of social science in which social reality is reduced to those phenomena that can be observed (Sayer, 1984, 2000). Instead, critical realists distinguish different levels of reality, notably (1) the ontological level of the objects and their generative power, and (2) the level of the actual on which these causal forces are (or are not) actualized and put to work and made to interact in specific contexts to produce social reality in all its complexity, messiness, and open-endedness. From this perspective, eventful temporality can combine a penchant towards "historical contextualization" with a focus on (multiple) causal analyses, developing an interpretation of urban change that draws on the strengths of both history and sociology and "avoid[s] highlighting one static causal factor" (Corfield, 2013, p. 828). Thus, urban theories reducing the process of sub/urbanization to a single universalizing process – for instance, the dynamics of agglomeration (see Scott & Storper, 2015) – will from a historical/historian's perspective always be perceived as unsatisfactory and needlessly restrictive, since it negates place- and time-specific historical contextualization and the chronological sequencing of multiple causal mechanisms at the same time (see, for instance, Harris in this volume; Walker, 2016).

But how to operationalize a research strategy that combines historical contextualization with multiple causal analyses? We want to argue here that the notion of path dependency, when it is lifted out of the context of neoclassical economic equilibrium thinking, is one promising way to do this. Path dependency explicitly puts time central to causal analysis, is sensitive to the contextually specific interaction between multiple causal factors, and operates on a notion of eventful temporality by seeing time as fateful, contingent, and complex. The notion of path dependence was initially developed to explain stability (Schneiberg,

2007) and is associated with the mechanism of "lock in" (Martin, 2010). Lock in refers to an evolutionary economic process in which an initial event "becomes progressively locked in" (Martin, 2010, p. 4) because of network effects or increasing returns around which stable patterns of technological and industrial development, including a locational geography, are established. However, it is unhelpful to reduce path dependency to the idea of lock in, which is predicated on conventional economic notions of equilibrium thinking in which stability can only be punctured by external events and shocks. To take history seriously implies leaving open the possibility – and even the likelihood – of endogenously generated change, continuous evolution, and even transformation. As Sewell (2005) noted, historians see time as highly complex. In every particular historical event "social processes with very different temporalities" (Sewell, 2005, p. 9) are at work, resulting in a mix of historical continuity, newness, and change. To briefly illustrate this point, consider the following: While residential suburbanization in early-modern Europe could take the form of the bourgeois *villa suburbana* described by Fishman (1987) – influenced by Classical and Renaissance ideas of anti-urbanism and the uplifting function attributed to the countryside – this has never been in history the dominant, first, or unchanging universal form of development. The suburban phenomenon was never "locked" in a path of one-single expression but was always taking place in different constellations and modalities *at the same time*. Some of these paths continued and persisted, while others transformed and changed through time, but all were connected to a complexity of human motives and societal processes (Jauhiainen, 2013).

In this context, historian Penelope Corfield's criticism of a linear understanding of urban growth is of particular relevance. Written as a conclusion to the over 800 page volume *The Oxford Handbook of Cities in World History*, Corfield (2013) starts by explaining how a linear or teleological understanding of urban growth – which we already criticized earlier in this chapter as leading to an inadequate understanding of the suburban phenomenon as formerly rural places on the (inevitable) path of becoming urban – has its roots in "Christian teaching [that] viewed history as a progressive upwards journey towards a 'shining city on the hill'" (p. 831). Such a Christian viewpoint later mutated into a secular version in which global urbanization since the eighteenth century was seen as a cumulative process of liberation and progress (see also Van Damme & De Munck, 2017). The problem with a linear history of urbanization, however, is that it "underestimates the diversity of outcomes ... [and] erases medium- and short-term fluctuations, rendering change too smooth and unidirectional" (Corfield, 2013, p. 832). Corfield concludes that the history of urbanization can only be understood by analysing it in terms of the interaction between "persistence, micro-change and revolution" (p. 834). A historically adequate conception of path dependency thus needs to be able to explain continuity or stability on the one hand, as well as incremental change and revolution on the other.

To illustrate Corfield's plea to study the long-term forces of both deep continuity and change in history, we can construct a multi-causal model explaining both path-dependent persistence and change in the sub/urbanization process. First, to account for stability, sub/urban research has to dig deeper in "location," "capital," and "community." Location refers to the initial time- and place-specific geographic (and other) advantages that attract people to sub/urban land (for instance, more green space, less noise, and so on), which is then reinforced and stabilized in path-dependent ways by time through the construction of transport, communication, and other networks. These infrastructures effectively bind people to place, leading to a material build-up of houses, amenities, and further investments; in other words, the construction of "huge amounts of stored overhead capital" (Corfield, 2013, p. 837; see also Walker, 2016). Invested capital eventually acts as ballast, restricting the possibilities of historical actors over time, but it can also be renewed and reformed, attracting new investors and developers. Stability of location and capital, however, cannot be taken for granted, since initial land advantages tend to change in the *longue durée*, and capital in the past has been as mobile as it is today. Thus, a final causal factor of stabilization has to be sought in the strengths or weaknesses of community ties among people themselves, especially "social traditions [that] regularly help to buttress the known and settled" (Corfield, 2013, p. 838). Sub/urbanites build families, kinships, neighborhoods, meeting places, and organize themselves in governmental bodies, and so on. People tend to "stay put" as long as they can find the means to survive (i.e., jobs), but equally important are social and cultural ties (for instance, nostalgia for birthplaces), and access to – or at least influence on – governing political levels.

Applying such multi-causal ways of thinking shifts the analytical focus to the interaction between causal factors in concrete spatio-temporal contexts and, hence, tends to be more sensitive to place specificity and changes through time. If we look, for instance, at the long-term evolution of Antwerp, Belgium – a city both authors have studied – we recognize similar, constitutive elements of stability popping up in the long-term sub/urban development of the city (De Smedt et al., 2010; Van den Broeck et al., 2015). Focusing on elements of stability entails pointing out the importance of the river Scheldt as a key international gateway and economic thoroughfare that defined the long-term harbour and commercial functions of Antwerp. Also important are the city's old infrastructural connections with its densely populated hinterland that was drawn into the orbit of the city from at least the fifteenth century onwards, leading to a shifting and constantly changing contact zone with the countryside that has been instrumentalized for suburban food production, residential purposes, and industry (beer brewing, craft production), as well as for externalizing unwanted public services (waste disposal, gallows field, leper hospitals, and so on). Moreover, since at least the sixteenth century, urban elites have been buying and selling, parcelling out, and building upon sub/urban land as capitalist investment or for surplus-oriented sub/urban development, drawing in a constant stream of newcomers and developers even in times of "urban crisis," shrinkage, and contraction. Finally, the city has long had an active and mobile middle class that has formed long-standing civil and community ties (such as guilds and urban societies) and made successful trans-generational claims to live and work in the city or its surrounding suburban belt. Thus, "after suburbia" from the long-term historical perspective of Antwerp is a narrative of the shifting contact zones between city and urbanizing countryside (for instance, in the suburb of Hoboken), of the maintenance of "rural identities" (as in Borsbeek), and of the mobilization of community ties around "suburban distinctiveness" (as in Brasschaat) (see May et al., 2022).

Whereas "continuity in history tends to be unsung and under-analysed" (Corfield, 2013, p. 836), urban historians and theorists are more on familiar ground when linking incremental change and revolution in the path-dependent sub/urbanization process to shifts in power and political agency. Indeed, to integrate the possibility of endogenous and incremental change and revolution within the notion of path dependency we need to adopt an agency-based perspective – a perspective that brings us right back to Sewell's eventful temporality. "Paths," or historical trajectories, are to varying degrees internally diverse and heterogenous. They are not devoid of alternatives but contain within them a multiplicity of historically generated resources that can be mobilized to open up other possibilities and create alternative paths (Schneiberg, 2007). Path dependency is hence not seen as a structural causal factor "acting itself out" but as a historically produced set of constraints and possibilities that needs actors – who are themselves temporally and spatially situated – to be "actualized." In this context, Stark and Brustz (2001) argue that path dependency is not "*past* dependency – because it is not from a generalized (and hence ahistorical, because uneventful) 'past' that actors reenact, redefine, and recombine resources" (p. 1133). It is these resources that are the result of past actions that actors can use to overcome "obstacles to change" and can recombine in order to engage in socio-spatial innovation. According to Stark and Brustz (2001), "this exploitation of existing institutionalized resources is a principal component of the paradox that even instances of transformation are marked by path dependence" (p. 1132).

Any historically attuned analysis of the suburban phenomenon, then, needs to pay explicit attention to the spatio-temporally situated actors pursuing sub/urbanization in particular places, which, as Hamel and Keil (2019) indicate, requires explicit attention to power relations. For Hamel and Keil, analysing space "implies the recognition of multiplicity with the presence of the other – the social question – or how actors are going to live and make choices, introducing as a consequence the 'geography of power.' Power relations – defining and contesting time-space relations – are thus inevitable" (p. 56). The history of the suburban phenomenon is, indeed, full of examples of struggling political and economic stakeholders ascribing different and sometimes conflicting functionalities and outcomes to the contact zone between city and countryside described above. The specific urban–land nexus of suburbia – in general, there is more and cheaper land available in the hybrid suburban zone than in the city, but the suburban zone's proximity to urban amenities still makes it different from rural land – will influence the outcome of such power struggles and eventually influence "after suburbia." Certain types of development, shapes and forms are believed to be more feasible, desirable, and cost effective in the suburban passageway. Throughout history, we see *both* negative and positive functionalities being externalized to suburbia, sometimes existing at the same time in the same suburban moment.

If we return again to the history of Antwerp by way of example, we notice a path-dependent power struggle between, on the one hand, a central government clearly demarcating the shifting contact zones between city and countryside for military reasons from at least the middle of the sixteenth-century onwards, effectively walling in and controlling the mobility flows of inhabitants and newcomers and, on the other hand, local and political urban stakeholders pushing for suburbanization out of a complexity of political, industrial, and residential motives. Within this long-running power struggle, which had a path-dependent influence on the development of the city of Antwerp until well into the twentieth century, the agency of the suburbanites – the people living in the peripheral edges of the city and moving through time in the direction of the "city" or, rather, back again to the encircling hamlets and villages in the "countryside" – should also be taken into consideration. Whereas at the end of the sixteenth century, due to political warfare and a deep economic crisis, "after suburbia" came to mean a rapid depopulation and loss of connections between the city and its hinterland, a strong demographic boom in the Antwerp countryside from the 1730s onwards effectively meant a reversal in evolutions. "After suburbia" came to imply – to a greater or lesser extent, depending on place – urbanization, suburban distinctiveness, and an almost anachronistic attachment to a rurality gone by. These attachments became "imagined traditions," materialized in the build-up of the housing stock and expressed in cultural festivities and local identities, as well as the subject of local boosterism and development coalitions, projecting the future outlook of their suburban communities in more or less antagonistic relations with the central city Antwerp.

Conclusion

We started this chapter by pointing to the current upsurge in urban theorizing related to rapid urban developments. Already at the start of the new millennium, the late Edward Soja (2000) described the present-day "urban age" in Dickensian terms as "the best of times and the worst of times to be studying cities, for while there is so much that is new and challenging to respond to, there is much less agreement than ever before as to how best to make sense, practically and theoretically, of the new urban worlds being created" (p. xii). Understanding the nature of "extended urbanization, 'disjunct fragments' and global suburbanisms" (Keil, 2018) might be one of the most important challenges defining the field of urban studies in the decade, or even decades, to come.

   In our collective search for better concepts and theories, however, we have proposed how – paradoxically enough – looking backwards might also bring a better understanding of our speculative urban futures. Historical approaches, or at least a historical sensitivity, might overcome a widespread "recentism" and lead to better sub/urban theory-building. As a way of fully integrating history in urban studies, we have proposed two interventions. The first of these proposes to mobilize the urban past in a comparative way to overcome overly restrictive definitions of what counts as sub/urbanization in urban theory. The past, similar to geography, is a way to understand urban diversity, shaping trans-historical processes such as the suburban phenomenon in reaction to different structural conditions and cultural frameworks that were dominant in history. Thus, "after suburbia" from a historical perspective becomes a category of the past – an emergent and unstable phase defining the contact zone between "city" and "countryside" throughout history.

Yet, the past cannot be separated from the present day; both are entangled in "path-dependent" ways. In a second intervention, we have therefore proposed to consider time seriously when theorizing about how the past shapes the present. Drawing on the writings of Sewell (2005) and Corfield (2013), and on historical institutionalist approaches to path dependency more generally, we have proposed a more historically attuned analysis of the suburban phenomenon, in which a historically contextualized analysis of a multiplicity of causal factors leads to the singling out of "eventful temporalities" or chronologically sequenced paths of continuity and change. Stabilizing factors that have been pointed out – but which in and of themselves are in no way unchanging, devoid of historical context, or abstract – are "location," "capital," and "community." In this context, we do not approach "path dependency" as a structural factor explaining the "lock in" of certain urbanization trajectories (although this is one of the possibilities) but rather suggest the adoption of an agency-based perspective. Historical (de-)surbanization trajectories generate resources that can be mobilized by specific actors to pursue particular paths of (de-)surbanization, alternative or not, eventually defining the material outlook and imagined character of their suburban community in a dialectic relationship with the central city. Seeing through the darkness of future "after suburbia" implies that these important paths of change and revolution will be balanced and set-off by just as important paths of continuity and stability. With such a complex framework, a linear and one-dimensional explanation will never suffice to build a historically attuned theory of the sub/urban.

NOTE

1 The full quote from the episode of *Twin Peaks* (1990–1) can be found here: https://www.youtube.com/watch?v=OWv3qfschlE.

REFERENCES

Angelo, H., & Wachsmuth, D. (2015). Urbanizing urban political ecology: A critique of methodological cityism. *International Journal of Urban and Regional Research, 39*(1), 16–27. https://doi.org/10.1111/1468-2427.12105.

Brenner, N. (Ed.). (2014). *Implosions/explosions: Towards a study of planetary urbanization.* Jovis.

Callon, M., & Latour, B. (1981). Unscrewing the big Leviathan: How actors macro-structure reality and how sociologists help them to do so. In

K. Knorr-Cetina & A.V. Cicourel (Eds.), *Advances in social theory and methodology: Toward an integration of micro- and macro-sociologies* (pp. 277–303). Routledge.

Clark, P. (2009). *European cities and towns: 400–2000.* Oxford University Press.

Clark, P. (Ed.). (2013). *The Oxford handbook of cities in world history.* Oxford University Press.

Corfield, P.J. (2013). Conclusion: Cities in time. In P. Clark (Ed.), *The Oxford handbook of cities in world history* (pp. 828–46). Oxford University Press.

De Smedt, H., Stabel, P., & Van Damme, I. (2010). Zilt succes: Functieverschuivingen van een stedelijke economie. In I. Bertels, B. De Munck, & H. Van Goethem (Eds.), *Antwerpen: Biografie van een stad* (pp. 109–44). Meulenhoff/Manteau.

Epstein, S.R. (Ed.). (2001). *Town and country, 1300–1800.* Cambridge University Press.

Ewen, S. (2016). *What is urban history?* Polity Press.

Fishman, R. (1987). *Bourgeois utopias: The rise and fall of suburbia.* Basic Books.

Forsyth, A. (2019). Defining suburbs. In B. Hanlon & T.J. Vicino (Eds.), *The Routledge companion to the suburbs* (pp. 13–28). Routledge.

Guldi, J., & Armitage, D. (2014). *The history manifesto.* Cambridge University Press.

Hamel, P., & Keil, R. (2019). Toward a comparative global suburbanism. In B. Hanlon & T.J. Vicino (Eds.), *The Routledge companion to the suburbs* (pp. 51–61). Routledge.

Harding, A., & Blockland, T. (2014). *Urban theory: A critical introduction to power, cities and urbanism in the 21st century.* SAGE.

Harris, R. (2010). Meaningful types in a world of suburbs. In M. Clapson & R. Hutchinson (Eds.), *Suburbanization in global society* (pp. 15–47). Emerald Publishing.

Harris, R. (2019). Suburban stereotypes. In B. Hanlon, & T.J. Vicino (Eds.), *The Routledge companion to the suburbs* (pp. 29–38). Routledge.

Harris, R., & Smith, M.E. (2011). The history in urban studies: A comment. *The Journal of Urban Affairs, 33*(1), 99–105. https://doi.org/10.1111/j.1467-9906.2010.00547.x.

Harris, R., & Vorms, C. (Eds.). (2017). *What's in a name? Talking about urban peripheries.* University of Toronto Press.

Jauhiainen, H.S. (2013). Suburbs. In P. Clark (Ed.), *The Oxford handbook of cities in world history* (pp. 791–808). Oxford University Press.

Jayne, M., & Ward, K. (2017). *Urban theory: New critical perspectives.* Routledge.

Keil, R. (2018). Extended urbanization, "disjunct fragments" and global suburbanisms. *Environment and Planning D: Society and Space, 36*(3), 494–511. https://doi.org/10.1177/0263775817749594.

Keil, R., and Shields, R. (2013). "Suburban boundaries: Beyond greenbelts and edges." In R. Keil (Ed.), *Suburban constellations: Governance, land and infrastructure* (pp. 71–8). Jovis.

Madgin, R., & Kenny, N. (Eds.). (2015). *Comparative and transnational approaches to urban history: Cities beyond borders.* Ashgate.

Martin, R. (2010). Roepke lecture in economic geography: Rethinking regional path dependence: Beyond lock-in to evolution. *Economic Geography, 86*(1), 1–27. https://doi.org/10.1111/j.1944-8287.2009.01056.x.

May, L., Van Damme, I., & Oosterlynck, S. (2022). *Street-making is place-making: A political-economic perspective on suburbanization in Antwerp, Belgium (first half of the 20th century).* In progress.

McManus, R., & Ethington, P.J. (2007). Suburbs in transition: New approaches to suburban history. *Urban History, 34*(2), 317–37. https://doi.org/10.1017/S096392680700466X.

Molotch, H., Freudenberg, W., & Paulsen, K.E. (2000). History repeats itself, but how? City character, urban tradition and the accomplishment of place. *American Sociological Review, 65*(6), 791–823. https://doi.org/10.2307/2657514.

Oosterlynck, S. (2012). Path dependence: A political economy perspective. *International Journal of Urban and Regional Research, 36*(1), 158–65. https://doi.org/10.1111/j.1468-2427.2011.01088.x.

Peck, J. (2015). Cities beyond compare? *Regional Studies, 49*(1), 1–23. https://doi.org/10.1080/00343404.2014.980801.

Phelps, N.A. (2019). In what sense a post-suburban era? In B. Hanlon & T.J. Vicino (Eds.), *The Routledge companion to the suburbs* (pp. 39–47). Routledge.

Rickards, L., Gleeson, B., Boyle, M., & O'Callaghan, C. (2016). Urban studies after the age of the city. *Urban Studies, 53*(8), 1523–41. https://doi.org/10.1177/0042098016640640.

Robinson, J. (2013). The urban now: Theorizing cities beyond the new. *European Journal of Cultural Studies, 16*(6), 659–677. https://doi.org/10.1177/1367549413497696.

Sayer, A. (1984). *Method in social science: A realist approach.* Routledge.

Sayer, A. (2000). *Realism and social science.* SAGE.

Schneiberg, M. (2007). What's on the path? Path dependence, organizational diversity and the problem of institutional change in the US economy, 1900–1950. *Socioeconomic Review, 5*(1), 47–80. https://doi.org/10.1093/ser/mwl006.

Scott, A.J., & Storper, M. (2015). The nature of cities: The scope and limits of urban theory. *International Journal of Urban and Regional Research, 39*(1), 1–15. https://doi.org/10.1111/1468-2427.12134.

Sewell, W.J. (2005). *Logics of history: Social theory and social transformation.* University of Chicago Press.

Short, J.R. (2019). The end of the suburbs. In B. Hanlon & T.J. Vicino (Eds.), *The Routledge companion to the suburbs* (pp. 335–41). Routledge.

Smith, M.E. (2010). Sprawl, squatters and sustainable cities: Can archeological data shed light on modern urban issues? *Cambridge Archeological Journal, 20*(2), 229–53. https://doi.org/10.1017/S0959774310000259.

Soja, E. (2000). *Postmetropolis: Critical studies of cities and regions.* Blackwell.

Sorenson, A. (2015). Taking path dependence seriously: An historical institutionalist research agenda in planning history. *Planning Perspectives, 30*(1), 17–38. https://doi.org/10.1080/02665433.2013.874299.

Stabel, P. (2015). Town and countryside in medieval Europe: Beyond the divide. In A. Wilkin & J. Naylor (Eds.), *Town and country in medieval north western Europe: Dynamic interactions* (pp. 313–23). Brepols.

Stark, D., & Bruszt, L. (2001). One way or multiple paths: For a comparative sociology of East European capitalism. *American Journal of Sociology, 106*(4), 1129–37. https://doi.org/10.1086/320301.

Storper, M. (1999). The resurgence of regional economics, ten years later. In T.J. Barnes & M.S. Gertler (Eds.), *The new industrial geography: Regions, regulation and institutions* (pp. 23–53). Routledge.

Storper, M., & Scott, A.J. (2016). Current debates in urban theory: A critical assessment. *Urban Studies, 53*(6), 1114–36. https://doi.org/10.1177/0042098016634002.

Taylor, P.J., & Lang, R.E. (2004). The shock of the new: 100 concepts describing recent urban change. *Environment and Planning A, 36*(6), 951–8. https://doi.org/10.1068/a375.

Van Damme, I., & De Munck, B. (2017). Cities of a lesser God: Opening the black-box of creative cities and their agency. In I. Van Damme, B. De Munck, & A. Miles (Eds.), *Cities and creativity from the Renaissance to the present* (pp. 3–23). Routledge.

Van den Broeck, J., Vermeulen, P., Oosterlynck, S., & Albeda, Y. (2015). *Antwerpen, herwonnen stad? Maatschappij, ruimtelijke plannen en beleid, 1940–2012.* Die Keure.

Van de Walle, T. (2019). *Van twee wallen eten? De stadsrand als overgangszone tussen stad en platteland in de late 15de en 16de eeuw. Casus Oudenaarde* [Unpublished doctoral dissertation]. University of Antwerp.

Vaughan, L., Griffiths, S., Haklay, M., & Jones, C.E. (2009). Do the suburbs exist? Discovering complexity and specificity in the suburban built form. *Transactions of the Institute of British Geographers, 34*(4), 475–88. https://doi.org/10.1111/j.1475-5661.2009.00358.x.

Walker, R.A. (2016). Why cities? A response. *International Journal of Urban and Regional Research, 40*(1), 164–80. https://doi.org/10.1111/1468-2427.12335.

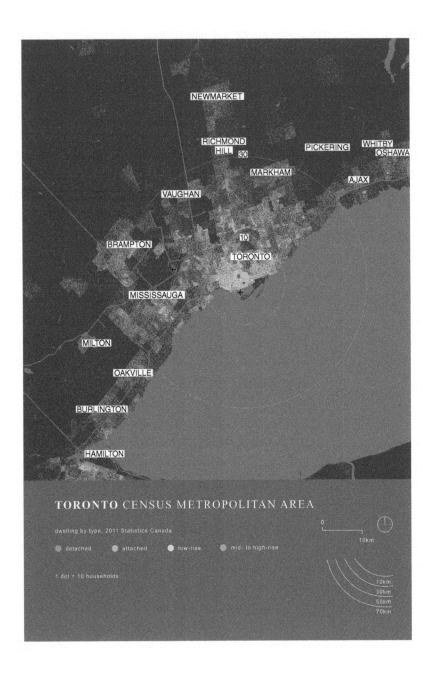

Image 15. Residential dwellings by type, Toronto

15 The Afterlives of "Suburbs": Methodological and Conceptual Innovations in Urban Studies

JENNIFER ROBINSON

It is time to argue for an intervention into urban theory on the basis of the suburban explosion.

– Keil, *Suburban Planet*

The initiative to think the urban from the sites that have been known as "suburbs" has provided many insights into the emerging form and challenges of urban areas across the globe. Shifting attention from the "inside-out" theorization of urbanization as spreading from a putative centre has definitively changed the perspective of urban studies. This represents an essential dislocation of urban studies to the "peripheries" – to the diverse territories where the impacts of the coming urban transitions in different parts of the world are being and will continue to be strongly felt. These are the territories where millions of men, women, and children find places to settle and to seek livelihoods or where enterprises seek to locate. Such sites, even in the poorest of urban contexts, are also often venues of value creation for global capital. Starting from "suburbs" or "suburbanization" has been an essential manoeuvre for developing the analytical resources to understand and address the future shape and challenges of urbanization.

It is clear, though, that if suburbs are to be drawn into analytical focus, taking a global approach is essential – from beyond the centre, from beyond the Global North, starting anywhere. Whatever suburbs are or might become, this ex-centric reference point embraces a very wide variety of urban forms. There are diverse sub/urbanization processes evident across different regions and diverse kinds of urban settlements. At the same time, ex-centric urbanisms in different places are also closely interconnected through a range of transnational processes, and thus they are also part of repeated but differentiated formations within wider circulations and circuits of urbanization and globalization

(Robinson, 2016b). In search of insights into suburbs, their afterlives, and their future trajectories, it is thus necessary to think from and with both *diversity* (i.e., outcomes are specific in different places perhaps because of long historical trajectories of development) and *differentiation* (i.e., interconnected processes associated with repetition and differentiation of urban form). The wider Global Suburbanisms research project, of which this book is one component, has inspired and enacted significant methodological and conceptual innovation in urban studies in order to grapple with this problematic of diverse and interconnected ex-centric urbanisms. This chapter seeks to assess these innovations, to show why they need to be clearly recognized, and to propose some implications for urban studies.

I approach this engagement with the Global Suburbanisms project through the rubric of the "afterlives" of sub/urbanisms and sub/urbanization using the instructive formulations brought together in Keil (2017) together with Benjamin's (2009) notion of "afterlives." The first sense of the afterlives of suburbs concerns its productivity as a conceptual and methodological object of concern, evident in the work of the wider project behind this book. Although there is a certain parochial trace to the concept of the suburb in its US- and UK-centric cultural resonances (Zhao, 2020), thinking with the diverse forms and afterlives of this concept and intersecting with a globalizing field of urban studies has yielded important theoretical and methodological insights that can support important new directions for urban studies. The second sense of the afterlives of suburbs considered here is material, in that specific suburban histories continue to be operative in shaping current and future configurations of the urban. Finally, the third sense of the afterlives of suburbs concerns Benjamin's (2009) figure of afterlives as a repetitive deployment without original form or reference. This helps us to capture the repetitive but differentiated production of ex-centric urban expansion as a product of the prolific globalized interconnections and circulations making the future city.

I consider these three aspects of the afterlives of suburbia (as concept/method, as ongoing material presence, and as vantage point for discerning the future of urbanization) from within an emerging vocabulary of comparative urbanism. This chapter therefore also tracks the methodological import of this focus on suburbs. Thus, we will note how the conceptualization of sub/urbanization has been reconfigured through an open engagement across highly diverse urban territories in search of emergent concepts. This operates as a form of generative urban comparativism. The grounded afterlives of specific suburbs in specific contexts indexes the potentiality of thinking beyond the single case. And tracing the interconnections shaping ex-centric urbanization across many different contexts exposes the form of genetic comparative practice, drawing connected cases into analytical proximity. The Global Suburbanisms project demonstrates important methodological principles for global urban studies and indicates potential trajectories for a critical reconfiguration of the meaning of sub/urbanisms and sub/urbanization, and more generally for inventing new urban concepts. In this chapter, I point to some of these achievements in reframing suburbs and draw out the implications for practices of theorizing or conceptualizing "the urban."

Approaching "Suburbs" Comparatively

Drawing on the important contribution of Thomas Sieverts, Keil (2017) has brought to our attention that the immediate challenge in thinking from "Zwischenstädte" is that it requires thinking across so many very different meanings and experiences of "suburb" – a world of suburbs – which Richard Harris very quickly began to explore in the context of the Global Suburbanisms initiative. As Keil (2017) puts it, the research group encountered "the Babylonian plurality in naming settlement away from the core" (p. 46). It was not just the names, though, as the collection of Harris and Vorms (2017) demonstrates; "suburb" and the myriad lexical terms used for urban peripheral territories index a wide range of experiences. Thus, any concept beginning with the US model, with features such as low density, peripherality, and exclusiveness, quickly breaks down in the face of the diversity of suburban forms, such as the French "banlieue" and South African townships, for example. After numerous such examples, in many ways only a remnant of the suburban remains: peripherality. In this way, a new and minimalist definition of the suburban emerged: "a combination of non-central population and economic growth with urban spatial expansion" (Ekers et al., 2012, p. 407). Harris (2010) further suggested a focus on "the ordinary urban periphery within a worldwide frame of reference" (p. 16), bringing together literatures across the Global North and South, which at that time he could say "had hardly communicated" (p. 20). But Global Suburbanisms has fundamentally changed that, bringing these diverse ex-centric urban contexts into conversation.

In the ensuing critical reflections, it has become clear that divergences from the putative "origin" model of the suburb abound. This origin has often been located, as Keil (2017) describes it, in the "marked term" indexing the "classic" US model of the suburb (p. 45). Certainly, the historical origins of the term "suburb" are more complex and much older than this and can be seen, for example, in the older "beyond the city walls" meaning developed within the experience and analysis of European urbanization (Phelps et al., 2017).[1] The distinctiveness of the US case is quickly apparent when bringing other contexts into consideration – for example, the ways in which suburban relocation has often been forced on residents and was not the classic individual consumer choice of the US suburb, or that many on the periphery are not escaping the city but arriving in it. Thus, the city has remained the focus of significant investment of mobility and daily effort to find livelihoods and to sustain life, often in modes of informality. As Phelps (2017) put it when reflecting on Southern and Eastern European experiences, suburbs often represent a base from which to secure "an approach to the city rather than an escape from it" (p. 7). This is even truer in poorer city contexts.

Opening up this conversation about assessing the meaning and utility of the suburb as a starting point for thinking about urbanization across such different experiences of the urban has required methodological innovation. Starting with an inherited and dominant concept relying on one primary contextual reference point, subsequently multiplied through investigations in many different contexts, research has pressed at this dominant meaning, confronted it with a great variety of experiences, and been most willing to disturb and dislocate the meanings inherited from traditional (northern) academic assumptions. This stream of research provides strong evidence for the value of "thinking with elsewhere" and of sustaining analytical conversations beyond the individual case (Robinson, 2016b). It has also, in my view, stimulated a distinctive form of comparative practice.

In other contributions (Robinson, 2011, 2016a, 2016b), I have called the kind of comparative practice driven by intellectual curiosity and the search for conceptual innovation "generative." In this vein, the particular style of analysis that Global Suburbanisms configures does not rest on a rigid practice of "variation finding," or a careful application of a formulaic research procedure. In such a revised formulation of (generative) comparative practice, reflecting on the diversity of suburban forms might rely on identifying some kind of "shared feature" that resonates across many different urban contexts and diverse outcomes, that draws our curiosity, and through which we can seek to expand our analyses to help us explain outcomes, to probe causalities, and to answer "why" questions. As Keil (2017) puts it, "the causalities, i.e. the 'why's' are beginning to reveal themselves through comparative, multi-site, multi-disciplinary, multi-method empirical and conceptual research and will lead to theoretical and conceptual insights that will guide further research on the topic" (p. 47). But while this sounds very proper and scientific, it glosses over the practices of this research project in a rather dated language, too quickly passing over what is innovative and unusual about the comparative approach adopted by this group.

Their starting point for analysis has been a certain kind of territory, in its multiple forms across the globe, with the shared feature of being relationally positioned in urban peripheries and a site of urban expansion. This comparative tactic has opened up the term "suburb" to a form of speculative concept formation: thinking the "suburban" from whatever is found in these kinds of territories. The potential for consequential emergent theorizing arises from the rich diversity of urban experiences across these territories. Inspired by Simone (2011), a methodological proposition emerges in which we can attend to the ways in which the urban world in all its diversity and potentiality presses on the conceptualizing practices of researchers, demanding of us to respond to the questions and challenges – the "problems" in a Deleuzian idiom – posed of us by urban worlds. Here, then, a form of generative comparative practice involves working in an open way with "urban territories" such that concepts are potentially emergent from the fullness of particular territories, or across a wide range of contexts, offering multiple starting points for conceptualization.

Urban studies has long had a commitment to the detailed analysis of individual cases. In the comparative practices I have outlined (Robinson, 2016a, 2016b), and those inspiring the Global Suburbanisms project, the distinctive urban outcome, or singularity, as starting point for conceptualization holds the potential for theorizing from anywhere, launching ideas into wider conversations about the urban from any particular case, insisting that theoretical innovation can begin anywhere in the urban world. The reconceptualization of suburbs puts this into practice, resting on the proposition that urban theory can begin anywhere.

A further methodological innovation is needed, though, to approach the world of sub/urbanization. Harris (2010) noted that common suburban forms were appearing repeatedly across a wide array of contemporary contexts in the guise of gated communities, satellite cities, enclaves, private cities, and "zones" – what Murray (2017) has collated as the "urbanism of exception." In an earlier era of suburbanization, we might also have noted garden cities, townships, social housing, and high-rise residential blocks. Now these kinds of repeated phenomena exemplify what I have been calling "genetic" grounds for thinking with elsewhere. This indicates an alternative route for bringing different contexts and experiences into comparative reflection, allowing the spatiality of the urban itself to draw one towards comparative reflections across different contexts and cases. Here, the profoundly interconnected nature of many different urban processes and outcomes proposes alternative ways of framing method as following the many urban phenomena that are repeated across different contexts. Thus, repeated instances are emergent from the vast array of interconnected processes that constitute, perhaps, the virtual urban – all the possibilities that might give rise to "any urban whatever," to paraphrase Simone (2011). This draws on Deleuze's (1994) analysis of how we can come to an understanding of phenomena through exploring their genesis in the entwined production of concepts and emergent entities, both marked by the proliferation of difference, rather than in (the failure of) mobile concepts that are thought to act as a more or less successful mirror on reality (Robinson, 2022, chap. 11).

Such reformatted comparative practices – generative and genetic – can provoke new conceptualizations, starting anywhere, with a strong orientation to revising existing concepts and building from the spatiality of the urban (the full social worlds of territories; the interconnections that compose urban outcomes, differently each time) and from our practices as engaged, embodied, and located researchers responding to the "problems" pressed on us by urban worlds (Robinson, 2022).

"Stretching" a Concept, Starting with Territories

At the heart of the texts associated with the Global Suburbanisms project are strongly generative comparative practices. Keil's *Suburban Planet* (2017) succinctly proposes a wider conversation across a range of highly diverse expanding, peripheral urbanities. France's banlieues, for example, join many other contexts in dispossessing the periphery of privilege and low density, as do the high rises at the edges of Istanbul, which signal removal of the urban poor from well-located settlements by a peremptory regime, often accompanied by loss of livelihoods and financial ruin (Karaman, 2013). Similarly financed and cognate developments in South Africa (see Caldeira, 2016) stand instead for a revolution won, a developmental state until recently committed to redistribution and a better life – if not for all, than at least for many. Likewise, Indonesian low-income high rises signal dispossession from dense, diverse inner-city neighbourhoods but also stand as sites for new associational projects (Simone, 2018). The hyper-privatized and gated suburbs of post-socialist Sofia, protecting personal freedoms longed for in the socialist era (Hirt, 2012), sit alongside the modest peripheral housing estates that keep intact an accessible and socially mixed urbanity in Prague (see also Ouředníček, 2016).

Within individual urban contexts, too, the periphery is clearly a multiplicity, and the extensions of many African cities highlight this. The diverse ex-centric urban settlement processes across Africa highlight more generally the multiplicity of processes and forms shaping urbanization: piecemeal, small-scale private developers (such as in Lagos; see Sawyer, 2014); segregated, modernizing, and linked to traditional authorities (Beall et al., 2015); extractive (Nairobi; see Murray, 2017); peremptory and historically dependent (Johannesburg; see Butcher, 2016); ambitious (Kigali; see Goodfellow, 2017); precarious (Lusaka; see Myers, 2011); diverse (Durban; see Todes, 2014b); shaped by concerns of rural settlements and extended, spatially dispersed families (Dar es Salaam; see Mercer, 2017) – these are some of the diverse sub/urbanisms across a continent that remains so neglected in urban studies and urban theory (Myers, 2011; Simone & Pieterse, 2017). The series of observations that the Global Suburbanisms project has stimulated on the African urban context has been hugely important in insisting on placing those experiences both conceptually and methodologically within the frame of the wider conversations of urban studies (Mabin et al., 2013). Thus, with so many diverse starting points for comparative reflections indexed through sub/urbanisms and sub/urbanization, the invitation to exceptionalize African cities is definitively cancelled. This is crucial – perhaps has never been more crucial than at this moment, when a "global" humanitarian project to "leave no cities behind" (Acuto & Parnell, 2016; Parnell, 2016) envisages rolling out infrastructure and housing across the world's poorest urban places, with a land value extraction model increasingly dominant, and in a situation where developmentalist imaginations of how African cities "should be" continue to occlude insights on emergent and transnational processes of urbanization. To understand the dynamics of contemporary African cities, these wider processes are an essential starting point. Certainly, as Bloch (2015) notes, "better urban governance, planning and management will be required if Africa's metropolitan areas, cities and towns are to reap the potential gains from economic growth and urbanization" (p. 258). But as he also observes, with the insertion of Africa's cities into the global economy, and in the context of increasing rates of urbanization, achieving this is far from assured. Such connections to wider processes, though, also indicates that building insights on global experiences of ex-centric urbanization needs to be informed by African experiences as much as by those of other contexts.

Thus, one effect of the Global Suburbanisms project has been to continue to deploy (rather than to jettison) the terms sub/urbanisms and sub/urbanization as referents of urban processes, but to simultaneously insist on their openness to new meanings, and to enable their widest relevance. Using the concept "suburb" in this way embraces variety and difference and has allowed significant conversations across the "world of cities." If you like, a "universal," which can name a phenomenon across different cases, has been assumed, but it has also been cut down to size and opened up to radical revision; a conversation has been enabled that is inclusive of any urban context. In the process, the initial concept of the suburb has been significantly mediated, transformed, thickened, and "clothed" in the variety of cases, perhaps until it looks like something else altogether. In this sense, the framing concept has been lightened sufficiently to enable its meaning to be transformed in its travels, to accumulate meanings, and to emerge rich and complex with potential to play a part in shaping distinctive understandings in different contexts. I have delighted in the insistence of the Global Suburbanisms project to change the meaning of the concept of suburb to definitively parochialize the analytically foundational US experience, and to make the "suburbanization process" stand for the diverse array of processes shaping the expansion and peripheralization of the urban.

Suburban Singularities: Thinking with the Afterlives of Distinctive Suburban Formations

While the call to hold the concept "suburb" as open and revisable is clearly foundational to this comparative project, and while we might well come to the end of the concept as such, as historical formations it is also clear that certain kinds of suburbs remain operative in many cities. This is one of the insights that I have drawn from Phelps and Wu's (2011) contributions on "post-suburbanization," and also from Keil's (2017) reminder that the processes of production and transformation of suburban and peripheral places continue long after the attention of developers, planners, and architects has withered.

Places known as "suburbs" – different in different contexts – thus condition and provide a scene for contemporary urbanization, even as they and the wider urban configurations of which they are a part are being profoundly remade. This brings into view another way to build an open and revisable conceptualization, a more targeted process of "thinking with variety" across a select number of distinctive cases moving beyond a causally narrow form of "variation-finding" as a methodological tactic. Thus, drawing on a recent comparative research project on Johannesburg, South Africa, and London, England,[2] we can consider how configurations of the suburban in particular urban contexts have shaped new rounds of development, demonstrating the persistence and "afterlives" of phenomena known as suburbs, even as the term may seem redundant given current forms of urbanization. Reflecting on this across different contexts provides an opportunity to draw insights from one case to thicken interpretations of another, while contributing again to enriching understandings of the diverse nature and potential meanings of "suburbs," and specifically of their afterlives in current processes of urbanization.

In Johannesburg, post-apartheid efforts to densify urban development and centralize low-income housing, seeking to combat the sprawling and divided form of the apartheid city, revealed the powerful alignment of interests and practices in a long-standing suburban model of development, albeit one divided between rich and poor residents (Butcher, 2016; Mabin et al., 2013; Todes, 2014b). These efforts follow in the tracks of strong path dependencies of sub/urban developments shaped by land-banking practices and relatively stable alliances of actors – developers, banks, and landowners. Together with the spatial organization of land prices, this has somewhat tragically led to the reproduction of apartheid-era residential formations, with not only continuing high-income suburban developments, but also large-scale state-led initiatives for poor housing located on the distant peripheries of the city (Todes, 2014a). Such a configuration of interests, committed to suburban formations, intensifies the challenges and dampens opportunities in the search for new alliances of actors to provide centrally located low-income housing along new transport corridors (Todes & Robinson, 2019).

By contrast, in London, the political intractability of resistance from residents to new development in the "outer suburbs" of London, as well as in the more distant areas in the wider region that might be considered the suburbs proper (the far-flung commuter settlements and small towns across the wider functional city-region), has instigated extreme densification in selected brownfield sites within the core urban area and mass displacement of working-class people as a result. The intensification of development in the core city is often normalized in imaginative references to high densities in Asian cities, but also through inter-referencing dense developments already achieved across London. Dependence on narrow and territorially confined financial instruments of planning gain to fund all aspects of developments including hard infrastructure and social facilities, together with secretive and informalized planning practices, permits and indeed necessitates extreme densities and heights (Robinson & Attuyer, 2020).

Massive changes are afoot in London, then, with substantial "expansion" and "growth" based on predicted population increases and a worsening housing crisis. But this case demonstrates that even the minimal definition of "suburbanization" can be questioned – as urban expansion is here primarily an involution, densifying and destroying not only social housing but also vital inner suburban economies and critical urban infrastructure. The new typologies are also placing in peril long-valued urban qualities of public space, tolerance, and accessibility (Hatherley, 2014). In addition to their mixture of "old Europe" and US suburban forms – as Phelps (2017) explains, a combination of social housing at high density and the "sprawl" of privatized individual homes – London suburbs now add a component of high and dense buildings, which closely match those of rapidly expanding Asian cities both in form and inspiration. Lacking the institutional format (land ownership and centralized state resources) for delivery, however, the outcomes are challenging for the future form of this city, especially in terms of sustaining levels of community and social infrastructure (Robinson & Attuyer, 2020; Shatkin, 2017).

While historical processes of suburbanization shape and condition future urban development in both Johannesburg and London, the form and meaning of "suburbs" are varied across the two contexts and are being differently reconfigured through globally circulating models and practices of urbanization. Whatever a "suburb" might be or become, it is precisely in exposing suburbs as differentiated and unpredictable across different contexts that the value of this ambitious project on global suburbanisms is to be found. To think with the variety of this term is to bring to urban studies a range and a richness, an attention to diversity from which it can never retreat. More than this, though, as the following section explores, the comparative and open perspective pursued by the Global Suburbanisms project suggests that there are productive conversations to be had across such diverse urban contexts.

Interconnections: Sub/urban Futures?

It is time to bid the (conceptual) white picket fence farewell.

– Keil, *Suburban Planet*

If not the afterlives of specific, historical processes of suburbanization, what comes after the suburban? Where are the starting points for thinking the future urban? Not a city, not a centre, not a suburb. Rather, I think, both methodologically and analytically, we need to deconstruct the urban to locate it again. There are potentially many issues to explore in this regard, but here I want to focus on the flows and interconnections that mediate distinctive territorializations of the urban, placing them as "difference" (produced in repetition and connection) rather than simply as "diversity." Keil (2017) brings together some of these concerns in evocative propositions around the idea of the "global suburb," which he suggests is about a "multifarious connectivity" that leads to "new assemblages of the global that surround our cities," replete with "new centralities" (pp. 47–8). He links this to emerging concepts of "extended urbanization" (Brenner & Schmid, 2015; Monte-Mor, 2014) and Lefebvre's (2003) famous hypothesis of the complete urbanization of society. In this context, Keil (2017) suggests that "it is time to bid the (conceptual) white picket fence farewell" (p. 52). "After suburbia," analysis and conceptual vocabularies need to engage with deterritorialized and extended urbanization processes and reconfigure what it means to think with and across the diversity of urban outcomes and experiences (Keil, 2020).

Many writers, including Keil (2017), suggest that understanding new phases of urbanization will involve looking not just from the sprawling (suburban) edges of the city, but from its exteriority too (see also Brenner & Schmid, 2014; Monte-Mor, 2014; Soja & Kanai, 2007). Here, I suggest we need to consider a few different things: (1) the potential for urbanization to be operative and defined at a planetary scale; (2) the many kinds of displaced urbanizations that stretch urban territories across vast areas, often along hard infrastructure corridors, calling for new analyses of the drivers of urbanization (Friedmann, 2020); and (3) the transnational actors and the prolific globalized circuits of ideas, materials, financing, and imaginative inter-referencing involved in shaping urbanization in many contexts. Each of these considerations asks us to take a view of whatever the urban might be from positions of ex-centrism, if not of exteriority. The first two give us a steer as to the future shape of urban areas, posing the question, "Where do we need to start looking for the urban?" As extended territories and operational landscapes of urbanization stretch across or influence potentially any part of the planet, the answer surely lies beyond the suburban (Brenner & Schmid, 2015). But it is the third consideration – the interconnections shaping urban outcomes – that brings into view both the dynamics shaping urbanization (as "concentrated" and "extended") and the nature of the urban territories that are emergent from these processes. From the vantage point of sub/urbanization, Keil (2017) identifies these interconnections as underpinning the future formation of urban territories, with the repetitive differentiation of urban forms produced through diverse transnational circuits expanding and even exceeding ex-centric urban areas.

The imagination of "extended" or "planetary" urbanization was initially linked for Lefebvre (2003) to the industrialized urbanization processes spreading "le tissu urbain" [the urban fabric] and "urban society" across wider areas, including beyond the patchwork metropolitan regions of centres and suburbs. Urbanists have long puzzled over the meaning of the influence of "urban" social and material life across diverse landscapes and contexts – the dam, for Louis Wirth; the urbanized villagers in Southern Africa for Max Gluckman. More recently, this analysis has worked well at a materialist level for South America's resource extraction economies (Monte-Mor, 2014), focusing on the spatial dynamics of economic processes: investment and extraction in areas remote from but closely entwined with urban areas. Generally, though, such a materialist analysis of urbanization requires reconfiguring to account for the many different forms of contemporary socio-economic globalizations drawing on a more rounded, Lefebvrian, heterodox, and non-reductionist Marxism attentive to the diversity of social processes constituting different forms of urbanization (Brenner & Schmid, 2015; Buckley & Strauss, 2016; Goonewardena, 2018; Schmid et al., 2018). Here the import of the Global Suburbanisms project is clear in its proposition, based on research from across the globe, that a wide range of dynamics and processes are shaping emerging ex-centric, extended urban territories. Thus, whatever "urbanization" might entail, it is certainly to be seen as diverse in both process and outcomes; simply focusing on globalizing political economic processes centred on formal capitalist systems will not reveal the diversity and mechanisms of urbanization.

Thus, circuits of finance, investment, and global capitalism are not the only circuits shaping urbanization, and we need to attend also to urbanization processes associated with informal, subaltern, or regional (rather than global) economies, developmental investments, and environmental policies (e.g., Bulkeley, 2010; Denis et al., 2012; Guyer, 2004; Simone, 2011; Parnell, 2016). In addition, the vast investments in infrastructures directly or indirectly driving urbanization, often linked to sovereign loans and aid as much as to private sector actors, form an important dynamic of contemporary urbanization – what Kanai and Schindler (2019) call a "scramble for infrastructure." Moreover, an emerging vocabulary of planetary urbanization would benefit enormously from paying attention to experiences of urbanization in many parts of the world that include circular or return migration and strong urban–rural links. Notably in Africa, a century-long process of extended urbanization – in which town and village formed, in the words of Max Gluckman, "one social field" – has long confounded a territorial view of the urban (see Fox et al., 2017; Mercer, 2017; Potts, 2010).

These dynamics not only register an analysis focused on diverse globalized processes of urbanization but also invite new methodological practices of thinking the urban across the differentiated outcomes they produce. To consider this further, we can begin in what are conventionally the most exceptionalized urban contexts, those in which Africa's approximately 500 million[3] urban dwellers have settled. As in many different regions, the production of new urban territories proceeds as much through informal, irregular, and self-built settlements as through the often widely dispersed satellite and eco-cities, new towns, and tech cities that propose themselves across the landscape (Cain, 2014; Murray, 2017; Shatkin, 2017; van Noorloos & Kloosterboer, 2017; Watson, 2014). In the wake of these and other emerging forms (such as the postcolonial suburb; see Mercer, 2017), Bloch (2015) asks us to reflect again on the meaning of the peripheries of cities in Africa:

> Africa's new suburbs tell a new story about African cities as places of growth and development, a story different from traditional developmentalist literature that commonly portrays African cities in a state of constant, ongoing, ineradicable urban crisis. (p. 274)

A developmentalist imaginary, however, continues to shadow and exceptionalize Africa's new ex-centric urban environments. Let us take that very African phenomenon, the "ghost city," the "(sub)urban fantasy," which is never likely to be built out, which so poorly matches with the identified needs of the many poor urban dwellers (van Noorloos & Kloosterboer, 2017; Watson, 2014). Grandiose and inappropriate, the spectrality of developments and their association with scarcely public governmental agendas seems perhaps typical of informalized and peremptory forms of government. However, by seeing these peripheral developments in Africa as part of wider circuits of investment and design, such exceptionalizing analytical practices need to be questioned.

The comparative research discussed earlier in this chapter, for example, brings into view many "ghost cities," spectral urban forms that can absorb vast resources of investors, states, and residents on the journey to producing nothing. Here, then, the potential of the yet-to-be-inhabited (Chinese) ghost city (Woodworth & Wallace, 2017) and the ghost city features of London's urban development landscape de-exceptionalize forms of African urbanization in relation to questions of spectrality and in terms of the informality and opacity of governance. Thinking with the prolific interconnections that shape urban governance and development can draw these cases into conversation and contribute different perspectives on each.

The "ghost city" is an important feature of London, where "buy to leave" apartments have become something of a scandal in the face of housing shortages – although as most foreign-owned apartments are in fact rented out, only a small percentage of the new build stock (around 3 per cent) is held in this way (De Verteuil & Manley, 2017; Scanlon et al., 2017). The more ghostly phenomenon for London is shared with the investment circuits in some African cities – where planning permission is obtained and itself becomes the asset to be traded, without any construction required to turn a significant profit. This then sometimes forms the basis for further negotiations with the planning system to reduce obligations and extend profitability. Also, the desire to receive maximum return on apartment sales slows down completion rates to stagger bringing apartments to the market. In March 2018, London was estimated to have a ghost city of 277,000 housing units in developments that had received planning permission but had not yet started with construction (Mayor of London, 2019). An additional concern is land banking; large numbers of major landowners are simply holding land, waiting for an opportune moment to even begin seeking planning consent. However, these practices are largely invisible in the urban landscape. Around many African cities, by contrast, empty fields destined for major new developments might be surrounded by barbed wire and highly visible, attracting a ready critique. With ambitious signs to signal the future city, these ghostly landscapes can seem to speak of opaque, perhaps shady deals and endemic uncertainty. However, behind the spectral veneer, this is often a future that is already being traded, shaping other outcomes and developments. Moreover, this future is already active in the present, shaping the desires and hopes of nearby residents (Kanai & Schindler, 2019; Van den Broeck, 2017; Watson, 2014). Returning to London, the hidden, spectral nature of urban development planning is striking; deals are done behind the scenes between planners and developers, and knowledge about the future shape of the city is buried deep within the highly documented and minutely regulated planning process, opaque to all but the well-informed.

Returning then to African examples on this basis, we might find ourselves posing new questions. In her discussion of what came to be a celebrated case of a Chinese-built ghost city on the outskirts of Luanda, Buire (2014) describes how one of the fantasy cities critically highlighted by Watson (2014) came to constitute a significant provision of middle-class housing (albeit reliant on state subsidization of the initially unaffordable price) in a city with extreme challenges of infrastructure and housing provision after a long civil war and extractive state practices. Far from a ghost city, its maintenance in pristine condition has become an important agenda for resident's organizations. Drawing on state input again, a now popular large-scale waterfront development in Luanda has fulfilled multiple functions for a city starved of public space and recreational amenities (Croese, 2018). From a characteristic South African perspective, interested in designing better policies to improve urban life for the poor, Turok (2016) anticipates that these new developments might be turned more systematically towards stronger public benefits. This would require attention to the complexities of transcalar urban governance and a search for lines of influence on key actors to ensure that the value produced through developments might return to improve urban life for existing residents (see also Cain, 2014).

Tracing the connections from ex-centric developments across Africa's urbanizing landscape as part of wider transnational circuits and flows can bring what seem to be distinctive African experiences (spectrality, stalled developments, informality) to enrich analytical repertoires in other contexts. Seeing London, for example, through the heightened informality of its planning process and its vast number of unfulfilled development plans can serve to de-exceptionalize urbanization in Africa. In turn, placing Africa's large-scale developments outside the usual developmentalist perspective on cities across that continent and inside the wider, properly urban politics of negotiating planning gain and governing urban outcomes might in fact open up more developmental possibilities from ambitious new "suburban" developments.

Thinking the Urban "After Suburbia"

Theoretical innovation in understanding the urban might therefore emerge through tracing the genesis of emergent urban forms, practices, and imaginations – the multiple interconnections that make up whatever the urban might become. Thus, the urban is emergent from a multiplicity of interconnected circuits, and from the territories in which urbanization unfolds. The territorial outcomes of urbanization are diverse, if often characterized by repeated elements. Diversity is shaped by historic distributions of land-use, opportunities for value creation and extraction, as well as the lived realities of social and economic practices. To think the urban, then, is to be willing to trace the genesis of distinctive territorial outcomes (if perhaps composed of repeated/differentiated instances) emergent in and through an interconnected set of socio-economic flows and relations.

Here, I want to return to the effect of the suburban as a conceptual innovator. We have been invited by the Global Suburbanisms project to start thinking the urban from (different) sub/urban spaces and sub/urbanization processes. I have explored in this chapter how on this basis we have been encouraged to consider the urban in its full diversity, following the tracks of extension, connection, and exteriority, as well as to build insights from distinctive territorial outcomes or singularities to incite new and different starting points for thinking the urban. These conceptual and methodological innovations do not prejudge the content of urban outcomes, urbanisms, or urbanization processes, nor the territorial forms of the urban, the relative significance of the political economy in shaping urban outcomes, or the ways in which diverse characteristics of urban life are constituted in and through particular lived realities and social or political identities. Rather, this initiative has insisted that while conversations about the urban remain vital and necessary, the starting points for thinking the urban or tracing urbanization processes might be anywhere, including the expanding peripheries, edges, and extended forms of the urban – indeed, they might be anywhere on the planet (Brenner & Schmid, 2015; Keil, 2017). This does not foreclose on a deep awareness of the incapacity of theory or concepts to exhaust the fullness of the world, nor does it suggest that there might not be other intellectual resources that could be brought to bear on urbanization processes and urban territories.

In this approach, then, singularities, or distinctive outcomes, are not stopping points for conceptualization but starting points for the kinds of modest universals – ideas, concepts – that can open themselves to necessary difference, as we have seen here with the concept of the suburb. Concepts certainly have ending points – moments when they are no longer useful – and this is vital to make way for new ways of thinking and engaging with a changing, diverse, and differentiated urban world. The invitation of the Global Suburbanisms project has been to profoundly recast the meaning of the term "suburb." Certainly, there is scope to go further than that and invent new concepts (Robinson, 2022; Zhao, 2020). Inspired by Deleuze (1994), a researcher confronting the multiplicity of an emergent (urban) world would initially turn to available concepts to seek to make sense of their (multiplicity of) observations. However, following Nietzsche, Deleuze imagines the moment of determination ("What is the phenomenon I am observing?") as a moment of conceptual innovation.

A metaphor illuminates this: The spectacular throw of the dice against the sky of chance dramatizes the moment of conceptual innovation. Each throw, each determination, opens to the possibility of all of chance – in our case, all that it might be possible to think about the urban. This symbolizes the virtual, here portrayed in Deleuze's (1994) reading of the Nietzschean "eternal return" – each return is different in this game of chance. In a post-representational world, certainly, the dice falls back each time distinctively, but it has many falling stars to keep it company. In terms of our engagement with urban outcomes and concepts, then, each instance, each conceptualization, has many closely related or relevant phenomena or analyses across which it is worth having a conversation. This is not to invite a purely micro-descriptive approach: To explore "the urban" is to potentially seek to understand some significantly "big" or "extensive" processes, and requires attention to practices and methods that speak beyond specific cases or territories as pure diversity (Jacobs, 2006, 2012).

If urban studies is to realize the potential of such conceptual genera-tivity, though, it will need to address the uneven presence of different urban situations and differently located urban scholars in theorization. This will require a renewal and transformation of institutional cultures and practices. Generating concepts of the urban, starting anywhere, should be characterized by respectful, open conversations able to engage across the insights of writers from a very wide range of theoretical tradi-tions and geographical contexts. But this means attending to the chal-lenges of working in a wide range of languages, and addressing extreme inequalities in the institutional and personal resources of scholars in dif-ferent contexts (Ferenčuhová, 2016; Parnell & Pieterse, 2016). The Global Suburbanisms project has demonstrated that this is both a priority for taking urban studies forward, and a feasible, productive enterprise.

NOTES

1 See also Topalov's (2017) fascinating account of the emergence and circulation of urban terminology.
2 I acknowledge funding from the Economic and Social Research Council (ESRC) for an Urban Transformations Grant (no. ES/N006070/1), "Governing the Future City: A Comparative Analysis of Governance Innovations in Large-Scale Urban Developments in Shanghai, London, Johannesburg," with Fulong Wu (University College London), Phil Harrison (University of the Witwatersrand), Ning Yuemin (East China Normal University), Allan Cochrane (The Open University), Jie Shen (Fudan University), and Alison Todes (University of the Witwatersrand).
3 This estimate is based on data from 2018. According to the UN Department of Economic and Social Affairs (2019), this number is projected to reach 1.48 billion by 2050.

REFERENCES

Acuto, M., & Parnell, S. (2016). Leave no city behind. *Science, 352*(6288), 873. http://doi.org/10.1126/science.aag1385.

Beall, J., Parnell, S., & Albertyn, C. (2015). Elite compacts in Africa: The role of area-based management in the new governmentality of the Durban city-region. *International Journal of Urban and Regional Research, 39*(2), 390–406. https://doi.org/10.1111/1468-2427.12178.

Benjamin, W. (2009). *The origin of German tragic drama.* Verso.

Bloch, R. (2015). Africa's new suburbs. In P. Hamel & R. Keil (Eds.), *Suburban governance: A global view* (pp. 253–77). University of Toronto Press.

Brenner, N., & Schmid, C. (2014). The "urban age" in question. *International Journal of Urban and Regional Research, 38*(3), 731–55. http://doi.org/10.1111/1468-2427.12115.

Brenner, N., & Schmid, C. (2015). Towards a new epistemology of the urban? *City, 19*(2–3), 151–82. http://doi.org/10.1080/13604813.2015.1014712.

Buckley, M., & Strauss, K. (2016). With, against and beyond Lefebvre: Planetary urbanization and epistemic plurality. *Environment and Planning D: Society and Space, 34*(4), 617–36. https://doi.org/10.1177/0263775816628872.

Buire, C. (2014). The dream and the ordinary: An ethnographic investigation of suburbanisation in Luanda. *African Studies, 73*(2), 290–312. https://doi.org/10.1080/00020184.2014.925229.

Bulkeley, H. (2010). Cities and the governing of climate change. *Annual Review of Environment and Resources, 35*, 229–53. https://doi.org/10.1146/annurev-environ-072809-101747.

Butcher, S. (2016). *Infrastructures of property and debt: Making affordable housing, race and place in Johannesburg* [Doctoral dissertation, University of Minnesota]. University of Minnesota Digital Conservancy. https://hdl.handle.net/11299/182191.

Cain, A. (2014). African urban fantasies: Past lessons and emerging realities. *Environment and Urbanization, 26*(2), 561–7. https://doi.org/10.1177/0956247814526544.

Caldeira, T. (2016). Peripheral urbanization: Autoconstruction, transversal logics, and politics in cities of the global south. *Environment and Planning C, Society and Space, 35*(1), 3–20. https://doi.org/10.1177/0263775816658479.

Croese, S. (2018). Global urban policymaking in Africa: A view from Angola through the redevelopment of the Bay of Luanda. *International Journal of Urban Regional Research, 42*(1), 198–209. https://doi.org/10.1111/1468-2427.12591.

Deleuze, G. (1994). *Difference and repetition.* Continuum.

Denis, E., Mukhopadhyay, P., & Zérah, M-H. (2012). Subaltern urbanisation in India. *Economic and Political Weekly, 47*(30), 52–62. https://halshs.archives-ouvertes.fr/halshs-00743051.

De Verteuil, G., & Manley, D. (2017). Overseas investment into London: Imprint, impact and pied-à-terre urbanism. *Environment and Planning A, 49*(6), 1308–23. https://doi.org/10.1177/0308518X17694361.

Ekers, M., Hamel, P., & Keil, R. (2012). Governing suburbia: Modalities and mechanisms of suburban governance. *Regional Studies, 46*(3), 405–22. https://doi.org/10.1080/00343404.2012.658036.

Ferenčuhová, S. (2016). Accounts from behind the curtain: History and geography in the critical analysis of urban theory. *International Journal of Urban and Regional Research, 40*(1), 113–31. https://doi.org/10.1111/1468-2427.12332.

Fox, S., Bloch, R., & Monroy, J. (2017). Understanding the dynamics of Nigeria's urban transition: A refutation of the "stalled urbanisation" hypothesis. *Urban Studies, 55*(5), 947–64. http://doi.org/10.1177/0042098017712688.

Friedmann, J. (2020). Thinking about mega-conurbations and planning. In D. Labbé and A. Sorensen (Eds.), *Handbook of megacities and megacity-regions* (pp. 21–32). Edward Elgar Publishing.

Goodfellow, T. (2017). Urban fortunes and skeleton cityscapes: Real estate and late urbanization in Kigali and Addis Ababa. *International Journal of Urban and Regional Research, 41*(5), 786–803. https://doi.org/10.1111/1468-2427.12550.

Goonewardena, K. (2018). Planetary urbanization and totality. *Environment and Planning D: Society and Space, 36*(3), 456–73. https://doi.org/10.1177/0263775818761890.

Guyer, J. (2004). *Marginal gains: Monetary transactions in Atlantic Africa.* University of Chicago Press.

Harris, R. (2010). Meaningful types in a world of suburbs. In M. Clapson & R. Hutchison (Eds.), *Suburbanization in global society* (pp. 15–47). Emerald.

Harris, R., & Vorms, C. (Eds.). (2017). *What's in a name? Talking about urban peripheries.* University of Toronto Press

Hatherley, O. (2014, August 21). London's new typology: The tasteful modernist non-dom investment. *Dezeen.* https://www.dezeen.com/2014/08/21/owen-hatherley-london-housing-typology-yuppie-flats-tasteful-modernist-investment/.

Hirt, S. (2012). *Iron curtains: Gates, suburbs and privatization of space in the post-socialist city.* Wiley-Blackwell.

Jacobs, J. (2006). A geography of big things. *Cultural Geographies, 13*(1), 1–27. https://doi.org/10.1191/1474474006eu354oa.

Jacobs, J. (2012). Commentary: Comparing comparative urbanisms. *Urban Geography, 33*(6), 904–14. https://doi.org/10.2747/0272-3638.33.6.904.

Kanai, J. M., & Schindler, S. (2019). Peri-urban promises of connectivity: Linking project-led polycentrism to the infrastructure scramble. *Environment and Planning A: Economy and Space, 51*(2), 302–22. https://doi.org/10.1177/0308518X18763370.

Karaman, O. (2013). Urban neoliberalism with Islamic characteristics. *Urban Studies, 50*(16), 3412–27. https://doi.org/10.1177/0042098013482505.

Keil, R. (2017). *Suburban planet: Making the world urban from the outside in.* Polity Press.

Keil, R. (2020) After suburbia: Research and action in the suburban century. *Urban Geography, 41*(1), 1–20. https://doi.org/10.1080/02723638 .2018.1548828.

Lefebvre, H. 2003. *The urban revolution.* University of Minnesota Press.

Mabin, A., Butcher, S., & Bloch, R. (2013). Peripheries, suburbanisms and change in sub-Saharan African cities. *Social Dynamics, 39*(2), 167–90. https://doi.org/10.1080/02533952.2013.796124.

Mayor of London. (2019). *Housing in London 2019: The evidence base for the Mayor's Housing Stategy.* Greater London Authority. https://data.london .gov.uk/dataset/housing-london.

Mercer, C. (2017). Landscapes of extended ruralisation: Postcolonial suburbs in Dar es Salaam, Tanzania. *Transactions of the Institite of British Geographers, 42*(1), 72–83. https://doi.org/10.1111/tran.12150.

Monte-Mor, R.L. (2014). Extended urbanization and settlement patterns in Brazil: An environmental approach. In N. Brenner (Ed.), *Implosions/ explosions: Towards a study of planetary urbanization* (pp. 109–20). Jovis.

Murray, M. (2017). *The urbanism of exception: The dynamics of global city building in the twenty-first century.* Cambridge University Press.

Myers, G. (2011). *African cities: Alternative visions of urban theory and practice.* Zed Books.

Ouředníček, M. (2016). The relevance of "Western" theoretical concepts for investigations of the margins of post-socialist cities: The case of Prague. *Eurasian Geography and Economics, 57*(4–5), 545–64. https://doi.org /10.1080/15387216.2016.1256786.

Parnell, S. (2016). Defining a global urban development agenda. *World Development, 78,* 529–40. https://doi.org/10.1016/j.worlddev.2015.10.028.

Parnell, S., & Pieterse, E. (2016). Translational global praxis: Rethinking methods and modes of African urban research. *International Journal of Urban and Regional Research, 40*(1), 236–46. https://doi.org/10.1111 /1468-2427.12278.

Phelps, N. (Ed.). (2017). *Old Europe, new suburbanization? Governance, land and infrastructure in European suburbanization.* University of Toronto Press.

Phelps, N., Mace, A., & Rodieri, R. (2017). City of villages? Stasis and change in London's suburbs. In N.A. Phelps (Ed.), *Old Europe, new suburbanization? Governance, land and infrastructure in European suburbanization* (pp. 183–206). University of Toronto Press.

Phelps, N., & Wu, F. (2011). *International perspectives on suburbanization: A post-suburban world?* Palgrave Macmillan.

Potts, D. (2010). *Circular migration in Zimbabwe and contemporary sub-Saharan Africa.* James Curry.

Robinson, J. (2011). Cities in a world of cities: The comparative gesture. *International Journal of Urban and Regional Research, 35*(1), 1–23. https://doi.org/10.1111/j.1468-2427.2010.00982.x.

Robinson, J. (2016a). Comparative urbanism: New geographies and cultures of theorizing the urban. *International Journal of Urban and Regional Research, 40*(1), 187–99. https://doi.org/10.1111/1468-2427.12273.

Robinson, J. (2016b). Thinking cities through elsewhere: Comparative tactics for a more global urban studies. *Progress in Human Geography, 40*(1), 3–29. https://doi.org/10.1177/0309132515598025.

Robinson, J. (2022). *Comparative urbanism: Tactics for global urban studies.* John Wiley & Sons.

Robinson, J., & Attuyer, K. (2020). Extracting value, London style: Revisiting the role of the state in urban development. *International Journal of Urban and Regional Research, 45*(2), 303–31. https://doi.org/10.1111/1468-2427.12962.

Sawyer, L. (2014). Piecemeal urbanisation at the peripheries of Lagos. *African Studies Quarterly, 73*(2): 271–89. https://doi.org/10.1080/00020184.2014.925207.

Scanlon, K., Whitehead, C., Blanc, F., & Moreno-Tabarez, U. (2017). *The role of overseas investors in the London new-build residential market: Final report for Homes for London.* LSE London. https://www.lse.ac.uk/business/consulting/assets/documents/the-role-of-overseas-investors-in-the-london-new-build-residential-market.pdf.

Schmid, C., Karaman, O., Hanakata, N., Kallenberger, P., Kockelkorn, A., Sawyer, L., Streule, M., & Wong, K.P. (2018). Towards new vocabularies of urbanization processes: A comparative approach. *Urban Studies, 55*(1), 19–52. https://doi.org/10.1177/0042098017739750.

Shatkin, M. (2017). *Cities for profit: The real estate turn in Asia's urban politics.* Cornell University Press.

Simone, A. (2011). The surfacing of urban life. *City, 15*(3–4), 355–64. https://doi.org/10.1080/13604813.2011.595108.

Simone, A. (2018). *Improvised lives.* Polity Press.

Simone, A., & Pieterse, E. (2017). *New urban worlds: Inhabiting dissonant times.* Polity Press.

Soja, E., & Kanai, M. (2007) The urbanization of the world. In R. Burdett & D. Sudjic (Eds.), *The endless city* (pp. 54–69). Phaidon.

Todes, A. (2014a). The impact of policy and strategic spatial planning. In P. Harrison, G. Gotz, A. Todes, & C. Wray (Eds.), *Changing space, changing city: Johannesburg after apartheid* (pp. 83–100). Wits University Press.

Todes, A. (2014b). New African suburbanisation? Exploring the growth of the northern corridor of eThekwini/KwaDakuza. *African Studies*, 73(2), 245–70. https://doi.org/10.1080/00020184.2014.925188.

Todes, A., & Robinson, J. (2019). Re-directing developers: New models of rental housing to re-shape the post-apartheid city? *Environment and Planning A: Economy and Space*, 52(2), 297–317. https://doi.org/10.1177/0308518X19871069.

Topalov, C. (2017). The naming process. In R. Harris & C. Vorms (Eds.), *What's in a name? Talking about urban peripheries* (pp. 36–67). University of Toronto Press.

Turok, I. (2016). Getting urbanization to work in Africa: The role of the urban land-infrastructure-finance nexus. *Area Development and Policy*, 1(1), 30–47. http://doi.org/10.1080/23792949.2016.1166444.

UN Department of Economic and Social Affairs. (2019). *World urbanization prospects 2018: Highlights*. https://population.un.org/wup/Publications/Files/WUP2018-Highlights.pdf.

Van den Broeck, J. (2017). "We are analogue in a digital world": An anthropological exploration of ontologies and uncertainties around the proposed Konza Techno City near Nairobi, Kenya. *Critical African Studies*, 9(2), 210–25. https://doi.org/10.1080/21681392.2017.1323302.

van Noorloos, F., & Kloosterboer, M. (2018). Africa's new cities: The contested future of urbanization. *Urban Studies*, 55(6), 1223–41. https://doi.org/10.1177/0042098017700574.

Watson, V. (2014). African urban fantasies: Dreams or nightmares? *Environment and Urbanization*, 26(1), 1–17. https://doi.org/10.1177/0956247813513705.

Woodworth, M., & Wallace, J. (2017). Seeing ghosts: Parsing China's "ghost city" controversy. *Urban Geography*, 38(8), 1270–81. https://doi.org/10.1080/02723638.2017.1288009.

Zhao, Y. (2020). *Jiehebu* or suburb? Towards a translational turn in urban studies. *Cambridge Journal of Regions, Economy and Society*, 13(3), 527–42. https://doi.org/10.1093/cjres/rsaa032.

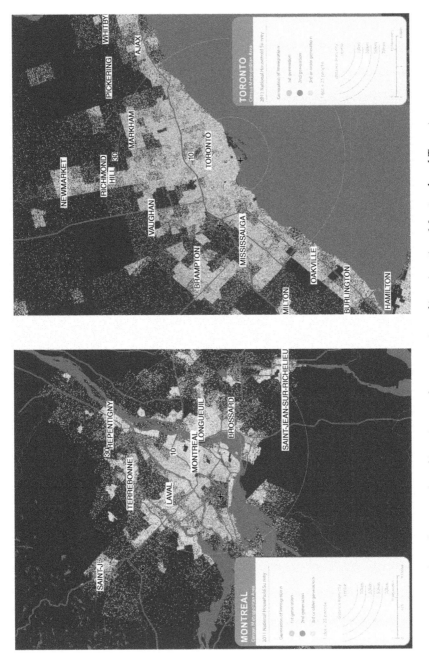

Image 16. Residential geography of immigrants by generation of immigration, Montreal and Toronto

16 Africa's Suburban Constellations

ROBIN BLOCH, ALAN MABIN, AND ALISON TODES

Introduction

Burgeoning interest in African "suburbanisms" and growth on urban peripheries has been sparked by a recognition of the remarkable scale and speed at which urbanization and urban spatial expansion have occurred across the continent this century (Bloch, 2015; Lall et al., 2017; Mabin et al., 2013). In broad terms, sub-Saharan Africa's (SSA) urban population has doubled since the mid-1990s and reached almost 400 million in 2016. The share of the urban population (or urbanization level) rose from 31 per cent in 2000 to 40 per cent in 2017.[1] By 2040, it is projected that this urban population will more than double to 1 billion.

Urban population growth has been accompanied by rapid spatial expansion. African cities are typically growing faster in size (that is, in spatial extent) than in population. One recent study found that in the period from 1990 to 2014, a population increase of 4 per cent per annum was outpaced by a 5 per cent growth of the built-up area in a sample of twenty-five African cities (Xu et al., 2019). The Organisation for Economic Co-operation and Development's (OECD) database on built-up area change reveals a doubling or even tripling of the built-up area for the same period in most fast-urbanizing African countries – an increase from 5,402 to 12,218 square kilometres in Nigeria, for instance, and from 727 to 2,045 square kilometres in Ethiopia (OECD, n.d.).[2]

Most population growth and new habitation is located on urban peripheries (or on what were once urban edges). Africa is now characterized by the growth of very large cities and their regions, and by urban primacy.[3] In consequence, immense, extended, often multi-centred urban regions and corridors are emerging across the continent, which combine metropolitan, city, town, and rural settlements into novel spatial formations. These vast, continuous, fragmented, and often uncontained urban fields typically manifest low income levels, high inequalities, and informal social, residential, and economic conditions (Bloch, 2015, 2016).[4] But it is also important to emphasize that population growth and expansion are not confined to the largest cities: Nearly three-quarters of the region's overall urban population lives in urban areas outside the largest city of each country (Hommann & Lall, 2019). Moreover, smaller cities can often feature even lower densities and more dispersed urban forms than medium-sized and large cities, which are themselves marked by relatively less compact urban forms than their comparators in other world regions (Xu et al, 2019).

Studies in several African cities point to the significance of rapid urbanization, economic growth, and the expansion of the urban middle classes, moving beyond an analysis of peripheral growth linked simply to poverty and informality (Bloch, 2015). Major flows of investment into property and the built environment as a source of profitability have also been key in some contexts (Goodfellow, 2017; Todes, 2014). Forms of growth and their underlying dynamics vary. For instance, in the case of Nigeria, Sawyer (2014) argues that both middle- and low-income residents are moving out of the increasingly expensive core areas of Lagos, where space is limited, in search of more affordable accommodation. "Piecemeal urbanization" in the form of single-family homes is being developed by owners on unplanned and un-serviced land, without access to mortgage finance. Developers also build rental housing for a range of income groups. In addition, there are some private estate developments sanctioned by the state offering varied levels of infrastructure and services. These forms of housing, however, are only accessible to those who have dependable employment and can manage the long and costly commutes (around three hours each way) to central areas, and in some cases the maintenance of rented sleeping spaces there (Sawyer, 2014).

Some similar dynamics are evident in Andreason et al.'s (2017) study of Dar es Salaam, where urban peripheries offer the possibility of building single owner–occupied homes (as well as other types of homes) affordable to a range of income groups. New occupiers mostly come from central areas of the city, and breadwinners frequently undertake long commutes to the city centre to work. Although services are often limited or lacking, the possibility of home ownership, and of building incrementally in a context where mortgage finance is absent, makes this type of development attractive. While Sawyer (2014) argues that the "suburban" concept has little popular purchase in Lagos, Andreason et al. (2017) argue that, although processes in Dar es Salaam are certainly not always guided by planners and are occurring informally – a tendency seen in many other countries and cities in Africa and in the world – suburbanization better describes the dynamics of peripheral expansion than peri-urbanization, which focuses on the rural–urban interface.

By contrast, Buire (2014) documents the development of new state-led housing for low-income (Panguila) and middle-class groups (Kilamba) some twenty to thirty kilometres from the city centre of Luanda, Angola. These forms of "tabula rasa urbanism" have different histories, with Panguila initially set up to accommodate those relocated from informal housing in the city centre to make space for state and private developments, but now accommodating a much broader demographic, while Kilamba, intended to develop as a new centrality, largely offers subsidized rent-to-buy apartments for public servants and the formally employed who cannot access the formal market. Poor roads and traffic mean very long commuting times, with work still concentrated in central areas, so some commute weekly or sublet. However, both developments are increasingly seen as desirable since they offer more space, homeownership, and better services. Buire (2014) argues that an "orderly" suburban way of life is emerging, reflecting social status and aspirations of belonging to a new middle class, which is very different from older urban practices in Angola.

New Meanings for African Suburbs and Suburbanisms

These developments call for a conceptual response. This swiftly changing urban terrain has framed and stimulated the recent upsurge of research on African suburbs (Andreason et al., 2017; Bloch, 2015; Buire, 2014; Sawyer, 2014; Todes, 2014). Researchers have begun to ask the following key questions: In what ways does contemporary Africa have suburbs? What kinds of suburbanisms can be identified in Africa? Or are these terms simply impositions from the Global North and Western world?

The indirect answer is that in cities that grow quickly, most expansion will be in places that are new relative to original sites and centres. Older areas will also experience many changes, among them increasing densities and changing uses of space. The term "suburban" applies both to many "recent" areas of expansion and to many "older" areas simply conceptualized as "in-between" old centres and new peripheries. Both, of course, exist in every African city, so again, Africa's cities have "suburbs" and "suburbanisms" associated with them. Some of them may even look like or feel like "suburbs" on other continents. But the diversity, not to mention recent alterations of so much African city space and the growth of the population, as seen above, means that many areas of African cities are unlike traditional notions of "suburbs," and that Africa now reveals both global as well as locally specific forms.

When reviewing the literature on African suburbanisms several years ago, Mabin et al. (2013) concluded that, "underpinned by wider economic and social dynamics such as economic growth and the rise of a consuming middle class, varied forms of suburbanisms are now combining to provide distinctively new terrain for urban life in Africa" (p. 16; see also Bloch, 2015). The literature suggested that the concepts of "suburb" and "suburbanisms" might be deployed as researchers grapple with multiple conceptualizations and experiences of "suburb," "periphery," and "peri-urban" across languages and places in SSA cities.

Three socio-spatial processes were emphasized in Mabin et al.'s 2013 review: (1) *a moving edge* in relation to political, social, and economic dynamics and pressures; (2) *the accelerating densification/retrofitting* of existing suburbs; and (3) *the emergence of new centralities* away from the traditional cores that related to new performances of city life in African cities. As the literature has rapidly expanded over the past decade, some authors have worked with a more fluid notion of the relationship between formal and informal habitats – and at the same time have refused African cities any exceptional status (Myers, 2011; Parnell & Pieterse, 2014).

Indeed, as Africa's cities have quickly changed, they have exhibited new urbanisms and new forms. Rapid change is especially concentrated in three kinds of places (cf. Mabin et al., 2013). First, in most accounts, are city peripheries, where the constant movement of urban frontiers is not merely extending the existing city but creating new configurations and spaces for different urbanisms. Second, older areas between "original" centres and current edges reveal very rapid changes of populations, densities of people and buildings, and activities. And third, new centralities have emerged, since both peripheral expansion and redevelopments in older areas destabilized former centralities as they remade patterns of urban life and movement.

Movement of people appears to be constantly ebbing and flowing between multiple spaces – between country and city – and cities, too, reveal an internal ebb and flow as they continue to grow. This complexity of movement is especially well-illustrated at the peripheries of cities. Some residents find themselves incorporated there as cities expand horizontally and their villages or customary settlements become part of a continuous fabric. Some arrive there from parts of the countryside and seek places to stay around the weakly defined urban edges (Harris, 2010). But many coming from the countryside seek urban insertion not at the periphery but within the older quarters (as in Kariakoo in Dar es

Salaam, central informal settlements like Nairobi's Kibera, or inner-city Johannesburg). At the same time, the periphery is a site of opportunity for longer-term urban residents, such as was the case of Ndogbassi, a new "planned slum" built on the outskirts of Douala in the mid-1970s. Elate (2004) explains how residents from the city of Douala, not the rural areas, relocated to Ndogbassi to gain for themselves ownership of a house – and this is but one of many such cases.

Spaces "inside" the cities also offer various opportunities, where newer arrivals of various origins compete and collaborate with those who are longer established, producing growing numbers of centralities. Some centralities are older. Some are very new. Within this flux, words such as "suburb," peri-urban, *subúrbio*, township, informal settlement, camp, and others in many more languages than English jostle for meaning (as the 2017 collection of Harris and Vorms has shown in a variety of contexts).

Indeed, it is provocative to observe uses of terms close to "suburb" or those given sometimes as equivalents in other languages. English approaches the "suburb" from the centre, but there are other linguistic trajectories. Areas for which "suburb" is used in English are often approached in African expression from *outside* the city rather than from within. The parts that come *before* the city replace the English implications embedded in "suburb" of moving *away from* the core city and, indeed, *towards* the countryside (Mabin, 2013). In isiZulu, one might hear about the "iphethelo ledolobha" as the borders and outskirts of the city, a term also used in general to denote the suburbs – although it is also common in some places for Zulu-speakers to talk of "lisabhebe" (obviously derived from the English word).

There is also a tangled web of meanings associated with French terms such as *cité* versus *ville, banlieue*, and *faubourg*, amongst others (Topalov et al., 2010). Likewise, the Portuguese word *subúrbio* has diverse histories and uses on at least three continents; it has had implications of periphery, the non-urban, and a hierarchy of social division of space. According to da Silva Pereira (2010), in Brazilian usage (sometimes influential in shaping language in Angola and Mozambique), *subúrbio* has come to be an "imprecise expression used everywhere to indicate the quarters [*bairros*] which do not appear on maps and are usually forgotten by public authorities" (p. 1203, our translation). During the colonial period, the word *subúrbios* emerged in reference not to the "white," planned "cement city," but rather to the informal areas of African residence on the periphery, where land was rented from elite settler landholders (Jenkins, 2009) and allocated through chiefly systems.

This "model of the colonial town" has persisted in some of the older spaces, while simultaneously another model of the urban is conceived "at its margins," where, as Roque (2009) has shown in the case of Luanda and its *musseques*, or informal settlements, material conditions can be precarious and spatial organization has not followed any specific urban plan (see also Morais & Raposo, 2005; Oppenheimer & Raposo, 2007).

"Suburbs" and peripheries take as many forms in various African contexts as they do elsewhere, perhaps even more. Older versions include townships and *bidonvilles*, housing estates for colonial officials as well as all kinds of self-built city zones. They include spaces that concentrate new economic activities and zones of middle- and upper-income residence (some of them gated), and they reveal diverse meanings of informality in building, land markets, and social activity.

Suburban growth has many drivers, shaped, as it is, by policy and institutional mechanisms that try to direct urban growth (and the reality of what happens in practice – which often shows very little impact of planning at all). The key actors involved in African suburban growth are, of course, also diverse – property developers, landowners, traditional authorities, administrators, households, associations, politicians – with perhaps a larger influence in many cases from far away (in the past, European colonial offices, nowadays China and India, for example) than is associated with local suburban development in some other parts of the world.

African city spaces now contribute an expanding part of global suburbanisms, as the number of city dwellers increases. These spaces are also increasingly complex. There is often no longer one centre, nor one periphery. What is interesting about the more central quarters is that they are changing, sometimes very fast – for example, low-density residential buildings that are fast becoming high density (such as eight-floor buildings where simple Swahili houses once stood in many areas of Dar es Salaam, or office buildings in former bungalow areas of Durban).

The term "suburban" can be negatively applied, however, to mean "less than" urban. Statements like this would potentially denigrate most space in Africa's cities as not-yet, not-fully, or not-adequately urban. Scholarly authors addressing suburbanisms in Africa thus avoid creating a morality of urbanisms and suburbanisms – there is a sense of searching for "the city yet to come" (Simone, 2004). But the suburb has also come to mean something about the *new* in acts of claiming the periphery. The probability is that "peripheral" spaces and quarters are inhabited by those most creative in music, art, and new forms of expression and being (da Silva Pereira, 2010). And the energy of the newer, outer areas may be an important sign of changes in spatialities of creativity and innovation, as various suburbanisms coalesce to provide a distinctively new arena for urban life in Africa.

New Perspectives

The research challenge has been taken up with enthusiasm in recent years by a range of scholars and analysts. A special 2014 issue of *African Studies* on African suburbanisms, as well as several reports, book chapters, and journal articles linked to the Global Suburbanisms project have explored forms of growth on the urban edge in African cities (e.g., Bloch, 2015; Buire, 2014; Rubin & Charlton, 2019; Sawyer, 2014; Todes, 2014). Questions of peripheral urban expansion also attracted academic attention from other directions (e.g., Melo, 2016), with some responding to debates over the usefulness of the suburbanization concept itself (Andreason et al., 2017).

The South African context, with its history of apartheid "townships" to which most Black people were confined on the urban edge, at some distance from economic centres, and with a growing number of gated communities, has provided fruitful material for studies of African suburbanisms and urban peripheries, although rates of urbanization and urban growth are relatively low compared to many other African cities. Recent studies have explored how and why large tracts of free, state-provided, mostly detached housing for the poor are largely continuing to occur on the periphery, despite policy aimed at densification and compaction (Charlton, 2014); the growth of forms of "auto-construction" (Caldeira, 2017) through backyard housing, especially in better located low-income settlements (Rubin & Charlton, 2019); the emergence of new mixed-income housing developments led by the private sector, but involving public–private collaboration in terms of finance and land access (Rubin & Harris, 2018); new, very large gated estates for the wealthy taking the form of "privatized urbanism" (Herbert & Murray, 2015); and major developments driven initially by large landowners in the north of Durban to realize higher values in urban land over the past two decades, and later by provincial initiatives to develop a new airport and major low-income housing projects in the area (Todes, 2014, 2017).

Research on peri-urbanization continues to develop new insights, particularly with regard to development on traditional lands that are now part of towns and cities in several African cities. Bartels (2020), for instance, argues that these are not simply spaces of poverty but places of middle-class development, such as in Ghana, for example. Mbatha and Mchunu (2016) make a similar point in the South African context, and Potts (2017) argues that the incorporation of existing villages and denser peri-urban areas into existing settlements forms a major part of urban growth.

There is also increasing interest in understanding how "transformation in the spatial peripheries of … African city-regions is shaped, governed and experienced" (Meth et al., 2021, p. 30. In a contemporary project focused on five different types of urban periphery in Gauteng (the city-region centred on Johannesburg) and Durban, South Africa, and two in Addis Ababa, Ethiopia, research sites reveal much about contemporary African urban peripheries. These sites include places of older and newer lower- and middle-income areas close to new spaces of infrastructure investment; traditional authority areas close to formal commercial and residential development; a "mega human settlement" on the edge of a former Black township; places with histories of relocation, "displaced urbanization," and industrial decentralization in the South African context; and both more established centres on the urban periphery and new areas of condominium development close to infrastructure development in Addis Ababa. The research explores the drivers and dynamics of change and governance in these contexts, and peoples' experiences of these places through social surveys, diaries, and interviews. The project aims to extend conceptualizations of urban peripheries and the different types of peripheries and change (for more, see Meth et al., 2021).

Planning for "new cities" in African contexts, many of which take the form of satellite cities, has also been a growing focus of research (Cote-Roy & Moser, 2019; Van Noorloos & Kloosterboer, 2018; Watson, 2013), with over seventy new city projects noted in a recent study (Falt, 2019). Inspired, in part, by new cities emerging in the Middle East and Asia, and overlaid with discourses of "smart cities" and "eco-cities," such planning and the developments that have resulted are linked to a broader resurgence in interest in new cities as a way to manage urban growth, escaping the complexity of responding to the challenges of existing dense, congested, and poorly serviced cities. Researchers argue that the "new city turn" reflects an "Africa rising" narrative based on the "(perceived) recent revival of African economies, and the assumption that African markets are poised for unprecedented growth" (Cote-Roy & Moser, 2019, p. 2395). New cities are being promoted by entrepreneurial states to attract international investment by re-imaging their countries as modern progressive places where "high tech and eco-cities are possible," as is the case in Kenya (p. 2402). Authors also speculate on the links to international processes of financialization, with investors seeking potentially profitable real estate markets following the 2008 global financial crisis; Chinese and Russian investments have also been important in some contexts, and some developments also include significant local funding (Van Noorloos & Kloosterboer, 2018).

The relative importance and roles of states and private developers vary contextually (Bloch, 2015). Private developments predominate, but state actors often play key roles in promoting new city ideas and facilitating development, and sometimes the state's role is even more direct, as, for instance, in both Luanda (Buire, 2014) and Addis Ababa, where major condominium developments have been developed by their respective states. While many proposals might be merely speculative, some are being realized, such as in Appolonia City outside of Accra (Falt, 2019). Most developments of this sort are focused on the wealthy, or at least the middle classes, despite discourses of greater inclusion. Developments often intend to be mixed use and multi-functional, but most are still largely residential and associated with long commutes to more central areas (Van Noorloos & Kloosterboer, 2018).

The idea of "new cities" has increasingly seized the imagination of policymakers in South Africa. In Gauteng, "mega human settlements" – developments of over 16,000 housing units, with a mix of incomes – have been proposed since 2015, in part as an alternative to the failure of policy to restructure apartheid cities (see Ballard & Rubin, 2017). In practice, developments of this sort have been wanting. Charlton (2017) shows that Lufhereng, a Johannesburg development in this mould, has been slow to develop, includes a narrow range of largely low- and lower middle-income housing (while high-income "new cities" exist in distant separate spaces), and still has limited social infrastructure and transport. Moreover, planned economic development has not been realized, resulting in a reliance on commuting. Despite extensive criticism of the concept (see, for example, the 2017 special issue of *Transformation: Critical Perspectives on Southern Africa*), "new cities" formed a central part of the June 2019 presidential State of the Nation address in South Africa (Ramaphosa, 2019).

The motivations that underlie the planning and development of new satellite cities can also be approached with the benefit of the findings of significant research conducted in recent years on urbanization and the spatial development of African cities. This work, as an aspect of a global program on urbanization in developing countries, was conducted primarily by researchers from or affiliated with the London School of Economics and Oxford University and was funded via the World Bank by the UK's Department of International Development. The research associated with this program has produced the most comprehensive and detailed analysis of the development of urban hierarchies and of urban forms in a number of African countries and cities to date. Further, it has provided insights to the consequences of the patterns and dynamics of urban growth and expansion for city-level economic development, urban and land use planning, housing provision, transportation, and the provision of infrastructure services including water, sanitation, and energy (for a sample, see Bird et al., 2017; Henderson et al., 2016; Michaels et al., 2019).

Key conclusions from this research program have been recently published by the World Bank Group in *Africa's Cities: Opening Doors to the World*. One such conclusion is that the weakness of African urban economies – characterized as a "low development trap" – is strongly related to their typically fragmented and dispersed urban forms and infrastructural (notably transportation) deficiencies. African cities are, in the words of the authors, "crowded, rather than economically dense, and they are physically disconnected; as a result, they are costly" (Lall et al., 2017, p. 28). These qualities – or the perceptions of them – and the dysfunctions and inefficiencies that follow can rationalize and impel new or satellite city developments.

The strongly data-driven spatial and economic analyses emerging from the urbanization and spatial development research program briefly referred to here – especially the insights they provide on emerging urban forms – can usefully complement and augment the research that is specifically addressing African suburbs and suburbanisms, which is arguably limited in its economic coverage. There is much that can be mutually learned from both streams of research.

Conclusion

This chapter has provided an overview of the current state of research on African suburbanization and suburbanisms. The Global Suburbanisms project played a creative role in encouraging new work. This research is, we would argue, increasingly important, as well as invigorating, for urban studies and practice. The continent of Africa now contains some 13 per cent of the world's urban population – the same proportion as Europe. The divergence from the "Old Continent," however, is marked: urbanization and spatial expansion will continue at a rapid pace and on a very large scale in Africa, and at all spatial levels. Between 2018 and 2035, the UN predicts that twenty-one of the world's thirty fastest-growing cities will be in Africa, including Bamako, Abuja, Ouagadougou, Kampala, and Dar es Salaam, which are expected to be among the top ten. Nigeria alone is projected to see an increase of some

189 million urban residents, and by 2050, the entire continent will be 60 per cent urbanized, harbouring at least nine megacities with populations exceeding ten million people, and around twenty-five cities of five million people or more.

Research on African suburbanisms to this point has explored their diverse dynamics, drivers, and forms (Mabin et al., 2013) but has, in particular, drawn attention to new forms of urban expansion on city edges linked to the economic growth and the rise of the middle classes (Bloch, 2015), moving beyond earlier discourses of poverty (see also Robinson in this volume). As we have noted, the applicability of the concepts of "suburbanization" and "suburbanism" to African contexts has been debated, yet, as Robinson (in this volume) suggests, these concepts have ultimately been productive in generating enquiry into the field of urban studies, and they have been used in the Global Suburbanisms project in an open-ended way (rather than in a narrow comparative sense) to invite exploration and reflection on existing and emerging urban forms. The search for better terms for and understandings of African suburbs and suburbanisms will continue, and indeed grow, as continental-wide urbanization proceeds, and African cities and their hinterlands – or, better, what were once their hinterlands – take new and unprecedented forms. New suburban constellations will continue to take shape, come alive, and supersede the city–suburb divide (Keil, 2018), with enormous significance for how we understand, theorize, and plan African cities and their regions.

NOTES

1 According to UN Population Division figures, 4.2 billion people, or 55 per cent of the world population, were residents in urban areas in 2018 (Angel et al., 2018).

2 Angel et al. (2011) estimated that, even at the same levels of density, urban land cover in 2050 in Africa will be four times higher than it was in 2000.

3 We define *urban primacy* as the share of total urban population in a country's largest city, with all SSA countries except South Africa characterized by primacy rates of above 30 per cent, with several above 50 per cent.

4 SSA is now 40 per cent urbanized, with an average per capita gross domestic product (GDP) of USD$1,100; at the same level of urbanization, Asia's GDP per capita is USD$3,500 (Lall et al., 2017). According to one UN-Habitat estimate, some 60 per cent of urban residents live in informal circumstances.

REFERENCES

Andreason, M., Agegaard, J., & Moller-Jensen, L. (2017). Suburbanisation, homeownership aspirations and urban housing: Exploring urban expansion in Dar es Salaam. *Urban Studies, 54*(10), 2342–59. https://doi.org/10.1177/0042098016643303.

Angel, S., Lamson-Hall, P., Guerra, B., Liu, Y., Galarza, N., & Blei, A. (2018, August). *Our not-so-urban world* [Working paper No. 42]. The Marron Institute of Urban Management, New York University. https://marroninstitute.nyu.edu/papers/our-not-so-urban-world.

Angel, S., Parent, J., Civco, D.L., & Blei, A.M. (2011). *Making room for a planet of cities*. Lincoln Institute of Land Policy.

Ballard, R., & Rubin, M. (2017). A "marshall plan" for human settlements: How megaprojects became South Africa's housing policy. *Transformation, 95*, 19–48. https://doi.org.10.1353/trn.2017.0020.

Bartels, L. (2020). Peri-urbanization as "quiet encroachment" by the middle class: The case of P&T in Greater Accra. *Urban Geography, 41*(4), 524–59. https://doi.org/10.1080/02723638.2019.1664810.

Bird, J., Montebruno, P., & Regan, T. (2017). Life in a slum: Understanding living conditions in Nairobi's slums across time and space. *Oxford Review of Economic Policy, 33*(3), 496–520. https://doi.org/10.1093/oxrep/grx036.

Bloch, R. (2015). Africa's new suburbs. In P. Hamel & R. Keil (Eds.), *Suburban governance: A global view* (pp. 253–77). University of Toronto Press.

Bloch, R. (2016). City-regions, urban fields and urban frontiers: Friedmann's legacy. In H. Rangan, M. Ng, L. Porter, & J. Chase (Eds.), *Insurgencies and revolutions: Reflections on John Friedmann's contributions to planning theory and practice* (pp. 61–72). Routledge.

Buire, C. (2014). The dream and the ordinary: An ethnographic investigation of suburbanisation in Luanda. *African Studies, 73*(2), 290–312. https://doi.org/10.1080/00020184.2014.925229.

Caldeira, T. (2017). Peripheral urbanization: Autoconstruction, transversal logics, and politics in cities of the Global South. *Society and Space, 35*(1), 3–20. https://doi.org/10.1177/0263775816658479.

Charlton, S. (2014). Public housing in Johannesburg. In P. Harrison, G. Gotz, A. Todes, & C. Wray (Eds.), *Changing space, changing city: Johannesburg after apartheid* (pp. 176–93). Wits University Press.

Charlton, S. (2017). Poverty, subsidised housing and Lufhereng as a prototype megaproject in Gauteng. *Transformation: Critical Perspectives on Southern Africa, 95*(1), 85–110. http://doi.org/10.1353/trn.2017.0023.

Cote-Roy, L., & Moser, S. (2019). "Does Africa not deserve shiny new cities?" The power of seductive rhetoric around new cities in Africa. *Urban Studies, 56*(12), 2391–407. https://doi.org/10.1177/0042098018793032.

da Silva Pereira, M. (2010). Suburbio. In C. Topalov, L. Coudroy de Lille, J.C. Depaule, & B. Marin (Eds.), *L'aventure des mots de la ville: À travers le temps, les langues, les sociétés* [The adventure of words of the city: Through time, languages, societies] (pp. 1197–206). Robert Laffont.

Elate, S. (2004). African urban history in the future. In T. Falola & S. Salm (Eds.), *Globalization and urbanization in Africa* (pp. 51–66). Africa World Press.

Falt, L. (2019). New cites and the emergence of "privatised urbanism" in Ghana. *Built Environment, 44*(4), 438–60. https://doi.org/10.2148 /benv.44.4.438.

Goodfellow, T. (2017). Urban fortunes and skeleton cityscapes: Real estate and late urbanization in Kigali and Addis Ababa. *International Journal of Urban and Regional Research, 41*(5), 786–803. https://doi.org /10.1111/1468-2427.12550.

Harris, R. (2010). Meaningful types in a world of suburbs. *Research in Urban Sociology, 10*, 15–47. https://doi.org/10.1108/S1047-0042(2010)0000010004.

Harris, R., & Vorms, C. (Eds.) (2017). *What's in a name? Talking about urban peripheries*. University of Toronto Press.

Henderson, V., Venables, A., Regan, T., & Samsonov, I. (2016). Building functional cities. *Science, 352*(6288), 946–7. https://doi.org/10.1126/science .aaf7150.

Herbert, C., & Murray, M. (2015). Building from scratch: New cities, privatized urbanism and the spatial restructuring of Johannesburg after apartheid. *International Journal of Urban and Regional Research, 39*(3), 471–94. https://doi.org/10.1111/1468-2427.12180.

Hommann, K., & Lall, S. (2019). *Which way to livable and productive cities? A road map for sub-Saharan Africa*. World Bank Group.

Jenkins, P. (2009). Maputo and Luanda. In S. Bekker and G. Therborn (Eds.), *Capital cities in Africa: Power and powerlessness* (pp. 142–67). HSRC Press.

Keil, R. (2018). *Suburban planet: Making the world urban from the outside in*. Polity Press.

Lall, S., Henderson, J., & Venables, A. (2017). *Africa's cities: Opening doors to the world*. World Bank Group.

Mabin, A. (2013). Suburbanisms in Africa? In R. Keil (Ed.), *Suburban constellation: Governance, land and infrastructure in the 21st century* (pp. 153–9). Jovis.

Mabin, A., Butcher, S., & Bloch, R. (2013). Peripheries, suburbanisms and change in sub-Saharan African cities. *Social Dynamics, 39*(2), 167–90. https://doi.org/10.1080/02533952.2013.796124.

Mbatha, S., & Mchunu, K. (2016). Tracking peri-urban changes in eThekwini Municipality – Beyond the "poor–dich" dichotomy. *Urban Research and Practice, 9*(3), 275–89. https://doi.org/10.1080/17535069.2016.1143960.

Melo, V. (2016). The production of urban peripheries for and by low-income populations at the turn of the millennium: Maputo, Luanda and Johannesburg. *Journal of Southern African Studies*, 42(4), 619–41. https://doi.org/10.1080/03057070.2016.1196955.

Meth, P., Todes, A., Charlton, S., Mukwedeya, T., Houghton, J., Goodfellow, T., Sileshi Belihu, M., Huang, Z., Asafo, D., Buthelezi, S., & Masikane, F. (2021). At the city's edge: Situating peripheries research in South Africa and Ethiopia. In M. Keith & A. De Souza Santos (Eds.), *African cities and collaborative futures: Urban platforms and metropolitan logistics* (pp. 30–52). Manchester University Press.

Michaels, G., Nigmatulina, D., Rauch, F., Regan, T., Baruah, N., & Dahlstrand-Rudin, A. (2019). Planning ahead for better neighborhoods: Long run evidence from Tanzania [Working paper No. 222]. Spatial Economics Research Centre, London School of Economics and Political Science. http://eprints.lse.ac.uk/id/eprint/86570.

Morais, J., & Raposo, I. (2005). Da cidade colonial às novas urbes Africanas: Notas exploratórias [From the colonial city to the new African cities: Exploratory notes]. In *Cidades Africanas: Cadernos da Faculdade de Arquitectura da Universidade Técnica de Lisboa* [African cities: Notebooks from the Faculty of Architecture of the Technical University of Lisbon] (pp. 88–91). Universidade Técnica de Lisboa.

Myers, G. (2011). *African cities: Alternative visions of urban theory and practice*. Zed Books.

Oppenheimer, J., & Raposo, I. (Eds.). (2007). *Subúrbios de Luanda e Maputo*. Edições Colibri.

Organisation for Economic Co-operation and Development (OECD). (n.d.). *Built-up area and built-up area change in countries and regions*. OECD.Stat. Retrieved 20 November 2019, from https://stats.oecd.org/Index.aspx?DataSetCode=BUILT_UP#.

Parnell, S., & Pieterse, E. (Eds.) (2014). *Africa's urban revolution*. Zed Books.

Potts, D. (2017). Urban data and definitions in sub-Saharan Africa: Mismatches between the pace of urbanisation and employment and livelihood change. *Urban Studies*, 55(5), 965–86. https://doi.org/10.1177/0042098017712689.

Ramaphosa, C. (2019, June 20). *President Cyril Ramaphosa: State of the nation address 2019* [Speech]. Government of South Africa. https://www.gov.za/speeches/2SONA2019.

Roque, S. (2009). *Ambitions of Cidade: War-displacement and concepts of the urban among bairro residents in Benguela, Angola* [Unpublished doctoral dissertation]. University of Cape Town.

Rubin, M., & Charlton, S. (2019). State-led housing provision twenty years on: Change, evolution and agency on Johannesburg's edge. In R. Keil, M.

Üçoğlu, & M. Guney (Eds.), *Massive suburbanization: (Re)building the global periphery one large scale housing project at a time* (pp. 241–66). University of Toronto Press.

Rubin, M., & Harris, R. (2018). An effective public partnership for suburban land development: Fleurhof, Johannesburg. In U. Lehrer & R. Harris (Eds.), *Suburban land question: A global survey* (pp. 258–79). University of Toronto Press.

Sawyer, L. (2014). Piecemeal urbanisation at the peripheries of Lagos. *African Studies, 73*(2), 271–89. https://doi.org/10.1080/00020184.2014.925207.

Simone, A. (2004). *For the city yet to come: Changing African life in four cities.* Duke University Press.

Todes, A. (2014). New African suburbanisation? Exploring the growth of the northern corridor of eThekwini/KwaDakuza. *African Studies, 73*(2), 245–70. https://doi.org/10.1080/00020184.2014.925188.

Todes, A. (2017). Shaping peripheral growth: Strategic spatial planning in a South African city-region. *Habitat International, 67*, 129–36. https://doi.org/10.1016/j.habitatint.2017.07.008.

Topalov, C., Coudroy de Lille, L., Depaule, J., & Marin, B. (Eds.). (2010). *L'aventure des mots de la ville: À travers le temps, les langues, les sociétés* [The adventure of city words: Through time, languages, societies]. Robert Laffont.

Van Noorloos, F., & Kloosterboer, M. (2018). Africa's new cities: The contested future of urbanization. *Urban Studies, 55*(6), 1223–41. https://doi.org/10.1177/0042098017700574.

Watson, V. (2013). African urban fantasies: Dreams or nightmares? *Environment and Urbanization, 26*(1), 215–31. https://doi.org/10.1177/0956247813513705.

Xu, G., Dong, T., Cobbinah, P., Jiao, L., Sumari, N., Chai, B., & Liu, Y. (2019). Urban expansion and form changes across African cities with a global outlook: Spatiotemporal analysis of urban land densities. *Journal of Cleaner Production, 224*, 802–10. https://doi.org/10.1016/j.jclepro.2019.03.276.

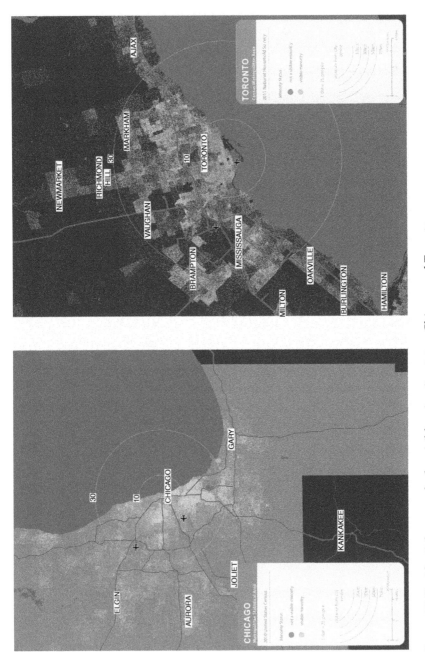

Image 17. Residential geography by visible minority status, Chicago and Toronto

17 Outside the Outside: Alienation, Fidelity, and New Sub-urbanizations

MATT HERN

Introduction

Surrey – part of the Metro Vancouver area and British Columbia's biggest suburb – has more than half a million people. This is just a little less than the City of Vancouver, but like so many suburban communities on the metropolitan fringes of big cities, Surrey is growing so fast that its population will surpass that of Vancouver's in a decade or two. Surrey already sits on a giant physical footprint: Built into the exurbs and mostly on fertile agricultural land, it has had little reason to constrict sprawl, and so sprawl it has. And like any good suburb, Surrey has placed few restrictions on development. Desperate for investment, it frantically courts capital and green-lights every half-witted development proposal and, as a result, is an immense jumble of metastasizing low-density housing tracts punctured by seemingly random sproutings of low-end condo towers, mismatched shopping malls, mini-malls, and strip malls, with no real downtown and so much traffic. Surrey is a classic North American suburb. You know this place. You've been here. Grown up here, maybe. Sneered at it. Used it as the butt of jokes. Maybe even moved here.

Suburbs are used as a proxy for every ill that besets the contemporary urban landscape. A synonym for bad urban development. Geographies of nowhere. Ecological catastrophes. Cultural wastelands. Shorthand for ugliness and banality and homogeneity and anomie. Most everyone who writes about, sings about, or researches suburbs wants to discipline or fix or bomb them. Suburbs are awful. The absolute worst. What am I doing here?

I know we need to start thinking more seriously about suburbs. The immense population growth in Surrey is hardly unique: In cities across the globe, poor, working-class, and migrant populations are being expelled from city cores and reordered further and further away on the margins of metropolitan regions. When I was young, the term "inner-city neighbourhood" was universally coded as "Black, immigrant, and poor." Now when my kids hear "inner city" they think "sanitized, high-end, condoization."

I have lived in East Vancouver for all of my adult life and have always worked closely with low-income kids and families. My own family rents in one central neighbourhood that has historically been the urban landing spot for immigrants, Indigenous people, working-class families, and everyone who needs cheap shelter. The Eastside has always been, in the indelicate words of a friend, the "dumping ground" for the rest of the city. But that is changing swiftly. As that same friend told me, "Matt, you need to get out to the suburbs. The people you think you work with aren't in East Vancouver anymore. They're in Surrey." I resisted strongly, but he was – and is – right. The East Vancouver that I know and love, the East Vancouver where I raised my family, does not really exist anymore. Like so many inner-city neighbourhoods, most of its residents have been displaced and replaced.

Twenty years ago, when a young, single mother on social assistance would show up at our door, interested in one of the community programs I ran, one of the first questions I would ask is where she lived. The answer was always one of the nearby neighbourhoods: the Drive, Mt. Pleasant, Strathcona, Downtown Eastside. When I get that sort of query email today, however, she lives in a place I have barely heard of and have probably never visited. It's a twenty-minute bus ride to the Brentwood Mall. It's a half-hour walk to the Guildford Town Centre. It's in some suburban zone that doesn't even really have a name and that I have to look up online.

This is not much of a surprise to anyone. As gentrification has gripped Vancouver, like it has so many other cities, my family's neighbourhood has been among the most vulnerable. Close to the downtown core with some preserved housing stock, a walkable high street, good public transit, and a vibrant, diverse reputation, this neighbourhood is being corroded by voracious capital, rising rents, property speculation, and a wave of swanky craft beer bars, cafes, toy stores, and weird hipster places that no one I know frequents. This community isn't helpless; its transformation remains uneven, halting, fraught and resisted, but the plight of this place will be immediately recognizable to most any urban dweller. It is fast becoming exclusive: wealthier, blander, and a whole lot whiter (see Nair & Carman, 2017).

In city after city across the globe, these ongoing cycles of displacement have become so chronic that ritualistic cleansings of local residents, businesses, and community ventures are now expected – normalized as just the everyday workings of the urban marketplace. We are all inured to it, numbed by the losses around us as our friends, families, and neighbours are pushed out of the neighbourhood, family businesses are closed, and buildings are torn down and replaced with things made of glass and brushed steel. Any attractive community in any city is exposed, especially if it is close to the urban core. All across the planet, new waves of upscale residents and residences, investment properties, spectacularist touristic forays, and all the social/cultural/architectural infrastructures that serve them are invading inner cities (Lees et al., 2016).

We are living in the midst of an exhaustively documented, historic, and historically dislocating global rush to cities. New urbanist planning and recently carved circuits of capital are driven by financialization and servicization to reshape central districts, making them more attractive, more livable, and more vibrant. And those armed with financial firepower are being convinced en masse to embrace revanchist urban living: Agile real estate, developer and marketing interests, and new occupying forms of capital – encouraged and greased by progressive urban planners – are reclaiming the city with startling ferocity.

The sheer speed of this urban occupation is aggressively pushing poorer residents out of city cores to places where their social marginalization is exacerbated by physical isolation. In cities from Seattle to Seoul to Sofia, and everywhere in between, middle-, working-class, and poor people are finding it increasingly difficult to secure viable housing and commercial opportunities in premium inner-city neighbourhoods. It is not happening, of course, at the same velocity or in the same patterns everywhere – every city evinces its own peculiarities and tendencies – but throughout the Global North and in so many cities across the planet there is a startling phenomenon unfolding.

In the wake of dramatic urban restructuring triggered by neoliberalism and globalizing economies, world and global city theory emerged in the 1980s as perhaps the dominant analytical vehicle for urban research. Since then, wide-ranging discussions on "planetary urbanization" have equally been scrutinized by feminist, postcolonial, and queer theorists (for a critique, see the 2018 special issue of *Society and Space* on planetary urbanization) and challenged by small waves of bold new scholarship. It is clear to so many of us that *something* is happening, and something perhaps unprecedented, in cities across the globe. But how to speak of it? Is it one major process, a collection of processes, or particular things in particular places? Is there a thread to pull that can offer insight into various kinds of peripheries? I am curious about new vocabularies that can describe existing places but can also be generalizably useful. I am not sure if the particularities of Surrey have anything to say to *barrios*, tent cities, slums, shanty towns, *borgatas*, *favelas*, *chabolas*, squatter villages, *banlieues*, or sub-urbs anywhere else – but my guess is that there are some common structural processes and common narratives at play.

It is true (as we have heard said metronomically) that for the first time in history, more than half the world's population finds itself in cities, but really it is more accurate to say that the vast majority of people live and work in *sub*-urban areas (Gordon & Shirokoff, 2014). North American suburbs, exurbs, periburbs, and ethnoburbs are all growing faster than central cities, and this holds true across the globe in *barrios*, *banlieues*, new towns, shantytowns, *gecekondus*, *favelas*, and a thousand different variations. As the *Economist* (2014) put it: "In developed and developing worlds, outskirts are growing faster than cores. This is not the great urbanisation. It is the great suburbanization."

The radical reshaping of inner cities means that we are also witnessing a related suburbanization of poverty (Badger, 2013; Kneebone, 2010) across North America and elsewhere (in many parts of the globe this has been true for a very long time) and thus, of course, a commensurate racialization of much of suburbia (Sullivan, 2011). People still tend to think about suburbs and the peripheral zones around major cities using specific and often calcified analytical lenses. Increasingly, however, those perceptions are being challenged by multiple considerations, not the least of which is the economic contours of both inner and outer suburbs.

The complex character of suburbia has always been evident to anyone who cares to look closely, and observers have been messing with tidy suburban clichés for generations. But today, new patterns of settlement – especially in the contexts of racial capitalisms, new orderings of poverty, and urban displacements – are presenting a novel set of circumstances. We need to find some new languages, some new vocabularies for thinking about what is happening around us, because what we have on hand currently is clearly insufficient.

The Remaking of the Urban Outside in Theory and Practice

I have several entwined goals here. The first is to join a growing number of (often tremendous) theorists who are working to cast a different kind of attention on urban peripheries. I would like to help to recast the sub-urbanization of newcomer, working-class, and poor populations in a more fulsome and generous light. Traditional urban theory cares little for sub-urbs except as a distasteful foil, exoticized fantasia, or technical problem, and I suspect this creates significant theoretical and political blind spots. I am entirely unclear if these new patterns of peripheralization are a transient set of phenomena, but my instincts suggest that what we are currently witnessing signals a long-term, fundamental reorganization of urban space triggered by the confluences of neoliberal remakings of city cores, transnational migrations and respondent racisms, global warming, economic restructuring, and revanchist urbanism.

These new forms of the sub-urbanization of poverty throw up a profound set of challenges for organizers, residents, activists, urbanists, scholars, and planners of all kinds – and all of us need to think hard on our work in light of these new currents. As suburban theorist Roger Keil put it to me, "The bourgeois Left really only feels comfortable in urban environments, and its political program revolves around a built landscape of coffee shops, public spaces, walkable streets, and public transit. We don't really know how to live otherwise" (personal communication, 3 January 2018). I think this comment is both insightful and impactful, and certainly speaks straight to me.

As I have become increasingly interested in sub-urbs everywhere, but most especially in Surrey where I now work and organize full-time, I have found myself in all kinds of surprising situations and places, totally out of my element, unclear on how to act and unsure what I am looking for. Over the last several years, I have been grateful to have been invited into mosques, temples, churches, and gurdwaras. I have had countless meetings in mall food courts. I have taken buses that I was convinced were never going to arrive. I have been lost more times than I like to admit. I have tried to walk to many places that were not within walking distance. I have gaped at new planning initiatives and housing tracts that do not fit my idea of the good life. I lost my sense of humour in many traffic jams.

What I have found, and keep finding, are new (or at least unfamiliar to me) and creative forms of social organization that are emerging on urban peripheries in every city I visit. New patterns of solidarities, new ways to confront social marginalization, new kinds of gathering spaces, and new ways of living are thriving within suburban landscapes, many of which do not adhere to my easy leftist imaginations. Over the last few years, I have been in evangelical churches above auto body shops, seen organizing campaigns conducted on buses, patronized pizza shops in garages, and boutiques in industrial parks. I have done my share of afterhours drinking in shopping plaza parking lots, bought food at fresh fruit and vegetable stands in mall food courts, played mini-golf in decommissioned suburban school gymnasiums, had a better-than-expected time at parties in sprawling chain restaurants, and attended protests and demonstrations taking over mall concourses. All of these experiences have defied my easy stereotypes and have helped me notice some of what I was not seeing.

For at least a couple of decades now, a (small) number of scholars have been using the term *post-suburbia* in a nod to the fractured, confused, confusing, and recursively shifting patterns of contemporary peripheries (e.g., Helbich & Leitner, 2009; Kling et al., 1995; Musil, 2007; Phelps & Wood, 2011; Phelps et al., 2010; Soja, 2000; Teaford, 1997). *Post-suburbia* is not necessarily meant as a binary departure/evolution from *suburbia*, but might be described as emerging as classical suburbia is "partly converted, inverted or subverted into a process that involves densification, complexification and diversification of the suburbanization process ... Post-suburbanization also entails a profound re-scaling of the relations and modes of governance that have traditionally regulated the relationships between centre and periphery in the suburban model" (Charmes & Keil, 2015, p. 3). I am frankly a little ambivalent about the term – as much as I am about its close cousins "edge cities," "technoburbs," "in-between-cities," and "fringe cities" – but all this points in the right direction: that sub-urbanization "after suburbia" does not look all that suburban anymore.

I am not – and do not want to become – any kind of apologist for extant suburbia.[1] Many, perhaps most, of the well-trod critiques stand: the grim isolations of the suburban built environment, the deep fetishization of the automobile, the fragmentation, the consumptiveness, the waste, the ecological perils – all of it pisses me off as much as anyone. I am not asking that anyone appreciate any particular existing suburban form. But I do want to challenge the condescending ease with which so much of progressive and leftist thought writes off massive areas of metropolitan regions where most of the urban population lives. Far too much of existing radical urban theory acknowledges gentrifications but cares little where displaced people end up. And, frankly speaking, I am absolutely guilty of this too.

Historically, the city has been conflated with modernity and democracy: the public milieus, the strangers, the difference, the shared institutions, the walkable scales – all of it supposedly nurtures ancient Athenian ideals of the *polis*. I think that prescription is correct in some ways, but does that then mean that people who live in sub-urban regions are relegated to a less-robust modernity, are in a democratic deficit, are less capable of civic life? This is exactly the kind of inference that induces so many suburbanites to feel so patronized.

This is the legacy of the Jane Jacobs, Andres Duany, and Richard Florida school of urbanism: an aggressively aestheticized neoliberal vision of the city that prioritizes a very particular kind of "vibrancy" as the scenery for bourgeois fantasies of public life. The New Urbanist worldview cares little for class or race, lionizing sanitized landscapes of "alive" shopping districts, public spaces, and mixed-use spaces. Much of Jacobs's work – and that of the New Urbanists since – reads like a gentrification how-to manual. It has laid the intellectual foundations for the reoccupation of inner cities by those who are seeking "authentic" urban experiences and have the money to buy them. Progressive urbanist fascination with Paris, for example, seems to come just short of overtly pining for a revived pan-Haussmannization that can sweep clear the poor with mass evictions and remake any city with a cleaner, more-ordered, nostalgically bourgeois façade. For the New Urbanists, the sub-urbs are just an icky laughingstock – a sneering punchline to jokes told with little curiosity about the social forces that create them, and with little generosity for the social milieus that flourish within them.[2]

The only substantive engagement most progressive urbanists today have with sub-urbs is to offer boilerplate templates for walkable high streets, densification, TOD (transit-oriented development) or POD (pedestrian-oriented development) planning, compact communities, and bike lanes. I find it particularly galling to hear the confident prescriptions for "fixing" the suburbs coming from the exact same mouths that have eviscerated inner-city neighbourhoods and now want to foist the same "solutions" onto the only affordable areas left in the region.

I hear smug calls for a "Vancouverist" remaking of Surrey echoed by exactly the same people who have fucked over Vancouver by creating what is perhaps the world's least affordable city. I see the same kinds of dynamics in every city I visit – the same blithely ignorant pseudo-progressivism trashing poor and working-class communities with a shrug. There is so much that needs changing in Surrey, just like in every suburb everywhere, but we cannot defer to the same tired formulas that are disfiguring inner cities across the globe (Peck et al., 2014).

These prescriptions for "fixing the suburbs" are further hamstrung by the misplaced confidence that easy divisions between the city and suburb can be drawn. Certain place names are code for certain built forms (alongside all kinds of racialized clichés). Here in Metro Vancouver, "Surrey" means endless cheap residential housing tracts and trashy malls, while "Vancouver" means glassy condo towers and bike lanes. But there are "town centres" in Surrey that are denser than most of Vancouver, and new towers are springing up by suburban SkyTrain hubs. Downtown Vancouver is very dense, but the vast bulk of the city is actually made up of single-family housing on large lots where everyone drives. The clean distinctions between sprawling suburbs and dense cities rarely hold easily, and the edges of where the city stops and the suburbs begin are unclear, permeable, and unfixed.

This is true in every city, where territorial borders are far more performative than material. Consider density, for example. Most of us are pretty sure we know what it looks and what it feels like, but the least-sprawling city in America is not New York, as everyone assumes, but Los Angeles (Romero, 2015). I have spent enough time in both places to find that claim confusing, but it underlines the problem with trying to assert easy binary distinctions and draw fixed borders. In every urban area, cities and suburbs are bound together, reflecting and constructing one another, and it is hard to say where one stops and the other starts. Imagining that it is somehow possible to talk about Surrey as separate from Vancouver, or Orange County as distinct from Los Angeles, is to ignore the existing economic and social patterns of centring and peripheralizing.

Understanding the kinds of development being given permission, the types of wealth and poverty being created, and where urban power resides necessarily means seeing cities and suburbs in relational contexts. Consider Paris's *banlieues*, for example, or *favelas* in Brazil, new towns in India, "lost cities" in Mexico, *gecekondus* in Turkey, commuter towns in China, shanty towns in Pakistan, slums in Nigeria, and all the wildly divergent patterns of peripheralization across the planet. Each of these examples requires an examination of specific logics of social and spatial segregations that are generating new forms of accumulation alongside new forms of poverty. But we can pull on some common threads here. The globalized phenomena of urban expulsions and banishments are not universalizable; they are articulated very differently in every city, but the processes that are driving these new spatial logics can be generalizably theorized. I am not interested in mapping or taxonomizing the so-called uncharted sub-urbs, but in understanding how sub-urban places fall off the map and become rendered *outside*.

This is my first goal in this chapter: I want to push urban theory towards a more convivial and generous approach to suburbs and urban peripheries of all kinds. I am most curious here about the forms of displacement and dispossession that are buffeting sub-urbs from multiple directions. My instinct is that this project will have to travel through property, which necessarily means thinking suburbia through whiteness and settler-colonialism.

Centre–Periphery Narratives Reconsidered

My second goal in this chapter is to trouble the ontological underpinnings of centre–periphery narratives. Unfortunately, the vocabularies I have at hand are inadequate – or even distorting and regulating. I am really struggling for a grammar. So far, I have been using words like suburbs, peripheries, edges, fringes, margins – pinballing between terms without much clarity. In part that is because I am wholly dissatisfied with all of these. This is language that implicitly relegates certain parts of the urban to the "outside," and designates other parts – with pretty flimsy rationales – as the "centre."

The languages of suburbs, peripheries, and margins echoes the logics of empires – placing colonial metropoles at the centre of world, with all the rest undeveloped, backward, dependent, and subservient provinces. Those are not descriptive efforts; those languages actively constitute the places they designate, and our urban vocabularies are performing the same kinds of functions. For most urbanists and urban theorists, the suburbs are highly "Orientalized" – ugly, dark, mysterious, exoticized places where poor people now go to disappear into inscrutably car-infested placelessness.

You may have noted that in the last few pages I have started to use *sub-urbia* instead of *suburbia*. It's a bit cutesy, I know, but I am trying to recast/reframe peripheral urban communities outside the confines of conventional suburban discourses. I am going to toggle between those terms on occasion, but I am using the hyphenated *sub-urb* to refer to a constellation of settlements outside of city-wall borders that are believed (both within and from afar) to be less-than, the sub to the urb. I am using sub-urb here as opposed to the *suburbs*, which is a larger set of modes of living and ways of being that include the built form, but much more as well.

These terms overlap. Most settlements on peripheries around the world and, increasingly, most North American (and North American-style) suburbs are included in sub-urbia. I am not much interested in suburbs – they have been dissected and theorized sufficiently – but I am highly interested in sub-urbia, what that might mean, and what possibilities it makes available to all of us. One the core threads of this chapter is a struggle for terminology – not in a taxonomic sense, but in a search for useful ways to speak of new forms of banishments, relegations, and displacements. How we speak of these reorderings is not just a descriptive exercise; it is always prescriptive.

The categories, codes, and conventions of most urban studies are actively sub-urbanizing and peripheralizing as we go, and dominant theory actively centres its presumptive "centre" without much question. The languages (certainly including most of what I have been able to deploy thus far) are very explicitly performing marginalization as they purport to describe it. As Roy (2016) asks, "For whom is the city a coherent concept? Whose urban experience is stable and coherent? Who is able to see the city as a unified whole? By contrast, for whom is the city a geography of shards and fragments?" (p. 206). The question of where a city starts and stops, in whose interests is it so pressing to demarcate clear boundaries (and what lies outside of them and who is expelled) leaps out here – even as I grasp for terms that do not reinscribe those same kinds of languages that cannot think past centre/hinterlands, metropoles/colonies, empires/provinces, cities/peripheries.

It strikes me that there is never a "periphery." That is just an idea that is leveraged in the interests of power. The idea of what is "outside" describes that which does not count as part of the "centre," and thus what does not really matter. The periphery is an idea that is actively constituted as a function of legitimizing accumulation and control. And this mode of thinking continues with the willing compliance of urbanists: The notion of the urban periphery performs a critical centring function. Why are the glassy condo towers, the exercises of high-end recreational shopping, and the architectures of financialization that mark downtown Vancouver the "centre" of this city? What are they the centre *of* exactly, and whose politics are driving those definitions? In whose interests is it to insist on these kinds of taxonomies?

The notion of the "outside," of the *periphery*, has very real and very material consequences. Some years ago, I was giving a lecture in Surrey to an audience of planners, city officials, social service organizations, and businesspeople. The event was one of those civic "visioning" exercises – a day-long series of events claiming to regenerate Surrey. The stated intent was to develop some new social and cultural energy as Surrey struggled to emerge as a fully-fledged *city*. I offered some ideas around civic participation, grassroots organizing, and planning from below. It was essentially an anti-creative-class kind of talk.

After the lecture, a woman waited to speak with me. She was a city councillor with ambitious ideas about her own future, and the question she asked kind of hit me. She said, "OK, everything you said is fine and good, I agree with some of it, but here's the real problem: Every young person with energy, ambition, or creativity desperately wants to leave Surrey. Most young people don't want to be here. They want to be in Vancouver, or Toronto, or Dubai, or Lahore, or … anywhere not stuck in the suburbs. What can we do about this?"

This is the same question that grips small towns, rural areas, and agricultural communities across the planet. The loss of ambitious young people is not the whole story, though, and it depends on how you ask that question. My everyday work is co-directing a project in Surrey building a cooperative of workers' cooperatives with youth from newcomer (refugee, racialized migrant – I don't like any of those terms, but don't know of any better ones right now) families. This work is the product and source of most of my ongoing interest and research in suburbs.[3] The kids we work with are all from newcomer backgrounds, and their families have taken convoluted and sometimes harrowing routes to land in Surrey. It is more than an hour via public transit to get to downtown Vancouver from where they live, and while all the kids get into the city sometimes – to shop, or whatever – none of them seems to be particularly drawn to Vancouver. For most of them, the centre of social gravity is their mosques, all of which are in Surrey or the next suburb over. They go to school locally, hang out in Surrey malls, play sports, and belong to various clubs, but it is manifestly clear that their "centre" is not downtown Vancouver. It makes me think about how imagined urban cartographies constantly orient and reorient political fidelities.

Centres and peripheries are not static ideas, and there is never any one centre. In any city, there are endless centres, big and small, centring various kinds of activities, forming and reforming, making and remaking. Some of these centres are in fact downtown – pivots of tourism, finance, and real estate investment – but there are many others, visible and obscured, and many of those exist on what are ostensibly "peripheries."

I think what I am most interested in is the *process* of centring, and by consequence, of peripheralizing. Contemporary urban restructuring is characterized by massive exclusions, displacements, and dispossessions, and I believe the right questions to ask are, Who is getting excluded from what, by whom, and on whose auspices? Who decides who is *inside* and who is *outside* – and inside or outside *what*? And what are the centrifugal forces that are ejecting low-income and racialized residents out from particular kinds of centres?

But here again, the vocabulary gets tricky. I just used the words "exclusions" and "ejecting," but are those the right descriptors? There is an expansive set of explanatory frameworks employed to describe how people are shifting and being shifted around in contemporary cities. Various theorists argue for expropriation, expulsion, dispossession, displacement, gentrification, banishment, evictions, exploitation, peripheralization, suburbanization, or marginalization (among others!), and each of these have value and inclarities. Trying to understand what is happening around us requires the right language, especially when we are trying to understand what and how to resist. This may seem overly punctilious: We can see urban crises of displacement all around us, so how does this analytical contortionism help?

Part of the answer is that we have to figure out how to fight the disfigurements that so many of the cities we love are suffering. We all want to start chucking some punches, but where to aim? Swinging wildly won't help. But there's more, too. How we speak of these processes and functions has repercussions and can groove lines of analyses that are often tricky ruts to escape. Take "dispossession," for example, which is a word I have been interested in for a long time. It is an evocative and powerful concept, but in the context of land battles, it is fraught. "Dispossession" implies an original "possession" that has been violated. *Dispossession* fixes certain kinds of property relations – the exact kinds of conceptions of property that many leftist claims (and many theorists from disparate traditions) are trying to unsettle – and subtly insists on property models of ownership and exclusive sovereignty. If we fight back against dispossession, is the implied answer a deepened commitment to possession? As Nichols (2018) puts it, "Since dispossession presupposes prior *possession*, recourse to it appears conservative and tends to reinforce the very proprietary and commoditized models of social relations that radical critics generally seek to undermine" (p. 3).

There is never one master process at play anywhere. There are always multiple, unstable, competing, and complementing forces at work in any place, but we can look closely, ask after specific experiences and specific examples, and theorize their origins and repercussions. I am happy to work both inductively and deductively here, and I want to propose some new analytical frames for thinking about sub-urbanisms. In thinking about generating new urban grammars, I want to pay particular attention to the (often very subtly) dismissive, degrading, or marginalizing implications of new conceptual apparatuses – and especially how they land for residents who live far from certain loci of power. In the same ways that the brilliant critiques of global city theory unsettle their implicit ordering work, it is useful to keep noticing how certain terms, such as "suburban," and certain kinds of analysis perform the same kinds of functions.

Conclusion: Out of the Settler Suburb

It's easy to fixate on suburban morphology, but that's missing the forest. American suburbs have defined the globalized American dream:

an owner-occupied, single-family lot with a grassy yard, a barbeque in the back, a recent-model vehicle on a sanitized cul-de-sac with a nice school within walking distance, and neighbours you can trust. We are all intimately familiar with this allusive and elusive imagery, but even more intimately we understand the economic, sexual, gendered, and racialized relationships it orders. Suburbanism demands fidelity to very specific and explicit ways of being.

There is an active constituency who want to describe this ordering as "market choice" – that this is how most people want to live and so developers, planners, and municipalities just deliver the goods. This, of course, is as absurd as it is offensive. Contemporary suburbanization has been structured and regulated as spatial expressions of white supremacy in precisely the same ways that segregated core-city neighbourhoods have been made and remade. Cities and municipalities have long used collusions of real estate and lending practices, racist grassroots movements, zoning policies, infrastructure spending, tax regimes, municipal land ordinances, and everyday threats of violence alongside micro-exertions of power, control, fear, and exclusion to create highly affective and effective suburban regimes.

Suburbanism is a land-allocation argument about how land should be distributed, used, and controlled (Harris & Lehrer, 2018). That argument is articulated by transportation, shopping, and public space design, but it is anchored by a deep and abiding commitment to individual home-ownership. More than highways, commutes, malls, or cul-de-sacs, the suburban fantasy is about property and homeownership. As Shields (in this volume) nicely frames it, "Suburbs have been a point of intersection between the land development economy, the reproduction of labour and the dispossession of resources, a lack of cultural recognition, and the inequalities of domestic arrangements within families." And in that, it is always an argument about whiteness. Land is the first commodity, and new suburban developments transform non-valuable and stagnant rural, agricultural, and un-used land immediately into property commodity. The explosion of post-war prosperity in the Global North, especially in America, was driven by the manufacturing of wealth out of "nothing land" – land that had to be rescued from what Hernando de Soto Polar famously called the "citadel of dead capital"[4] and transformed into property. This story is of course the story of settler colonialism, and suburban expansionism is properly understood as a re-enactment of the settler-colonial drive to capture land from its *terra nullius* torpor via occupation.

It is no surprise, then, that it is in the United States, Canada, Australia, and New Zealand where the suburban form was birthed and nurtured. Other places try to mimic suburbanism, and often succeed in patches and places, but it is in unreconstructed and unapologetic settler-colonial societies where real suburbanism exploded on the landscape. As Quandamooka scholar Aileen Moreton-Robinson (2015) puts it, "It takes a great deal of work to maintain Canada, the United States, Hawai'i, New Zealand, and Australia as white possessions" (p. xi), and suburbia is carrying a heavy workload.

Referencing Cheryl Harris's seminal 1993 *Harvard Law Review* article, "Whiteness as Property," Moreton-Robinson (2015) writes the following: "As a form of property, whiteness accumulates capital and social appreciation as white people are recognized within the law primarily as property-owning subjects" (p. xix). Property and whiteness entangle, and thus, as a utopian exercise, suburbia has to be exclusive. Australian scholar of settler-colonialism Lorenzo Veracini (2012) describes suburban exclusivity as re-enacting settler-colonialism. Both are producing "localized sovereign capacities," and it is easy to draw parallels: "Settlers and suburbanites are founders of political orders and are especially focused on exercising local control over local affairs" (Veracini, 2012, p. 7).

Like settler-colonial logics, suburban boosters are particularly fond of escape narratives. White fragility and anxiety are prominent features of such narratives, requiring both a fearful flight to the periphery from a dangerous, cluttered, and unstable centre and a nostalgic return to "the way things used to be": pastoral, secure, sanitary, and unencumbered by racialized newcomers. This is a world made anew in the service of white idealism and, like all "community" discourses, requires constant vigilance to defend borders, build moats, and ensure common-sense understandings of what and who is and is not acceptable. Much like settler-colonialism, there is now significant destabilization of the suburban ideal, a fraying and unraveling that is increasingly complicated to contain. Borders become permeable, whole chunks fall away, territory is relinquished, and doubt and fear creep in. White anxiety, like capitalist crises, is never "solved," it is only moved around.

The spatial fix of suburbia was built by spectacular post-war infusions of both public and private investment, precipitating the still-unfolding convergence of real estate and global financial markets. Beginning in the 1960s, as financialization emerged as the dominant mode of contemporary accumulation, every form of capital concentration, from investment banks to pension funds to insurance funds to a fantastic array of exotic investment vehicles, has poured resources into real estate and suburban industrial expansionism. That voracious consumption of land meant that sometime around 1990, the pre-eminent suburb-building countries – the United States, Australia, and Canada – had all become mostly suburban – although that designation is of course questionable and weird. No longer is it clear at all where cities start and stop, it has never been clear what exactly constitutes a suburb, and it is now clearer than ever that what is typically designated as such often contains multitudes of shifting forms and patterns. Just as what actually defines a city is ineffable and unstable, naming a suburb is mostly performative.

What is clear is that suburbs are a mess. There are fewer and fewer suburbs that adhere to easy clichés, so much of what we call "classical suburbia" is restructuring as sub-urbia. Urban peripheries across the globe are becoming increasingly fraught, complex, and destabilized. As non-white people acceleratingly breach suburban gates, the centre cannot hold, and white people are fleeing to new urban gated communities downtown – settlements that vertically segregate in condo towers – and displacing and banishing huge swaths of low-income residents who rearrange themselves on messy – and cheaper – urban fringes.

I contend that as suburbia fractures and reorders into sub-urbia, it opens up new possibilities for land and property. As the materiality of contemporary sub-urbia consistently fails to articulate the "white possessive," what kinds of fidelity might new sub-urbias nurture? Those are questions I ask myself over and over again as I take the train and then the bus from the now gentrified "inner-city" neighbourhood where my family has rented for so long. As I commute for an hour and a half, I keep wondering what the hell I am doing. What am I possibly thinking, heading out into deep suburbia, leaving behind the now tidy and well-ordered urban streets of East Vancouver? And my answer is always the same: that it feels exciting, like so much more is in play, like the certainties of this tightly ordered little city start giving way to something far more complicated, unfamiliar, unpredictable.

The sheer messiness and unplanned nature of Surrey is what makes so many urbanists/urbanites anxious – there is nothing even vaguely like a "smart city" in Surrey, nor in many sub-urbs. And it is exactly that messiness that offers so many opportunities to unravel narratives of property and ownership. All that Surrey sprawl is happening on unceded Indigenous territory, but that occupation is uncertain and complicated, and I wonder how the frothy diasporic layering of Surrey might offer new possibilities for a decolonial set of narratives. Can suburban patterns of displacement and banishment nurture a new politics from below? How can the uncertainties of the sub trouble, even dismantle, the accumulative power of the urb?

NOTES

This chapter is adapted with permission from my forthcoming book of the same name.

1 And I certainly have no interest in the Wendell Cox/Randal O'Toole schools of pro-sprawl, ugly, and (mostly) racist "urbanism."

2 I know it is an anathema to impugn Jane Jacobs. She is rightly celebrated for her resistance to modernist urban renewal brutes, for her scholarship, and for vision. But it is way past time to interrogate her legacy. New Urbanists willfully misunderstand causalities and disregard the politics of property, and their paths are littered with one decimated, sanitized neighbourhood after another.

3 This cooperative project is called Solid State Community Industries (see www.solidstate.coop). We describe our project as trying to articulate new narratives around newcomers and racialized migrants, especially youth, in an era of rising xenophobia, racism, and nationalism. We are combating notions from the Right that cast newcomers as a threat, and those on the Left that insist on charity for helpless arrivals. We believe both of these narratives are desubjectifying and need to be resisted, and we are thus trying to build a more constructive set of stories. We do this by working through the cooperative movement, especially workers' co-ops, with youth. Cooperatives in this part of the world are often (incorrectly) coded as primarily middle-class, old, white people, and we are hoping to add to that story by building a cooperative of youth-owned co-ops. Finally, Solid State Community Industries works to develop a constructive approach to addressing community safety and issues of youth violence that builds on the creativity, skills, and energy of youth.

4 Slow clap for Hernando de Soto Polar.

REFERENCES

Badger, E. (2013, May 20). The suburbanization of poverty. *Bloomberg CityLab*. http://www.citylab.com/work/2013/05/suburbanization-poverty/5633/.

Charmes, E., & Keil, R. (2015). The politics of post-suburban densification in Canada and France. *International Journal of Urban and Regional Research*, *39*(3): 581–602. https://doi.org/10.1111/1468-2427.12194.

The Economist. (2014, December 6). A planet of suburbs: Places apart. *The Economist*. https://www.economist.com/essay/2014/12/06/places-apart.

Gordon, D., & Shirokoff, I. (2014, July). Suburban nation? Population growth in Canadian suburbs, 2006–2011 [Working paper]. School of Urban and Regional Planning, Queen's University. https://canadiansuburbs.ca/files/Gordon_Shirokoff_2011.pdf.

Harris, C.I. (1993). Whiteness as property. *Harvard Law Review, 106*(8), 1707–91. https://harvardlawreview.org/wp-content/uploads/1993/06/1707-1791_Online.pdf.

Harris, R., & Lehrer, U. (Eds.). (2018). *The suburban land question: A global survey*. University of Toronto Press.

Helbich, M., & Leitner, M. (2009). Spatial analysis of the urban-to-rural migration determinants in the Viennese Metropolitan Area: A transition from suburbia to postsuburbia? *Applied Spatial Analysis and Policy, 2*(3), 237–60. https://doi.org/10.1007/s12061-009-9026-8.

Kling, R., Olin, S., & Poster, M. (1995). *Postsuburban California: The transformation of Orange County since World War II*. University of California Press.

Kneebone, E. (2010, January 20). *The suburbanization of poverty: Trends in metropolitan America, 2000 to 2008*. Brookings Institute. https://www.brookings.edu/research/the-suburbanization-of-poverty-trends-in-metropolitan-america-2000-to-2008/.

Lees, L., Morales, E.L., & Shin, H.B. (2016). *Planetary gentrification*. Polity Press.

Moreton-Robinson, A. (2015). *The white possessive: Property, power, and Indigenous sovereignty*. University of Minnesota Press.

Musil, R. (2007). Globalized post-suburbia. *Belgeo, 1*, 147–62. https://doi.org/10.4000/belgeo.11718.

Nair, R., & Carman, T. (2017, October 28). East Vancouver becoming less diverse, census shows. *CBC News*. http://www.cbc.ca/news/canada/british-columbia/east-vancouver-becoming-less-diverse-census-shows-1.4373858.

Nichols, R. (2018). Theft is property! The recursive logic of dispossession. *Political Theory, 46*(1), 3–28. https://doi.org/10.1177/0090591717701709.

Peck, J., Siemiatycki, E., & Wyly, E. (2014). Vancouver's suburban involution. *City, 18*(4–5), 386–415 http://doi.org/10.1080/13604813.2014.939464.

Phelps, N.A., & Wood, A.M. (2011). The new post-suburban politics? *Urban Studies, 48*(12), 2591–610. https://doi.org/10.1177/0042098011411944.

Phelps, N.A, Wood, A.M., & Valler, D.C. (2010). A postsuburban world? An outline of a research agenda. *Environment and Planning A: Economy and Space, 42*(2), 366–83. https://doi.org/10.1068/a427.

Romero, D. (2015, February 18). L.A. is America's "least sprawling" city!? *LA Weekly*. https://www.laweekly.com/l-a-is-americas-least-sprawling-city/.

Roy, A. (2016). Who's afraid of postcolonial theory? *International Journal of Urban and Regional Research, 40*(1), 200–9. https://doi.org/10.1111/1468-2427.12274.

Soja, E.W. (2000). *Postmetropolis: Critical studies of cities and regions*. Basil Blackwell.

Sullivan, J. (2011, October 10). Black America is moving south and to the 'burbs. What's it mean? *Colorlines*. https://www.colorlines.com/articles/black-america-moving-south-and-burbs-whats-it-mean.

Teaford, J.C. (1997). *Post-suburbia: Government and politics in the edge cities.* Johns Hopkins University Press.

Veracini, L. (2012). Suburbia, settler colonialism and the world turned inside out. *Housing, Theory and Society, 29*(4), 339–57. https://doi.org/10.1080/14036096.2011.638316.

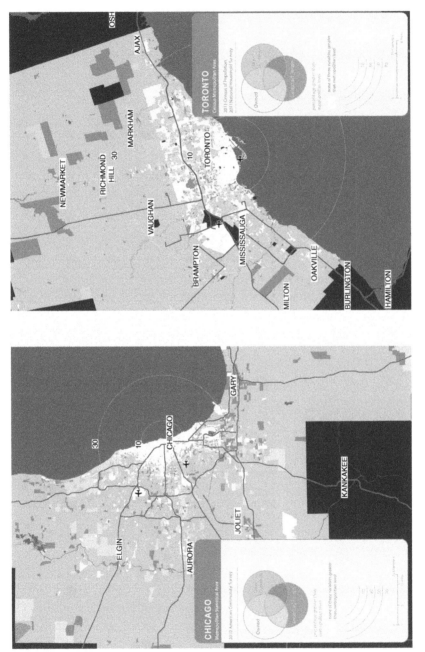

Image 18. Suburban ways of living as defined by homeownership, automobile use, and single-detached dwellings, Chicago and Toronto

18 (Sub)Urban Vibrations: The Suburbanization of Sex Shops and Sex Toys in Australia

PAUL J. MAGINN AND CHRISTINE STEINMETZ

Introduction

The suburbs have long been portrayed as a space of residential settlement by largely white, middle-class, home-owning, nuclear families (Keil, 2018). A space historically defined, (re)produced, and governed by patriarchy and capitalism. These structural forces, combined with scholarly analyses and popular culture representations of suburbia via television and film, have helped reinforce stereotypes about the geographic, physical, and aesthetic character of suburbia and, more crucially, within the context of this chapter, the gendered, sexual, and consumptive practices of suburbanites. Put simply, suburbia has been framed as a space of (1) domesticity, (2) conformity, (3) heteronormativity, and (4) conspicuous consumption (Harris & Larkham, 1999; Hubbard, 2012; Medhurst, 1997). The (suburban) dream of homeownership, single-family homes on a quarter-acre lot with a driveway and garage, front lawns with white picket fences, and separation from the city has offered suburbanites a deep sense of ontological security, territoriality, privacy, and protection from the ills and vices of the city (Saunders, 1990; Vaughan, 2015).

During the mid-twentieth century, as the economy boomed in Anglo-Saxon settler colonies like Canada, the US, and Australia, healthy rates of employment resulted in increased household income and expanding social mobility. Simultaneously, the costs of producing cars and constructing suburban homes fell as a result of Fordist mass production processes providing the freedom of open roads and homeownership to a newly defined, yet up-and-coming, middle class. The signs of upward mobility were also reflected in conspicuous consumption habits of suburbanites who were influenced by a new wave of advertising in the 1950s. As captured with stylistic effect in the hit television series *Mad Men*

(Weiner, 2007–15), suburban homes were bedecked with various domestic appliances (e.g., refrigerators, washing machines, vacuum cleaners, and lawn mower), luxury appliances (e.g., televisions, telephones, and hi-fi systems), and home furnishings for the living room, dining room, and master bedroom, and a car for the garage. *Mad Men* offered realistic insights into cultural representations and commodification (Knox, 2005) of the suburbs as bourgeois utopias (Fishman, 1987). In contemporary times, conspicuous consumption habits of suburbanites, as reflected in the material accoutrements that adorn their homes, are signifiers of taste, status, and identity (Belk, 1988; Holt, 1995; Veblen, 1899/1994).

This chapter, however, is about an increasingly commonplace range of *intimate* household accoutrements found within many suburban homes, typically hidden from plain sight. Specifically, it focuses on the commodification, mass consumption, regulation, and (sub)urbanization of sex shops (also referred to as "restricted premises" in Australia and "adult novelty stores" in the US) and sex toys (e.g., vibrators, dildos, fetish wear, lingerie, and lubricants). These retail spaces and products have often been viewed as deviant, immoral, or criminogenic by politicians and policymakers and thus subject to strict regulations (Kelly & Cooper, 2015; Seaman & Linz, 2014). Such political and policy attitudes seem to run counter to the market demand for sex toys and the growth and diversification of adult retailing over the last few decades. Nevertheless, adult retailing continues to generate vibrations and reverberations within suburbia.

The chapter has three parts. First, an overview of the representation of sex toys in film and television is presented to illustrate their mediated cultural association with the city and suburbia. Second, a review of the shifting geographies and typologies of sex shops with empirical emphasis given to Perth (Western Australia) and Sydney (New South Wales) points to the socio-spatial mainstreaming of sex shops. Third, the extent of the suburbanization of sex toys is illustrated via an analysis of a unique dataset of sex toy sales obtained from an online adult retailer based in Australia. This sales data provides insights into the spatial patterns of "sexual consumership" (Attwood, 2005; Wood, 2017) within Australia. Sexual consumership is indicative of an exploration in human sexuality, or expressions of "sexual dissidence" (Rubin, 1993) and "sexual citizenship" (Bell, 1995). Ultimately, this chapter breaks new ground in the scholarly analysis of the geographies of sexual consumership.

Sexing the Suburbs

Fishman (1987) observed that "[s]uburbia can never be understood solely in its own terms. It must always be defined in relation to its rejected opposite: the metropolis" (p. 27). This has resulted in a series of enduring (sub)urban myths about the supposedly unique physical, social, economic, demographic, cultural, and sexual character of the city and of suburbia. Put simply, "the suburbs" have been stereotyped as a space of family, femininity, domesticity, safety and morality, homogeneity, and heterosexuality and monogamy. Conversely, "the city" has been cast as a space of individuality, masculinity, work, risk and immorality, heterogeneity, and sexual diversity and promiscuity (Hubbard, 2012; Johnston & Longhurst, 2010; Silverstone, 1997). While this neat characterization of the city and suburbia might suggest they are distinctive spaces, this is too simplistic a proposition. The city and the suburbs are engaged in a complex and dynamic symbiotic relationship with one another in economic, social, cultural, and sexual terms.

The sexual mores of (middle-class) suburbanites have been framed as heteronormative, monogamous, and "vanilla," with sex confined to the master bedroom and pursued for the purposes of procreation as opposed to pleasure, thereby reifying the idea of the family home (Gorman-Murray, 2008; Hubbard, 2012). This long-standing, socially conservative view of the sexual character of suburbia has not always been the case, however. Again, as Fishman (1987) notes, "[i]n Shakespeare's London so many houses of prostitution had moved to these disreputable outskirts that a whore was called a 'suburb sinner'" (pp. 6–7). Ultimately, the suburbs have always been a highly complex and dynamic sexual space, with its inhabitants intimately involved in the production and consumption of all manner of sexual practices. As Hubbard (2012) explains, "if one pulls back the net curtains of the 'average' suburban home one can reveal promiscuity, perversion and parody aplenty" (p. 64).

Pop Cultural Representations of Sex Toys

Suburban Vibrations

In cultural terms, this "promiscuity, perversion and parody" is captured in the Hollywood movie *Parenthood* (Howard, 1989), which depicts the fraught relationship travails between the intergenerational members of Gil Buckman's family. Various scenes and characters in the movie capture the complex social realities of sexuality in suburban family life – for example, the introverted teenage son infatuated with pornography, and the rebellious senior high school daughter who challenges suburban sexual norms by taking risqué photos of herself and her boyfriend. One of the most memorable scenes depicting sex in suburban life is an incident involving a family dinner, a power outage, and the search for a flashlight wherein the main character, Gil Buckman, finds himself in possession of a large vibrator that belongs to his divorcee sister. The fact that Gil's sister is divorced, remains unmarried, and owns a vibrator conveniently serves to reinforce her presumed perverted sexuality amongst her family members. As various feminist scholars have noted, the celebration of female masturbation and the use of vibrators and/or dildos by women in the 1970s denoted expressions of sexual exploration and liberation and, moreover, a reclamation of female sexuality and power over their own bodies as well as the politics of gender and sexuality (Comella, 2017). The presence of the vibrator in *Parenthood* ultimately indicates an (albeit secret) expression of sexual dissidence/ citizenship (Rubin, 1993), in the sense that the divorcee has taken control of her body and sexual pleasure. Furthermore, the vibrator is also a marker of sexual consumership.

Ultimately, the sexual proclivities of the Buckman family as portrayed in *Parenthood* offer an insightful, if light-hearted, cultural representation of the realities of the socio-sexual "moral order" of suburbia (Baumgartner, 1988). In many ways the sexual scripts portrayed in *Parenthood* resonate with contemporary debates about porn consumption and amateur porn production (Dines, 2010; McKee et al., 2008; Ruberg, 2016) as well as female sexuality and sexual pleasure (Comella, 2017; Lieberman, 2017; Smith, 2007). While *Parenthood* was a successful movie, grossing USD$126 million at the box office, its mainstream success did not result in a cultural crossover in terms of the commodification of sex toys and a boost in sexual consumership of such products in the real world. This accolade belongs to two other cultural phenomena: (1) the HBO comedy-drama series *Sex and the City* (Star, 1998–2004), and (2) E.L. James's book trilogy *50 Shades of Grey* (2011, 2012a, 2012b). Both *Sex and the City* (*SATC*) and *50 Shades* have helped propel sex toys into mainstream commercial and public consciousness and have initiated scholarly and media discussions about (1) female sexuality and pleasure and (2) fetish/kink practices. Moreover, they have prompted public curiosity, experimentation and, of course, increased sexual consumership (Deller et al., 2013).

Both *SATC* and the *50 Shades* trilogy have been the subject of much scholarly scrutiny, especially within feminist, cultural, and media studies, with particular attention given to how gendered and sexual tropes are used and what they signify or reflect in social reality (Arthurs, 2003; Attwood, 2005; Comella, 2003, 2013, 2017; Wood, 2018). Interestingly, however, the underlying spatial (i.e., [sub]urban) contexts of representations of sexuality, sex toys, and sexual consumership within media such as *SATC*, *50 Shades*, and other popular films and/or TV shows has received tacit acknowledgment. The focus here will be on *SATC*, because although this show presented a counter-narrative to the domesticated suburban housewife, it was nonetheless influential on the sexual citizenship/consumership of suburban women.

Urban(e) Vibrations

The comedy-drama *SATC* centres on the "the dating rituals and sex lives of four single, [Caucasian, 30–40 something,] professional women in New York" (Comella, 2003, p. 109), presenting a mediated, post-feminist representation of these women who are economically independent, materialist consumers, and, crucially, sexually liberated. As Arthurs (2003) highlights, in earlier television comedy-dramas (e.g., *I Love Lucy*, *Leave It to Beaver*, *The Adventures of Ozzie and Harriet*, *Father Knows Best*, and *The Honeymooners*) women were generally prevented, unlike their male counterparts, from "having it all" because home and work were generally depicted as distinct gendered spaces. In *SATC*, however, this gendered trope is turned on its head as "the world of work largely disappears from view as a distinct space ... work is collapsed into the private sphere and becomes another form of self-expression, alongside consumption, thereby side-stepping the post-feminist problematic" (Arthurs, 2003, p. 84). The "have it all" attitude exuded by the four female protagonists is facilitated by the fact that *SATC* is set in New York.

New York arguably provided the ultimate city backdrop for the hypersexual and consumerist behaviours of the lead characters. New York has always had something of an enduring reputation as a highly sexualized space on account of Times Square, a renowned "vice district" since the late nineteenth century (Ryder, 2004) that emerged as the epicentre of all things XXX – theatres, peep shows, sex shops, and prostitution – during the 1970s and 1980s (McNamara, 1994, 1995). This seedy and sleazy aesthetic gradually gave way to a sanitized Disney-land of mass entertainment and corporate towers when the then mayor, Rudy Giuliani, led a crusade against the pornification of Times Square and surrounding neighbourhoods in the early 1990s. By the late 1990s, urban renewal and gentrification, underscored by neoliberal policies and the rise of Wall Street as a key node in global financial services in the early 1980s, had well and truly taken hold within New York (Fainstein et al., 1992; Sassen, 1991). If the 1980s were New York's crass "yuppie era," as exemplified in Tom Wolfe's *Bonfire of the Vanities* (1987), Bret Easton Ellis's *American Psycho* (1991), and director Oliver Stone's *Wall*

Street (1987), then the 1990s came to represent an era of bourgeois bohemia, which characterizes the overall aesthetic of *SATC* (Arthurs, 2003).

The bourgeois bohemians, or BoBos, as Brooks (2000) has referred to them, were essentially upwardly mobile creative and media types who lived in loft apartments in trendy neighbours such as SoHo, Greenwich Village, Tribeca, and the Meatpacking District. Furthermore, they exhibited progressive social attitudes, especially in relation to sex, and a predilection for materialist consumption (Brooks, 2000). *SATC* (re)presents a post-feminine interpretation of the bourgeois bohemia reflected in the material consumerism of the lead characters. Simultaneously, the bohemian id of *SATC*'s main protagonists is depicted in their sexual attitudes and behaviours, as evidenced in the frank and detailed conversations they have with one about their sexual fantasies, conquests, and mishaps. As Arthurs (2003) notes, "[*SATC*] publicly repudiates the shame of being [a single female] and sexually active in defiance of bourgeois codes that used to be demanded of respectable women" (p. 85). This is evident in terms of how female masturbation, sexual pleasure, and sex toys are depicted in the series, although there are fleeting signs of conservative suburban sensibilities expressed when the topic of vibrators is raised.

There are three episodes – "The Turtle and the Hare" (Fields et al., 1998), "Ghost Town" (Spiller et al., 2001), and "Critical Condition" (King et al., 2002) – where female sexuality and sex toys feature prominently in the storyline. "The Turtle and the Hare" is arguably the most culturally and commercially impactful of these episodes in that the featured Rampant Rabbit vibrator was responsible for a "dramatic rise in sales of this sex toy" (Attwood, 2005, p. 393). It is in this episode that Miranda takes Charlotte and Carrie to the Pleasure Chest – an adult store –in order to introduce them to the rabbit vibrator. Although it "reads like a cautionary tale warning woman not to masturbate with sex toys" (Liberman, 2017, p. 271), "The Turtle and the Hare" has played a critical role in provoking the cultural mainstreaming and (sub)urban legitimization of sex toys. Since this episode of *SATC*, there have been numerous feature articles about the rabbit vibrator and other sex toys, as well as female pleasure more broadly, in popular magazines such as *Cosmopolitan*, *Elle*, and *Glamour*. Relatedly, there has been a growth in feminine-oriented sex shops and erotic boutiques and a rise in online adult retailing, both of which have made it much easier, safer, and pleasant for women to shop for sex toys (Crewe & Martin, 2017; Maginn & Steinmetz, 2015; Martin, 2015, 2016). In the last decade, sex toys have increasingly featured in more TV shows, which have pushed and blurred the sociocultural boundaries of gender, sexuality, and sexual practices. These have included *Broad City* (Comedy Central, 2014–19);

Mad Men (Lionsgate, 2007–15); *The L Word* (Showtime, 2004–9); *Younger* (TV Land, 2015–21); *Sense8* (Netflix, 2015–18); *Love* (Netflix, 2016–18); *Girls* (HBO, 2012–17); *Orange Is the New Black* (Netflix, 2013–19); and *Grace and Frankie* (Netflix, 2015–22). This increased pop cultural presence of sex toys has also coincided with an increase in the number and type of adult stores and the consumption of sex toys.

The Geography/Typology of Adult Stores

Sex toys have become big business. A recent report by Bloomberg (2021) notes that the global adult toy market in 2021 was valued at almost USD$31 billion and was predicted to grow to USD$55.6 billion by 2028. In terms of global regional markets, the three largest markets are Asia-Pacific and China, Europe, and North America. In another survey, this time focusing on online searches for sex toys, the UK-based firm VoucherCloud (n.d.) found that the number of yearly sex toy searches per 1,000 internet users was highest in Denmark (118), Sweden (115), and Greenland (108). The remaining top ten countries, in rank order, were as follows:

> #4 – the US (104)
> #5 – the UK (96)
> #6 – the Netherlands (88)
> #7 – Russia (87)
> #8 – Bulgaria (86)
> #9 – Italy (84)
> #10 – Australia (82)

Based on an analysis of adult stores listed in the Yellow Pages, research conducted by the Australian data analytics firm ipData found that Adelaide, a predominantly suburban metropolitan region in South Australia, had the highest number of adult stores (5.22) per 100,000 people within Australia (Fleet, 2014). In short, then, the global scale of the sex toy market, social curiosity in sex toys, and the number of adult stores could be indicative of the mainstreaming of sexual consumership *and* sexual citizenship (Bell, 1995; Brents & Sanders, 2010; Maginn & Steinmetz, 2015).

From Out of Sight, Out of Mind to Online High Street

The spatial and typological contours of adult retailing in countries such as the UK (Martin, 2016), Australia (Maginn & Steinmetz, 2015) and the US (Comella, 2017; Kelly & Cooper, 2015; Lieberman, 2017) have evolved significantly over the past forty years. Adult stores have progressively moved from peripheral locations within the inner city to not only the high/main street but also shopping malls and even suburban homes via sex toy parties and online shopping. This dynamic geography points to an increasing diversification in the aesthetic design, consumer orientation, and philosophy of adult stores. Put simply, in the 1960s and 1970s, adult stores were primarily seen as "seedy and sleazy" hyper-masculine spaces that sold a mix of pornography and "marital aids," with some even offering live peep shows. These were generally located within "vice districts" such as Times Square (NYC), Soho (London), Kings Cross (Sydney), and other marginalized areas within the city.

By the 1980s, corporatized chain stores such as Ann Summers (UK), Adam and Eve (US), Beate Uhse (Germany), and ClubX (Australia) had become increasingly more visible. These stores occupied a mix of marginal and high street locations, but sought to broaden their consumer base, especially in terms of gender. While feminist-oriented stores (e.g., Good Vibrations and Eve's Garden) had also been established in the US during the 1970s, it was not until the 1990s that they became more prominent, with stores such as Toys in Babeland established in Seattle and New York, Sh! opening in London in 1992, and New Zealand's D.VICE in 1998. More upmarket adult stores, or "erotic boutiques," such as London-based Agent Provocateur, Myla, and Coco de Mer (see Kent & Berman Brown, 2006), plus Australian retailer Honey Birdette, underpinned by their own interpretations of sex-positive feminism, came to prominence in the early 2000s. These stores tended to occupy high-end retail precincts and/or (sub)urban shopping centres (Maginn & Steinmetz, 2015; Sanders-McDonagh & Peyrefitte, 2018).

Interestingly, these erotic boutiques have essentially evaded planning and licensing regulations long imposed against traditional sex shops (Crewe & Martin, 2017). This is largely because these newer types of stores have tended not to sell pornography and have devoted more floorspace to lingerie and clothing as opposed to sex toys. As such, they do not stand out as "sex shops" per se. They have nonetheless come in for criticism from some feminist advocacy groups for their advertising and window displays, which are seen as objectifying women and being inappropriate for public spaces inhabited by children and young people (Kalms, 2017).

Adult retailing is generally associated with brick-and-mortar stores. Notably, however, the sale of sex toys and related apparel have also occupied off-street locations as well. Before there were sex shops, vibrators were marketed as medical instruments or consumer appliances used for muscle and facial massage purposes and sold via mail order with adverts in magazines, catalogues, and newspapers (Juffer, 1998). As Lieberman (2017) notes,

> The earliest ad for a consumer electric vibrator in America appeared in 1899 in *McClure's Magazine*. This ad positioned the Vibratile vibrator as both a beauty and health device, suitable for "removing wrinkles" and "curing nervous headache." By 1909, Vibratile was joined by at least twenty other companies hawking similar vibrators. Vibrator ads were everywhere in the early 20th century; *Popular Mechanics, McCall's*, and even Christian publications featured ads for vibrators, as did regional newspapers in Racine, Wisconsin, and national papers like the *New York Times*. Vibrators were available for mail-order in Sears Roebuck catalogs, while department stores like Macy's always kept them in stock. (p. 34)

Mail order shopping facilitated the infiltration of sex toys into suburbia. Of course, this was not the only means by which to purchase these products. Using a business model dating back to the early 1950s, when mass suburbanization was in full swing, the Tupperware Corporation utilized a large network of suburban housewives to host demonstration parties of "a product which developed contemporaneously with postwar suburbia [and] wholeheartedly embraced domesticity and conspicuous consumption" (Clarke, 1997, p. 132) – Tupperware! Sex toys also made their way directly into suburban homes via these demonstration parties. This approach was adopted in the US in the late 1970s by sex toy company Doc Johnson (Lieberman, 2017); by the British sex shop retailer Ann Summers in the 1980s (Storr, 2003); and, in the early 2000s, by Australia retailers such as Honey Birdette and PASH (Jackman, 2010). Ultimately, this direct marketing and selling approach meant that women could avoid going to sex shops that were still largely unwelcoming, (hyper-)masculine spaces throughout the 1980s and 1990s.

Purchasing sex toys and avoiding being caught entering or leaving a sex shop has been made easier with the emergence of online retailing over the last two decades. Prospective consumers can now easily browse and purchase products via their PC, laptop, tablet, or smart phone and have them delivered under plain packaging to their suburban homes. Most traditional brick-and-mortar adult stores now have an online presence so that they can extend their retail threshold. Other adult retailers only have an online presence. Lovehoney, a UK-based firm established in 2002 and located on an industrial estate in the outer suburbs of Bath, England, is arguably one of the most notable examples of this type of adult retailer. A recent report highlights that Lovehoney saw its sales revenue across its UK, Europe, US, and Australia markets increase by 60 per cent between 2015/16 and 2017/18, from £58 million to £93 million (Rigby, 2017). They have also been instrumental in facilitating the crossover of sex toys from mediated cultural materiality to mass consumption via their licensing and distributing of branded sex toys. These brands include (1) the *50 Shades of Grey* collection; (2) sex toys inspired by the US television show *Broad City*; and (3) a series of branded "pleasure tools" by the rock 'n' roll bands Motörhead and Mötley Crüe, who both have large suburban fanbases.

Adult stores of various kinds – "seedy and sleazy," "corporate chains," and "erotic boutiques" (Maginn & Steinmetz, 2015) – remain a feature of the (sub)urban retail landscape. It would appear, however, that the "seedy and sleazy" type are in decline. While prospective consumers can engage in virtual window shopping via the internet, purchasing adult products still necessitates being able to inspect them in person to make an informed purchasing decision. The growth and diversification in the geography and typology of brick-and-mortar stores has made it relatively easier for prospective consumers to visit adult stores, particularly for women and sexual minorities wanting to explore their sexuality and assert their sexual citizenship. As Bell (1995) has noted, "[w]ithin a late capitalist economy, then, we can argue that spaces of sexual citizenship are to some extent constituted through consumption" (p. 141). The diversity of adult stores plus other sexualized spaces (e.g., strip clubs, brothels, and queer spaces) within the city act as a signifier of the city's *cosmo-sexuality* – that is, its recognition and acceptance of diverse sexualities, plus consensual commercial sexual practices and spaces (Maginn & Steinmetz, 2015).

Suburbanization of Adult Retailing: Perth and Sydney

The suburbanization of adult retailing is evident in the geography of adult stores within the Perth (Western Australia) and Sydney (New South Wales) metropolitan regions. Table 18.1 highlights the overall number of stores identified within each local government area (LGA) in metropolitan Perth (2009 and 2015) and Sydney (2011 and 2015) plus the distance of each LGA from its respective central business district (CBD). Both metropolitan regions experienced a net increase in the number of stores. Perth saw the number of stores increase from 45 to 48 (6.6 per cent), and Sydney's increased from 68 to 76 (11.8 per cent). Based on the estimated resident population (ERP) of both metropolitan

Table 18.1. Geography of adult stores in Perth (2009/2015) and Sydney (2011/2015) by LGA

	Perth Metropolitan Region					Sydney Metropolitan Region			
LGA	Distance from CBD (kms)	2009	2015	Change	LGA	Distance from CBD (kms)	2011	2015	Change
Perth	0	7	5	−2	Sydney	0	2	3	+1
Vincent	3	4	4	0	North Sydney	3	2	2	0
Subiaco	3	1	0	−1	Wolhara	5	0	1	+1
Victoria Park	5	5	3	−2	Randwick	6	0	1	+1
Bayswater	7	0	3	+3	Marrickville	7	7	7	0
Belmont	8	5	4	−1	Botany Bay	7	1	2	+1
Canning	10	1	4	+3	Waverley	7	1	1	0
Stirling	10	6	3	−3	Rockdale	12	3	4	+1
Fremantle	19	4	3	−1	Ryde	12	3	3	0
Swan	20	3	4	+1	Burwood	12	1	1	0
Gosnells	20	2	2	0	Canterbury	15	2	1	−1
Cockburn	24	3	3	0	Bankstown	17	3	3	0
Joondalup	25	2	2	0	Warringah	18	3	3	0
Wanneroo	25	1	2	+1	Hornsby	21	2	1	−1
Armadale	28	0	1	+1	Hurstville	23	2	1	−1
Rockingham	40	1	4	+3	Parramatta	25	3	4	+1
Mandurah	75	0	1	+1	Sutherland	26	1	2	+1
					Liverpool	27	3	3	0
					Fairfield	30	1	2	+1
					Blacktown	35	3	2	−1
					Penrith	50	3	5	+2
					Campbelltown	55	2	4	+2
TOTAL		45	48	+3	TOTAL		68	76	8

regions, Perth had 2.62 (ERP = 1,711,756) and 2.44 (ERP = 1,964,398) adult stores per 100,000 people in 2009 and 2015 respectively. Sydney, in comparison, had 1.60 (ERP = 4,240,340) and 1.67 (ERP = 4,547,327) adult stores per 100,000 people in 2011 and 2015 respectively. Since "the suburb" is the most common unit of geography within Australian cities, all sex shops are effectively suburban in terms of their spatial character. If, however, the capital city LGAs – the City of Perth and the City of Sydney – are deemed urban and all other LGAs are suburban, this means that the majority of adult stores in Perth – 2009 (n = 38, or 84.5 per cent) and 2011 (n = 43, or 89.5 per cent) – and Sydney – 2011 (n = 46, or 67.6 per cent) and 2015 (n = 53, or 69.7 per cent) – were located in suburban areas.

In terms of inner-ring (up to 5 kilometres), middle-ring (between 6 and 19 kilometres), and outer-ring (20+ kilometres) LGAs within Perth, there were 17, 16, and 12 adult shops respectively in 2009. By 2015, this had changed to 12, 17, and 19 stores, thereby indicating a more pronounced suburbanization of adult retailing. In Sydney, the distribution of adult stores across the different suburban rings increased by approximately the same volume, up from 24, 24, and 20 stores in 2011 to 27, 27, and 24 stores by 2015. The general increase in the number of stores, especially in middle- and outer-ring suburbs, points to a mix of market and planning forces at play. In terms of market forces, the establishment of new stores is indicative of an increase in real/perceived demand for sex toys and other adult-related products.

As discussed at the beginning of this chapter, there has been somewhat of a cultural mainstreaming of sex toys due, in part, to their presence in various television shows and films over the last decade or so. Notably, the increase in the number of adult stores in Perth and Sydney between 2009 and 2015 coincides with the release of the *50 Shades of Grey* book trilogy that sold 100 million copies globally by February 2014 (Bosman, 2014). The *50 Shades* books are argued to have stimulated wider social interest in sex toys and kink/fetishism, particularly amongst "suburban housewives," the primary readership base. According to Deller et al. (2013), sex toy manufacturers and adult retailers sought to exploit this interest via the commodification of kink/fetish practices:

> Retailers were quick to capitalize on the success of the [*50 Shades* books], with bookshops promoting the "next *Fifty Shades*" and adult stores using the novels as a selling point, offering *Fifty Shades* themed evenings and imploring people to buy the items featured in the novels in-store. Indeed, adult retailer Lovehoney released an official *Fifty Shades* range of merchandise including handcuffs and paddles. (p. 859)

Figures 18.1 and 18.2 highlight the geographical distribution of the key types of adult stores – "seedy and sleazy," "mainstream/chain," and "elite/specialist" – in 2015 in Perth and Sydney respectively. The central business district (CBD) in both metropolitan regions had the greatest concentration of adult stores. This can be explained by the historical fact that the CBD was the "natural area" (Park & Burgess, 1925) for this type of land-use, where the city permitted such activities due in large

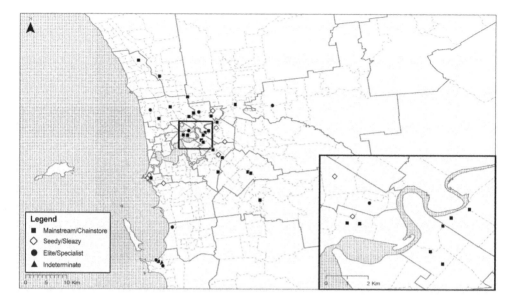

Figure 18.1. Geography of adult stores, Perth, 2015

part to market demand from the largely male working population. In Perth, the adult store landscapes within the inner city and the suburbs were dominated by mainstream/chain stores (e.g., @AdultShop.Com, ClubX, and Lovers Adult Stores) with a smaller number of seedy and sleazy stores (e.g., Vibrations, Taboo, and Vic Park Book Exchange). The elite/specialist adult stores were very much a minor aspect of the adult retail landscape in Perth. The erotic boutique chain Honey Birdette was the dominant player with five stores located in suburban shopping malls across Perth. Conversely, the Sydney CBD comprised an even mix of mainstream/chain stores such as Club X, Adult World, and The Pleasure Chest; seedy and sleazy stores like Sydney Adult Book Exchange and Everything Adult; and elite/specialist retailers such as Honey Birdette, Hapsari, and Toolshed. In suburban Sydney, the seedy and sleazy type of store was slightly more common than mainstream/chain stores.

The spatial distribution of adult shops in Perth and Sydney shows the clear suburbanization of this type of retailing. This is despite the opposition that adult stores often face from local councillors and restrictive planning regimes that seek to control the location, signage, and window displays of adult stores in an effort to render them invisible. In Perth

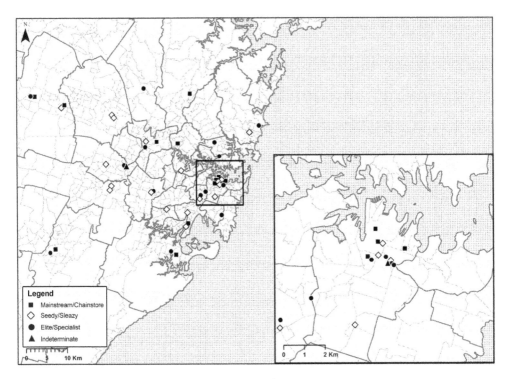

Figure 18.2. Geography of adult stores, Sydney, 2015

and Sydney, adult stores are not deemed a "normal" form of retailing within local planning schemes because of the types of products they sell. In Perth, adult stores do not have automatic planning approval within any land-use zones set out in local planning schemes. They are permitted in certain types of zones – town centres, commercial centres, industrial areas, and mixed-use areas – but planning approval is discretionary. This often means that approval may be subject to scrutiny from planners *and* local councillors on the council's planning sub-committee. In Sydney, adult stores are permitted in the same types of zones: local centres, commercial cores, and mixed-use areas. Although permitted in these zones, adult stores will not necessarily be granted development approval. While planning officers may grant approval because a development application for an adult store meets the various requirements of the local planning scheme, there is always the chance that any such application may be politicized by councillors, vocal local residents, or businesses who are opposed to adult retailing.

A Snapshot of Sexual Consumership

The increased number of adult stores in Perth and Sydney (and indeed in other Australian cities) is indicative of the wider social, cultural, and economic mainstreaming of sex toys. Interestingly, however, there has been no scholarly analysis on the geographies of the sale of sex toys at the city and/or suburban level. This chapter contributes to that research gap by providing a snapshot of the spatial distribution of the sales of an online adult retailer – SexToys247/Frisky.com.au – based in Melbourne. The focus here is on the geography of the volume of sales (i.e., 66,561 orders) over the course of a thirty-nine-month period (March 2012 to May 2015).

Unlike brick-and-mortar adult stores, online retailers have – at least theoretically – a much greater spatial threshold in terms of consumer footfall because prospective consumers can engage in virtual window shopping from anywhere in the world. Furthermore, although adult stores may be in prominent locations such as high streets or suburban shopping malls, there still appears to be some embarrassment amongst many people about entering/leaving such stores. Hence, there is a possibility that (many) people who frequent brick-and-mortar adult stores will visit one outside their normal retail activity zone in order to minimize the risk of being observed entering/leaving by someone they know. The stigma associated with frequenting an adult store is somewhat reduced by online shopping because purchases can be made anonymously from home.

Figure 18.3 provides an illustration of the geography of sales for SexToys247/Frisky.com.au at the postcode level. It is clear that their customer base extends throughout most postcode areas within metropolitan, regional, and even remote Australia. Although this online retailer is based in Melbourne, the state of Victoria was only its third biggest market as measured by volume of sales (i.e., 12,200 orders, or 18.3 per cent). A majority (57 per cent) of orders went to suburban postcode areas, 41 per cent of orders went to regional (i.e., rural) Victoria, and the remaining 2 per cent of orders were for customers within the Melbourne CBD postcode. New South Wales (NSW) was the largest state market with almost 22,200 orders (30.2 per cent of total orders). This was followed by Queensland with almost 16,700 orders (25 per cent), and then Western Australia (WA) with 8,400 orders (12.6 per cent).

In NSW, the majority of orders (56.4 per cent) were from customers located in regional areas. A total of 41.2 per cent of orders were sent to suburban customers in metropolitan Sydney. The remaining 2.4 per cent of orders were for customers located within the Sydney CBD, an

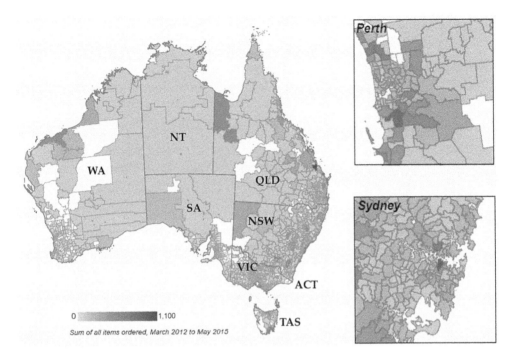

Figure 18.3. Total number of item orders by postcode. Data source: *SexToy247/Frisky.com.au* (Sales data March 2012–May 2015)

area with many brick-and-mortar adult stores. In WA, the majority of orders (67 per cent) went to customers in outer suburban postcodes. A small proportion of orders (1.5 per cent) went to customers within the CBD postcode area. The remaining share of orders (31.5 per cent) went to towns in regional WA, where brick-and-mortar adult stores are few and far between. That said, adult stores do exist in various regional towns across Australia. The presence of stores in these areas, plus the spatial distribution of orders from SexToys247/Frisky.com and other online adult stores within and outside Australia, clearly point to the suburbanization and ruralization of sex toys.

Conclusions

During the twentieth century, suburbia was stereotyped as a space of conformity, domesticity, and conspicuous consumption. Relatedly, the sexual mores of suburbanites have been portrayed as private, heteronormative, and monogamous, and their sexual proclivities characterized as "vanilla." In contrast, "the city" was a highly sexualized and risqué space, as this was the "natural area" for the proliferation of commercial sexual premises such as brothels, XXX theatres, strip clubs, peep shows, and sex shops. Of course, many, if not most, of the consumers of these sexualized spaces were male suburbanites who worked in the city. In other words, suburban sexual pleasure-seeking and consumership has always been expressed – albeit surreptitiously and predominantly by men.

The commercial sex shops that emerged in the late 1960s and early 1970s were also once the preserve of a predominantly male consumer base and often located in marginalized spaces within the city, far from the suburbs. Despite this spatial separation, sex toys/marital aids (and pornography) found their ways into suburban homes via various modes since the 1950s. These included (1) mail order catalogues, (2) home-based demonstration parties, (3) sex shops, and, more recently, (4) online adult retailing. By the 1990s, brick-and-mortar sex shops had become increasingly more diverse in terms of location (back street, high street, shopping mall, and online), type ("seedy and sleazy," "corporate chains," and "erotic boutiques"), and consumer base, with different types of stores catering to male/female/coupled and heterosexual/LGBT customers. The growth and diversification of adult retailing paralleled a pop cultural mainstreaming of sex toys (and sexuality more broadly) that began in the late 1990s. The hit US television series *Sex and the City* arguably lays claim to being responsible for the beginnings of the mass feminization, commodification, and consumption of sex toys in the late 1990s. It would be another fourteen years (i.e., 2011) before the first *50 Shades of Grey* book was released, when adult retailing and sex toys experienced its second major wave of mass commodification and consumption.

Both *SATC* and *50 Shades*, plus broader changes in sexual attitudes, social mobility, and the rise of the internet over the last twenty to thirty years, have helped transform the sexuality of and sexual consumership habits within the suburbs in Australia and globally. Sex toys are now commonplace in many suburban households – coupled and single – and emblematic of an increasingly sexually liberated and adventurous mindset. Anecdotal observations seem to suggest that brick-and-mortar adult stores, especially "seedy and sleazy" ones, are closing down as more and more adult retailing takes place online – the virtual high street/shopping mall. The sex toy industry continues to evolve with more female entrepreneurs entering the sextech industry and working hard to develop products that centre on female sexuality and pleasure. This is arguably what will underpin the next major wave in the mass commodification and consumption of sex toys.

ACKNOWLEDGMENTS

We would like to express our thanks to SexToys247/Frisky.com.au for providing us with the sales data that underpins part of the empirical analysis in this chapter. Thanks also to Dr. Alistair Sisson (University of Wollongong) for producing figures 18.1–18.3.

REFERENCES

Arthurs, J. (2003). *Sex and the City* and consumer culture: Remediating postfeminist drama. *Feminist Media Studies, 3*(1), 83–98. https://doi.org/10.1080/1468077032000080149.

Attwood, F. (2005). Fashion and passion: Marketing sex to women. *Sexualities, 8*(4), 392–406. https://doi.org/10.1177/1363460705056617.

Baumgartner, M.P. (1988). *The moral order of a suburb*. Oxford University Press.

Belk, R.W. (1988). Possessions and the extended self. *Journal of Consumer Research, 15*(2), 139–67. https://doi.org/10.1086/209154.

Bell, D. (1995). Pleasure and danger: The paradoxical spaces of sexual citizenship. *Political Geography, 14*(2), 138–53. https://doi.org/10.1016/0962-6298(95)91661-M.

Bloomberg. (2021, December 10). Sex toys market size worth $55,648.88 Mn, globally, by 2028 at 7.6% CAGR – exclusive report by the Insight Partners. *Bloomberg.* https://www.bloomberg.com/press-releases/2021-12-10/sex-toys-market-size-worth-55-648-88-mn-globally-by-2028-at-7-6-cagr-exclusive-report-by-the-insight-partners.

Bosman, J. (2014, February 26). For *Fifty Shades of Grey*, more than 100 million sold. *New York Times.* https://www.nytimes.com/2014/02/27/business/media/for-fifty-shades-of-grey-more-than-100-million-sold.html.

Brents, B., & Sanders, T. (2010). Mainstreaming the sex industry: Economic inclusion and social ambivalence. *Journal of Law and Society, 17*(1), 40–60. https://doi.org/10.1111/j.1467-6478.2010.00494.x.

Brooks, D. (2000). *Bobos in paradise: The new upper class and how they got there.* Simon & Schuster.

Business Wire. (2016). *Global adult toys market to exceed USD 29 billion by 2020, according to Technavio.* https://www.businesswire.com/news/home/20160412005747/en/Global-Adult-Toys-Market-Exceed-USD-29.

Clarke, A.J. (1997). Tupperware: Suburbia, sociality and mass consumption. In R. Silverstone (Ed.), *Visions of suburbia* (pp. 132–60). Routledge.

Comella, L. (2003). (Safe) *Sex and the city*: On vibrators, masturbation, and the myth of "real" sex. *Feminist Media Studies, 3*(1), 109–12.

Comella, L. (2013). *Fifty shades* of erotic stimulation. *Feminist Media Studies, 13*(3), 563–6. https://doi.org/10.1080/14680777.2013.786269.

Comella, L. (2017). *Vibrator nation: How feminist sex-toy stores changed the business of pleasure*. Duke University Press.

Crewe, L., & Martin, A. (2017). Sex and the city: Branding, gender and the commodification of sex consumption in contemporary retailing. *Urban Studies, 54*(3), 582–99. https://doi.org/10.1177/0042098016659615.

Deller, R.A., Harman, S., & Jones, B. (2013). Introduction to the special issue: Reading the *Fifty Shades* phenomena. *Sexualities, 16*(8), 859–63. https://doi.org/10.1177/1363460713508899.

Dines, G. (2010). *Pornland: How porn has hijacked our sexuality*. Beacon Press.

Fainstein, S.S., Gordon, I., & Harloe, M. (1992). *Divided cities: New York and London in the contemporary world*. Blackwell.

Fields, M. (Director), Star, D. (Writer), Avril, N. (Writer), & Kolinsky, S. (Writer). (1998, August 2). The turtle and the hare (Season 1, Episode 9) [TV series episode]. In D. Star (Creator), *Sex and the city*. HBO.

Fishman, R. (1987). *Bourgeois utopias: The rise and fall of suburbia*. Basic Books.

Fleet, N. (2014, February 26). Australiasbestcity.com.au says Adelaide is not the city of churches after all. *The Advertiser*. https://www.adelaidenow.com.au/travel/australiasbestcitycomau-says-adelaide-is-not-the-city-of-churches-after-all/news-story/c0cae1bf386fce08b9d2b857bfe0c790.

Gorman-Murray, A. (2008). Masculinity and the home: A critical review and conceptual framework. *Australian Geographer, 39*(3), 367–79. https://doi.org/10.1080/00049180802270556.

Harris, R., & Larkham, P.J. (Eds.). (1999). *Changing suburbs: Foundation, form and function*. E & FN Spon.

Holt, D.B. (1995). How consumers consume: A typology of consumption practices. *Journal of Consumer Research, 22*(1), 1–16. https://doi.org/10.1086/209431.

Howard, R. (Director). (1989). *Parenthood* [Film]. Universal Pictures.

Hubbard, P. (2012). *Cities and sexualities*. Routledge.

Jackman, C. (2010, February 6–7). Good vibrations. *The Weekend Australian Magazine*, 12–16.

James E.L. (2011). *Fifty shades of grey*. Vintage Books.

James, E.L. (2012a). *Fifty shades darker*. Vintage Books.

James, E.L. (2012b). *Fifty shades freed*. Vintage Books.

Johnston L., & Longhurst, R. (2010). *Space, place and sex: Geographies of sexualities*. Rowman & Littlefield.

Kalms, N. (2017). *Hypersexual city: The provocation of soft-core urbanism.* Routledge.

Keil, R. (2018). *Suburban planet: Making the world urban from the outside in.* Polity Press.

Kelly, E.D., & Cooper, C. (2015). From perception to reality: Negative secondary effects and effective regulation of sex businesses in the United States. In P.J. Maginn & C. Steinmetz (Eds.), *(Sub)urban sexscapes: Geographies and regulation of the sex industry* (pp. 214–60). Routledge.

Kent, T., & Berman Brown, R. (2006). Erotic retailing in the UK (1963–2003): The view from the marketing mix. *Journal of Management History, 12*(2), 199–211. https://doi.org/10.1108/13552520610654087.

King, M.P. (Director), Star, D. (Writer), & Junge, A. (Writer). (2002, August 25). Critical condition (Season 5, Episode 6) [TV series episode]. In D. Star (Creator), *Sex and the city.* HBO.

Knox, P. (2005). Vulgaria: The re-enchantment of suburbia. *Opolis, 1*(2), 33–46. https://escholarship.org/uc/item/5392f4vq.

Lieberman, H. (2017). *Buzz: A stimulating history of the sex toy.* Pegasus Books.

Maginn, P.J., & Steinmetz, C. (Eds.). (2015). *(Sub)urban sexscapes: Geographies and regulation of the sex industry.* Routledge.

Martin, A. (2015). Sex shops in England's cities: From the backstreets to the high streets. In P.J. Maginn & C. Steinmetz (Eds.), *(Sub)urban sexscapes: Geographies and regulation of the sex industry* (pp. 44–59). Routledge.

Martin, A. (2016). Plastic fantastic? Problematising post-feminism in erotic retailing in England. *Gender, Place & Culture, 23*(10), 1420–31. https://doi.org/10.1080/0966369X.2016.1204994.

McKee, A., Lumby, C., & Albury, K. (2008). *The porn report.* Melbourne University Press.

McNamara, R.P. (1994). *The Times Square hustler: Male prostitution in New York City.* Praeger.

McNamara, R.P. (1995). *Sex, scams and street life: The sociology of New York City's Times Square.* Praeger.

Medhurst, A. (1997). Negotiating the gnome zone: Versions of suburbs in British popular culture. In R. Silverstone (Ed.), *Visions of suburbia* (pp. 240–68). Routledge.

Park, R.E., & Burgess, E.W. (1925). *The city.* University of Chicago Press.

Rigby, C. (2017). *Lovehoney marks 15th birthday with 76% profits rise.* Internet Retailing. https://internetretailing.net/themes/themes/lovehoney-marks-15th-birthday-with-76-profits-rise-15449.

Ruberg, B. (2016). Doing it for free: Digital labour and the fantasy of amateur online pornography. *Porn Studies, 3*(2), 147–59. https://doi.org/10.1080/23268743.2016.1184477.

Rubin, G.S. (1993). Thinking sex: Notes for a radical theory on the politics of sexuality. In A. Habelove, M.A. Barale, & D. Halperin (Eds.), *Lesbian and gay studies reader* (pp. 3–44). Routledge.

Ryder, A. (2004). The changing nature of adult entertainment districts: Between a rock and a hard place or going from strength to strength? *Urban Studies, 41*(9), 1659–86. https://doi.org/10.1080/0042098042000243093.

Sanders-McDonagh, E., & Peyrefitte, M. (2018). Immoral geographies and Soho's sex shops: Exploring spaces of sexual diversity in London. *Gender, Place & Culture: A Journal of Feminist Geography, 25*(3), 351–67. https://doi.org/10.1080/0966369X.2018.1453487.

Sassen, S. (1991). *The global city.* Princeton University Press.

Saunders, P. (1990). *A nation of homeowners.* Routledge.

Seaman, C., & Linz, D. (2014). Are adult businesses crime hotspots? Comparing adult businesses to other locations in three cities. *Journal of Criminology, 2014,* Article 783461. https://doi.org/10.1155/2014/783461.

Silverstone, R. (Ed.). (1997). *Visions of suburbia.* Routledge.

Smith, C. (2007). Designed for pleasure: Style, indulgence and accessorized sex. *Journal of Cultural Studies, 10*(2), 167–84. https://doi.org/10.1177/1367549407075901.

Spiller, M. (Director), Star, D. (Writer), & Heinberg, A. (Writer). (2001, June 4). Ghost town (Season 4, Episode 5) [TV series episode]. In D. Star (Creator), *Sex and the city.* HBO.

Star, D. (Creator). (1998–2004). *Sex and the city* [TV series]. HBO.

Storr, M. (2003). *Latex and lingerie: Shopping for pleasure at Ann Summers.* Berg.

Vaughan, L. (Ed.). (2015). *Suburban urbanities: Suburbs and the life of the high street.* UCL Press.

Veblen, T. (1994). *The theory of the leisure class.* Penguin. (Original work published in 1899)

VoucherCloud. (n.d.). *Where's your country in the sex toy world rankings?* Retrieved 14 January 2022, from https://www.vouchercloud.com/resources/sex-toy-world-rankings.

Weiner, M. (Creator). (2007–15). *Mad men* [TV series]. Lionsgate Television.

Wood, R. (2017). *Consumer sexualities: Women and sex shopping.* Routledge.

Image 19. The suburban cul-de-sac reimagined

19 Manhattan in Orange County: Lippo and the Shenzhen of Indonesia

ABIDIN KUSNO

Meikarta surpasses anything this country has ever seen, epic in its scale and vision as a truly integrated city of the future. Not only does Meikarta redefine what a modern city should look like and feel like, it sets the new standard for a world city in Southeast Asia and beyond.

– Lippo Homes, "Meikarta"

On 17 August 2017, at the 72nd anniversary of Indonesian Independence, the Lippo Group, a family-run conglomerate and one of the largest property developers in Indonesia, launched its "largest project ever." The project, called "Meikarta" (named after the wife of the group's founder, Mochtar Riady), is located in Cikarang, Bekasi, about thirty-four kilometres east of Jakarta, where the Lippo Group promises to develop an area of 5,400 hectares into the "Shenzhen of Indonesia." The first phase of development, with a value exceeding $USD19 billion, involves 500 hectares (with 100 hectares of open green space), 250,000 units of prime residential property, and 1,500,000 square metres of prime commercial space. James Riady, the CEO of Lippo Group and son of the group's founder, told media that "we foresee that the future of the Indonesian economy is on the outskirts of Jakarta. Together with the government's massive infrastructure development plans – which aim to connect the capital with the Greater Jakarta area – we believe our project can create jobs for at least 6 to 8 million people in the near future" (as cited in Soegiarto, 2017). James Riady also suggests that Meikarta will outcompete Jakarta and become the new economic centre of the country. The timing of the launch coinciding with Indonesia's Independence Day, the allusion to the future, and the CEO's remark about reviving the national economy all suggest that the Lippo Group is leading the nation through a production of a new mega-city. The new city is not to be a suburb of either Jakarta or Bandung, the two closest major postcolonial cities. Instead, it is to be a new "world city."

The Meikarta project, however, carries with it malpractices. Before necessary permits were obtained, the Lippo Group was already selling the property to buyers. To fast track the project, the executive directors of the project bribed officers of the Bekasi regency to issue licences (Kahfi, 2018). Considering the unclear regulatory framework of local government, however, such a strategy is not unusual for any group pioneering development projects in the country. It is understood that developers of major development projects require good networks within government institutions to help ease or speed up the completion of their projects. Roy (2015) refers to such spatial-economic transactions between the state and developers as a legitimized practice of informality, even though it is a violation of planning and land-use regulations. Yet, today, this tradition of "elite informality" is being undermined by the law enforcement of the Indonesian Corruption Eradication Commission. Meikarta has thus been under a legal spotlight since the arrest of the Bekasi regent and certain directors of Lippo Cikarang for their roles in bribery connected to the issuance of property permits for the project. There is much to unpack here in considering this largest-ever project of the Lippo Group.

This chapter seeks to contribute to the discussion around the idea of "after suburbia" by teasing out components, discourses, and forces that constitute the property development of Lippo Group in which Meikarta is a part. It begins with a brief history of Indonesian property development across the Asian financial crisis in the late 1990s and how such a history has shaped the Lippo Group and its production of space. The narrative is organized around the agency of the Group, especially its founder, Mochtar Riady, and his family. It shows the interface of business, culture, and politics in the production of space "after suburbia."

(After) Suburbia

The concept of "suburbia" has undergone a series of changes over time. Its earlier reference to only residential development is now considered by many to be too narrow. Urban scholars, especially in the US, have recently identified new or different developments in the suburbs. The sprawling of Los Angeles, for instance, with its multiple centres, includes not only residential areas but also spaces for commercial and economic activities – spaces that have come to be called "post-suburbia" (Phelps et al., 2006). Others have developed other terms to refer to the same idea, such as "edge cities" (Garreau, 1991), "edgeless cities" (Lang, 2003), "post-metropolis" or "exurbia" (Soja, 2000), and "technoburbs" (Fishman, 1987). Concomitant to the "LA School" is a formulation by scholars working on the extended metropoles of Asia. Led by Terry McGee (1991), they observed a more diffuse and scattered form of urbanization characterized by a profound mix of urban and rural land uses in the expanding urban fabric of metropolitan areas such as of Jakarta and its surrounding region, known then as Ja-botabek (Jakarta and its surrounding Bogor, Tangerang, and Bekasi). From the urban–rural transactions at the fringes of major Southeast Asian cities, terminological innovations ensued, including "extended metropolitan region" (Douglass, 2002), "desakota" (McGee, 1991), and "mega-urban regions" (McGee & Robinson, 1995). All these terms share an agreement that the concept of "suburb" as a residential development has failed to capture the complexity or multiplicity of urbanization. They also generated interesting debates on whether Asian suburbanization is different from its American counterpart (see Dick & Rimmer, 1998; Friedmann & Sorenson, 2019).

In this chapter, I choose not to enter into the debate about whether the Asian and American post-suburbias are converging. I will simply show how the spaces that the Lippo Group produced cut across different cultures while being tied to the political economy of the place in which it is a part. By doing so, I seek to contribute to the discussion on the "layered sediment of contradictory temporalities and spatialities" of our suburban planet (Keil, 2018). I argue that the Lippo Group sought to build a private city (or town), but that its value is derived essentially from the ideology of suburbia. And yet, the group seeks to transcend the dependency of the suburb on the city by tapping into the ideology of autonomy, centrality, and (exclusive) community. What follows is a story of Lippo Group and the forces, processes, and agents that shaped its production of an Indonesian post-suburbia.

The World of Lippo

The mega-project of Meikarta stemmed in some ways from the period of the Suharto regime, when financial investment was increasingly shifted to the real estate sector (for accounts of this crucial period, see Cowherd, 2002; Dorleans, 2002; Firman, 2000; Kenichiro, 2015; Leaf, 1994; Shatkin, 2017). Lippo, however, has a distinctive experience of that era. But first, let us get acquainted with this family-run conglomerate. The Lippo Group owns a number of companies in different industries under the control of a single developer: Mochtar Riady (and his family). The group's promoter refers to the conglomerate as the server (if not saviour) of one's lifetime, from cradle to casket. The Lippo Group has a hospital in which to give birth or to recover when sick and it has a cemetery park for the dead to rest. In between this circle of life, the group provides education from junior kindergarten to university, offices for working, malls for shopping, venues for businesses, lifestyle, and entertainment, as well as residential housing for living with family. The group also runs high-tech communication and media industries to connect the whole community. In short, Lippo's diversified businesses promise a total environment for a generation and beyond. Much of this totality is represented by the group's slogans, which seek to control both time and space: "the World of Ours" and "the Future Is Here Today."

"Today," however, is inseparable from the past. And Lippo's past is best understood through its founder, Mochtar Riady, born Li Wen Jin, who turned ninety-two years old in 2021. Like many early wealthy Chinese Indonesians, Mochtar worked his way from poverty to undreamed riches, mostly through instinct, timing, connections, and, no less importantly, patronage. Yet unlike other early, wealthy, ethnic Chinese businessmen, Mochtar Riady commands Indonesian, English, and Chinese and possesses a degree from a university in Nanjing. These language skills, and later a connection to China and the US, proved beneficial in building his business empire in Indonesia.

Mochtar started his career by running a trading service office (which he called "Lippo," to mean "the treasure of power") in Hong Kong to help Indonesian importers obtain letters of credit (Riady, 2018b). He then built his fortune at home as a banker in the 1960s. After a few years of successfully fixing and upgrading some small local banks in Java, he accepted an offer in 1975 to run Bank Central Asia (BCA), which was owned by Liem Sioe Liong, also called Sudono Salim, a close friend of President Suharto, who was also an indirect shareholder of the bank. It is known that Salim and Suharto helped each other become immensely rich. By 1997 (a year before the collapse of the Suharto regime), the Salim Group was reportedly "the world's largest Chinese owned conglomerate, with $20 billion in assets and some 500 companies" (Harvey, 2005, p. 34). And the family of Suharto, by the time the president was forced to resign, was the largest landowner in Indonesia; they were estimated to have controlled over 3.5 million hectares of land with a value close to $USD57.3 billion (Properti Indonesia, 1999).

Yet, despite his closeness to the oligarchs, Mochtar always considered himself a "professional" as he worked to expand BCA from only one office to become the largest "private" bank in Indonesia with 600 branches by 1991. As he recalls, the asset value of BCA under his management "surged from $1 million to $3 billion in 1990" (Riady, 2018d). While Mochtar was credited for the rise of BCA, much of its success was due in no small measure to the bank's connection with Suharto, who had the power to funnel government or semi-government businesses through the bank. Backed by Suharto, BCA was perhaps more powerful than the state bank, as it was almost free from any regulatory constraints.

Only towards the end of 1980s, during what Mochtar called the "new era" of deregulation and privatization of banks, did he begin to think of owning his own bank (Riady, 2018d). By then, the government allowed foreign and domestic joint-banking ventures and encouraged the opening of branches by Indonesian banks. Mochtar took the opportunity to merge his trading office of Lippo with other local banks – but he did not forget to include Sudono Salim in his enterprise. That was how Lippo Bank was born. By the end of the 1980s, Mochtar (Riady, 2018b) recalls that he "had come to effectively oversee the management of two banks – BCA and Lippo Bank." Both were related in a complex way and they co-produced diversified conglomerates: "Sudono and I had become co-owners of many financial institutions through mergers and acquisitions" (Riady, 2018d). As Harvey (2005) noted, Mochtar (having also a share in BCA) can be seen as pioneering in Indonesia the fusion of ownership and management, which contributed to "the restoration of economic power to the same people" (p. 31).

The "new era" of the deregulation and privatization of banks that started in 1988 made the property industry a golden business. By 1995, Salim and Mochtar were co-listed in some major landownerships (Indonesia Property Report, 1995). Before we move on to discuss the shift of Lippo from banking to land development, however, it will be instructive to reveal a little bit more about Lippo's elite informality, which seems to be practicable everywhere in various forms. It also left a mark in the US, as the Riady family ventured to America and lived in a Californian suburb.

The American Adventure

As a businessman in a developing country, Mochtar would "lament," a privileged relationship to state power could be very beneficial, from getting protection to special treatment and bailout from crisis. Without such elite informality, it is hard to imagine how Lippo could do business in China when government ministries own companies that are Lippo's partners. Yet, as in Indonesia, Mochtar was also aware of the danger of getting too close to the source of power. In his memoir, he indicates his discomfort with Suharto's patronage: "I felt that Sudono and Suharto were too friendly with each other. Their personal relationship made me uneasy" (Riady, 2018d). This uneasiness, as real as it might be, was perhaps also compounded by the sense that while considered "success-ful" as executive director of BCA, Riady's share in the bank was smaller than that of Sudono Salim and the Suharto children (Lee, 2009c). The uneasiness was perhaps also prompted by the opportunity of the new era in finance wherein Lippo Bank would be better off as an indepen-dent bank.

Riady decided to leave BCA even though, as he claimed in his mem-oir, both Salim and Suharto objected to his resignation: "Seeing that I was determined to strike out on my own, Sudono finally agreed to let me go my own way. I exchanged my shares in BCA for Sudono's shares in Lippo Bank under a deal designed to make Lippo an inde-pendent bank" (Riady, 2018f). After that, Mochtar took steps to turn Lippo Bank into an international financial conglomerate. He partnered with groups beyond his usual Sino-sphere, which include Japanese and French counterparts, and opened a branch for Lippo Bank in California.

In the US, as early as the mid-1980s, Mochtar bought two banks. One was a small Chinese-owned bank in San Francisco, which he later renamed "Lippo Bank" (Lee, 2009c). The second was the Worthen Bank in Arkansas. Through the Worthen Bank, he befriended a young and ambitious politician named Bill Clinton, who wanted to become presi-dent of the United States. Mochtar immediately took part in financ-ing Clinton's campaign. This proved to be a good move, and Mochtar and his son James (who was sent to the US "to hone his management skills" by managing the two banks) were frequent visitors to the White House, and it was reported that a key Lippo official worked in the Clin-ton administration (LaFraniere et al., 1997). Through their contacts with Clinton, the Riadys set themselves up as lobbyists for both China and Indonesia, the two places where they located their major businesses.

To please their Chinese partners, they urged Clinton to renew Chi-na's trading privilege, reminding the US president that trade and human rights are two separate issues. For Indonesia, the Riadys sought to improve Indonesia–US ties at a time when the relation was strained by Indonesia's treatment of workers and human rights abuses (LaF-raniere et al., 1997). In this role, the Riadys proved themselves to be Indonesian nationalists (with a structure of feeling towards China). Certainly, they sought to impress upon Suharto that they were not only businessmen but also nationalists who worked to connect Suharto to the White House (something that none of Suharto's business cronies would be able to do). Other than that, it was all for their business so that it could grow or be rescued by the government when needed. As declared by the US court that charged James Riady for illegal campaign contributions to the Democratic Party in 1996, "Defendant James Riady generally believed campaign contributions to be good for the reputation and business of Lippo Group and various Lippo entities, including defendant Lippo Bank California" (as cited in Jackson, 2001).

Despite the bittersweet time with Clinton, Riady's banks in the US were caught several times for violating regulations due to the Riadys funneling the bank money to their private enterprises (LaFraniere et al., 1997). This was common in Indonesia during that time (as exemplified by BCA), where banks provide a ready pool of cash for their owners to finance their private businesses, but such a practice was untenable for US regulators. Not long after the 1997 Asian financial crisis, the Lippo Banks (in the US and various places in Asia, including those in Indonesia) were sold to partners.

Recalling the 1997 Asian financial crisis, Mochtar (Riady, 2018c) believes that it was created "by a combination of speculative investments by US funds and rumours of a looming currency crisis." For him, "in 1997, Indonesia was in a relatively good economic shape." He blames the rumours that had weakened the rupiah and crashed the currency. He acknowledges, nevertheless, that "Indonesian banks were left with mountains of bad loans as one corporate borrower after another defaulted." Almost all private banks were nationalized and put under state control. Some were listed for liquidation under a state-appointed asset management agency, others were taken over by the state. BCA, the Salim Group bank that was too big to fall, received an injection of public funds. Lippo Bank, according to Mochtar, was only "shaken" thanks to Lippo's diverse interests in China and Hong Kong. "Lippo Bank managed to avoid this (fate of nationalization)," but it too entered the Indonesian Bank Restructuring Agency (IBRA) for recapitalization of its asset. The state still had to inject about USD$430 million and took over 59 per cent of the shares in Lippo Bank before the bank could be sold to an investment agency called Swiss–Asia Group for about USD$90 million (Pramisti, 2016).

Since then, Lippo Bank has moved from one hand to another, but it never returned to Mochtar. The shock seemed to have discouraged Mochtar from relying on banking industries. As Mochtar recalls, by the end of 2004, "I had sold not only Lippo Bank but also most of my bank related businesses in Macau, Hong Kong and the US. I decided to focus Lippo's operation on two areas: land development, and information and telecommunications" (Riady, 2018c). In the end, Mochtar considered banking business "onerous" and "stressful" and not worth passing over to his children and grandchildren. He had seen how, after the financial crisis, "Lippo Bank was under the strict supervision of the Bank of Indonesia, the central bank, and we had to obtain permission for every new business activity. We were a commercial bank, yet we had no freedom to make management decisions" (Riady, 2018c). Yet he also acknowledges the role of the state as Lippo Bank "needed funds to deal with its huge pile of bad loans." Much of these non-performing loans came out of the bank's engagement with speculative property business. In addition, his bank problems were also related to cash flows issues tied to his property projects (Lee, 2019c).

Phoenix from the Ashes: The Birth of a Post-Suburbia

With Lippo Bank gone, it seemed a treasure had been lost. But also gone was all the endless trouble of the banking business, the ordeal of bad loans, and the stress of elite informality under the Suharto regime. Indeed, the loss of the Lippo Bank turned out to be a blessing in disguise. Mochtar (Riady, 2018f) explains that in the early 1990s, when loans and land were easy to give and take, one of Lippo Bank's corporate borrowers got into trouble due to an unexpected surge in interest rates. As a result, the borrowers left Lippo Bank with three tracts of land that were used as collateral for the loan. "These tracts of land, totaling 70 square kilometres in area, were located between 25 km and 59 km from central Jakarta." They were too far from the city that he doubted any investors would be interested in the plots. Furthermore, "it was a wasteland with no trees." Be that as it may, it came at the right moment, as it enabled Lippo to expand and switch its operation from finance to the development of land resources.

The "wasteland" also came early enough in the 1990s for the Riadys to turn it into a major asset. Lippo Bank was still around then to support the Riadys' newfound property projects. There was also more going on in the Riadys' existential life during that time: The Riadys became Christian. James once confess that God spoke to him personally: "I thought I was a good man. God said, 'You're a horrible man.' I cried and cried and cried. Since then I try to be a better person, and my life has changed" (as cited in Lee, 2009b). As for Mochtar (who by then had turned sixty), after a period of some reluctance, a pastor's sermon (in a church where attendees were top corporate executives) on how "all humans were born with a sinful nature" affected him deeply. He started to realize that all his good deeds – from building Buddhist temples, Muslim mosques, and Christian churches to giving jobs to many people and making countless donations – were not enough. "'Who had never cheated the government?' the pastor asked. I found that I could not swear that I had never done so. And 'who among you has never lied to your wife at some point?' came the second question. Tears flooded my eyes. I realized that I was indeed a sinful person" (Riady, 2018a).

The seized "wasteland" could then be seen as a blessing, divinely speaking – a chance to redeem the Riadys' sinful souls through the space they sought to build for a new community. But while Mochtar had accepted the lands as his and was determined to carry out the mission of developing it himself, he could not find any model to follow. These tracts of land were so far away from Jakarta. They could not be planned as a suburb. Mochtar went to different places to learn about suburbs and townships, but true inspirations only came when he travelled to Shenzhen. He was impressed with how this special economic zone, which is some distance away from both Hong Kong and Guangzhou, could become a metropolis of its own with manufacturing jobs and housing for millions. So came Mochtar's master plan, which he revealed in his memoir:

> One of three plots of land in Lippo's possession was in Karawaci, 25 km west of Jakarta. There were many plants operated by foreign companies in the area, but few houses suitable for foreigners. I decided to transform the land in Karawaci into a new housing area. I started building a new town complete with a school, hospital, supermarket and golf course. On the eastern wasteland of Cikarang, I decided to build an industrial park. This industrial park was divided into four areas for manufacturers from four Asian economies – Japan, South Korea, Taiwan and Indonesia. A housing area was also developed for the workers at these factories. The last plot of land, in the hilly Karawang area, was the biggest development challenge. It was too far from central Jakarta even for an industrial park, so I turned the land into a sprawling cemetery. These projects transformed three tracts of wasteland into valuable assets. (Riady, 2018f)

These three tracts of land, seized as collateral from a defaulted debt in the early 1990s, became what is today the biggest asset of the Riady family. After merging its eight different companies in 2004, a business empire was thus born, the Lippo Group. It was principally organized around these tracts of land, which were used for business while fulfilling the family's heart and soul. By the 2000s, with the tightening of rules, banking was no longer the most effective way to generate cash to support property development. There were other more productive financial instruments such as the real estate investment trusts (REITs), which just entered Singapore in 2002. The Riadys decided to monetarize their properties in Indonesia via this property trust in Singapore. Backed initially by their successful shopping malls in Greater Jakarta, the Riadys' illiquid properties were effectively turned into cash to fund their property projects thanks to REITs in Singapore. Lee (2009a) reported that "Lippo [was] among the most enthusiastic sponsors" (i.e., owners who inject their properties in the trust).

By 2012, ten years into REITs, Lippo, having built up a huge portfolio of properties in Indonesia, was considered "the largest property developer in Indonesia as measured by assets, land bank, revenue and net profit" (Oxford Business Group, 2012, p. 185). The Lippo Group was based in Indonesia, but Mochtar also set up property development headquarters in Singapore and put one of his sons, Stephen Riady, in charge of Lippo Group's overseas operations. He appointed James Riady for the Indonesian operation, and James became CEO of the Lippo Group. Today, it is all hands on deck for the family to develop Meikarta, the USD$19 billion investment and the biggest project in the history of the family-run conglomerate. As Mochtar (Riady, 2018e) reminds everyone, "I am pouring my heart and soul into creating a new city in Indonesia. Meikarta will integrate residential areas with academic research facilities to create a new kind of community." It is where the "future is here today," but as we have discussed so far, such today is all due in no small measure to taking advantage of the climate of the Suharto era. Before Lippo left banking, a huge chunk of Lippo's money had already been invested in the property business. So, before Meikarta, there is perhaps something to learn from Lippo's first two post-suburban projects in the 1990s: Lippo Village and Lippo's Orange County.

The World Lippo Seeks to Create

The development of Lippo Karawaci (formerly called Lippo Village) started in 1992 with the construction of a central 500 hectares of a total 2,800 hectares. Supported by Lippo Bank and aided by Hong Kong business magnate Li Ka-shing as a shareholder, Mochtar's vision of a new community was already crystallized there with the construction of as many "public" facilities as possible at the outset in the central area – a symbol of Lippo's dedication to serve the new community. By 1997, Lippo Karawaci already declared its 220,000-square-metre "Super Mall" to be the largest in Indonesia, and from an adjacent toll road, towers for offices and condominiums could already be seen. Lined up around the main entrance were an international hospital, a five-star international hotel, a fifty-two-storey residential tower and a forty-two-story residential tower, rows of shophouses, and a very expensive private school and university. All these represent the idea of world-class facilities.

Planned by English-born American golf course designer Desmond Muirhead, Lippo's residents live in a site organized hierarchically around a 6.5-hectare, 18-hole golf course called Imperial Golf, and a lake that serves as the centre of the residential complex. But with the home of James Riady and his family on the island, accessible only by private helicopter, the lake is more like a moat. Hogan and Houston (2002) think that the site of Riady's mansion (with soaring French windows and Greco-Roman columns) echoes "the traditional symbolism of Javanese Sultanate and Chinese emperor system of authority" (p. 239). Houses on and around the golf course are the most expensive and secured. Each cluster of houses is called *taman* ("garden") and all of them are gated with only one entrance for each cluster. Each *taman* refers to a place in the world (such as Taman Holland, Taman Boston, Taman Mediterranean, Taman Osaka, Taman Paris, as well as (the less expensive choice) Taman Ubud, Bali. The houses in each *taman* are built according to the stereotypical architectural style of the place after which it is named. All the residents (including expatriates) are housed in an array of such thematized estates.

Lippo Karawaci also offers complete infrastructure and assumes a self-sustained governmental power. It is proud to provide its own town management. It has a town manager of its own, so there is no need for city governor or local government to get involved. According to Gordon Benton, the town manager and planner of Lippo Karawaci, "it is the only township throughout the nation to have drinking water from the tap, municipally-treated central sewage system, all services underground, hierarchical and traffic-calmed street designs, and ... a Town Management Division" (as cited in Hogan & Houston, 2002, p. 249). The Town Management Division is effectively the city government. It promises full security for its residents who are not involved in the governance of their environment, as represented by the town's motto: "We take greater care of YOU!" (as cited in Leisch, 2002, p. 346). This seems to follow Mochtar's vision of a new community, but it also embodies the consolidated, diverse interests of the conglomerate. It represents all eight elements of life that Mochtar has sought to coordinate. The space of Lippo Karawaci is an inside-out of the Lippo Group:

> We placed a high priority on ensuring that the town would be able to provide residents with all of the services that are essential for daily life. A residential area must have the facilities to meet these needs. We thought that eight types of facilities were essential: shopping malls with various specialty stores, supermarkets, movie theaters, play areas for children, hotels and schools. This meant that Lippo had to expand its operations into these

eight areas. I decided to operate these businesses (through acquisitions or partnerships) as independent companies and grow them into nationwide chains. (Riady, 2018f)

California Transfer

The second property, about 3,000 hectares in Cikarang, Bekasi, has a different quality. It is about thirty-four kilometres east of Jakarta, way too far to allure Jakarta's middle class. It is, however, near the official industrial zone designated by the central government for multinational companies (largely from East Asia) as they moved to Indonesia for cheaper labour (Kenichiro, 2011). In this context, it was logical for Lippo Cikarang to develop properties that would attract foreign and local managers, as well as workers who could afford to buy houses.

Taking advantage of its proximity to private industrial estates, Mochtar thus established in the 1990s a subsidiary of Lippo Karawaci called Lippo Cikarang Tbk, with a focus on finding supports and developing partnerships to build an urban centre that would integrate real estate and industrial estate with the development of infrastructure. Planned by a "Master Planner" called Meng Ta-cheang, the aim was for Lippo Cikarang "to develop into a viable city of its own serving as a new centre to the surrounding industrial and suburban housing" (Lippo City Masterplan, as cited in Hogan & Houston, 2002, p. 251).

Two zones were set up: (1) "Delta Silicon Industrial Park" to accommodate the working needs of different manufacturing and trading enterprises and (2) "Orange County," a residential zone with a variety of themed housing styles (resembling those of Lippo Karawaci) and luxurious condominiums. Both zones were organized around a city centre that boasted a complete set of facilities such as a twelve-storey office tower, a seven-storey global trade centre, a four-star international hotel, a shopping mall and "water world" park, a hospital, schools, and a golf course with club facilities.

Names such as "Delta Silicon" and "Orange County" recall California, where James Riady once lived. Despite all the bitter memories of banking business in California and his US campaign contribution – which cost him an exorbitant fine, two years' probation, 400 hours of community service (to be carried out in Indonesia), and being banned from entering the US – James Riady could not forget California suburbs such as Orange County, Irvine, Beverly Hills, and Newport Beach. He arranged for Lippo's directors and planners to visit these places (Cowherd & Heikkila, 2002) and went on to hire architectural design firms from Orange County. As a result, as Hogan and Houston

(2002) indicate, what have been produced (in both Lippo Karawaci and Orange County) were nothing short of "direct clones of Southern California housing styles and estate themes with minor modifications to allow for different climatic conditions" (p. 258). Yet, while Californian in spirit, Orange County was essentially a project in partnership with Japanese trading company Mitsubishi Corporation (Tani, 2018). Japanese scholar Kenichiro (2015) goes as far as pointing out that Lippo Cikarang was essentially built by East Asian investors, such as Sumitomo (Japan) and Hyundai (Korea): "These foreign partners provided a large part of the capital and the actual construction work and Lippo itself functioned more as a facilitator for various permits, land consolidation and marketing" (p. 486).

The Eastern Connection

Despite all the extraordinary efforts to connect East and West, Orange County produced only partial results. Orange County was relatively sleepy from the beginning. There is a relatively well attended shopping mall and few other facilities, but other than these, most of the area in Orange County is still vacant. Initially, Orange County targeted expats working in the Indonesian industrial estate of foreign companies, but there is not enough demand from the estate, even though it is home to more than 1,000 plants of mostly East Asian manufacturers, including giants such as Toyota Motor, Honda Motor, and South Korea's Hankook Tire. After some years, it was clear that foreign expats and foreign workers alone are not enough to populate Orange County. Bartholomeus Toto, the director of Orange County, did what he could for the development. In 2016, Toto even declared Orange County to be a trendsetter for urban development and proposed a new master plan based on transit-oriented development (TOD). The plan was to take advantage of recent government plans to build a toll road and railway to link Jakarta and Bandung. Located between these two cities, the area of Orange County would immediately benefit from this future infrastructure development. In collaboration with Mitsubishi Jisho Sekkei Inc., Toto invited renowned Japanese architect Kengo Kuma, who designed Japan's National Stadium for the 2020 Tokyo Olympics, to serve as design advisor for the first phase of a new master plan for Orange County. According to Toto, "Kuma is an expert in combining expressions of traditional culture with modern materials and technology in a design that is artistic and phenomenal" (as cited in Himawan, 2016).

Koki Miyachi, general manager of the architectural design department of Mitsubishi Jisho Sekkei Inc., confirmed that the plan will be based on the concept of the "liveable city" to "improve the quality of life of everyone in the city." It will apply "green architecture" supported by artistic "pedestrian way." The city will be known for its "unique and impressive skyline." Through this collaboration with Japan, Lippo's Orange County is expected to become a "new California city" supported by a "super complete facilities of 32-in-1" (as cited in Himawan, 2016).

Toto's timing seems to be right. Just five years earlier, in 2011, the then president of Indonesia, Susilo Bambang Yudhoyono, launched an ambitious Master Plan for the Acceleration of Expansion of Indonesian Economic Development. The idea was to use infrastructure to "debottleneck" the flow of investment to Indonesia. Soon, forty-four regulations concerning investment were revised to accelerate the flow of capital. Such a flow relies on physical infrastructure, which is itself an object for investment (Ridha, 2018). Moving on, in 2012, a high-level meeting between Indonesia and Japan was held in Tokyo. A consensus was reached to carry out a program called Metropolitan Priority Areas, which was to bring connectivity to the scattered property development in Jabodetabek (Jakarta Metropolitan Area) though a coordinated transport infrastructure. The (close to USD$30 billion) project would be carried out with a public–private partnership (PPP) deal.

Due to Orange County's location, this government initiative was a big deal for Lippo. The state had created an opportunity for capital to flow in and through infrastructure. But the support of Japan, the old brother, alone would not be sufficient to boost and stretch Orange County so that it could take full advantage of the government's infrastructural plan. Fortunately, stars were in alignment for Lippo. In 2016, a business delegation from China's Shenzhen came to visit the Lippo Group. An ambitious deal with a value of USD$14.5 billion was struck (Siniwi, 2016). The Shenzhen Yantian Port Group and Country Garden Holdings agreed to develop the two wings of Lippo Cikarang: the industrial zone and the commercial property projects. The Chinese entry was big – bigger than its Japanese counterpart. With this Chinese deal, Orange County would no longer be the same. Toto, too, was immediately transformed. He was soon asked to cover a bigger-scale project by serving as executive director of Lippo Cikarang. The aim was to fold Orange County into the new ambitious project planned to kick-start in 2017. The time had come for Orange County to be rebranded. Japan's "new California city" would be recalibrated by a new marriage with China. Meikarta was thus born. The child was to be raised in Orange County, but she would look like Manhattan.

Meikarta: The World of Ours

In 2017, Meikarta was launched with the idea that Cikarang would be surrounded by a high-speed train that links Jakarta-Bekasi-Cikarang-Bandung, the Patimban Deep Sea Port, the Jakarta–Cikampek Elevated Highway, the automated people moved (APM) monorail, and Kertajati International Airport. It proposes an urban form that recalls Manhattan, included its famous Central Park. But unlike Manhattan, Meikarta is to be the largest manufacturing hub in Indonesia and at once a smart metropolis with a most sophisticated infrastructure for 1 million people. According to the promotional brochure, it will have "distinctive architecture for each residential tower with different styles from various eras and cultures" such as towers in an "Asian style," "European style," "American style," and "Modernist style."

The boss of Meikarta, James Riady, expects that Meikarta will immediately boost Orange County as "the CBD [central business district] of Meikarta" (as cited in Sugianto, 2018). Together with Delta Silicon, the "Silicon Valley of Indonesia," Orange County is expected to spur economic growth in the region, making Lippo Cikarang the centre of the Indonesian economy. There are already ten global institutions on board: Columbia University Medical Center, University College London, the University of North Carolina, Genesis Rehab Services, World Trade Centers Association, HTC Corporation, China Telecom Global, JM Eagle, Zhong Ying Finance, and Lausanne Hotel Management Institute. Together, they constitute "the truly integrated city of the future," which would set a new standard for a world city in Asia and beyond – or so the Meikarta brochure reads. The months leading up to the 2017 launch, Meikarta's advertising and promotional materials were everywhere in the media and malls owned by Lippo. Much of the success of the campaign was due to spending USD\$107 million, but it was also owing to a negative campaign blitz against the urban condition of Jakarta.

In *Bourgeois Utopia*, Fishman (1987) argued that "suburbia can never be understood solely in its own terms. It must always be defined in relation to its rejected opposite: the metropolis" (p. 27). All the Lippo projects have been defined in negative relation to Jakarta. In particular, the ad of Meikarta. It has made Jakarta essentially a city of garbage with flooding, pollution, poverty, crime, and traffic jams depicted as part of daily life. The unruly streets of Jakarta are contrasted with an interior of a condominium, surrounded by glass walls, that is equipped with interactive informational panels. The glass wall opens, not to the streets, but to the new skyline of technopoles as if all units are interconnected in a "milieu of innovation" by advanced informational and communication technology (Castells, 2010, p. 62). Such a presentation devalues Jakarta and, by extension, the state. Yet, while Meikarta can be seen as a prime example of a new informational global city above the state, its values are explicitly linked to the provision of transport infrastructure supported by the state.

In any case, the Meikarta launch was a hit, and the pre-construction sales targets were through the roof. Boosted a few months later by the topping-off ceremony of the first towers, which was attended by a powerful cabinet minister and former military officer, General Luhut Panjaitan, a top gun of President Jokowi, the success of Meikarta seemed to be all but guaranteed. Yet soon after that, a series of unfortunate events unfolded, and everything went downhill. Allegations that construction began before permits had been obtained were followed by Meikarta employee and contractor protests over unpaid salaries and bills due to a dwindling flow of Chinese funds (Tani, 2018). Then news circulated that China's tighter capital controls had led to the pulling out of Chinese partners including Shenzhen Yantian Port and Country Garden Holdings. Finally, there were the arrests of top Lippo executives, including Bartholomeus Toto, by Indonesia's anti-graft agency for paying off city officials to obtain building permits for Meikarta. These events, along with the raid of James Riady's home, caused Lippo's shares to plunge, and panic soon ensued, causing condo buyers to start demanding full refunds.

The future of Meikarta is no longer looking so bright. It must surely be hard for Mochtar to believe that Meikarta, a project of his heart and soul, built to accommodate the middle-class need for "affordable," world-class condos (unlike Lippo's other high-end projects) and a signification of his good deeds, would suffer from such a misfortune. But empowered by his talent for survival and his faith in Christianity, Mochtar has been quick to respond. To restore Lippo's scandal-tainted image, the Riady family immediately appointed a new CEO, Mochtar's thirty-three-year-old grandson John Riady, an MBA of the Wharton School of the University of Pennsylvania and a juris doctor of Columbia Law School. John is of the post-Suharto generation and a symbol of the future. If the wrongdoings of those from the Suharto era were just too many and unpardonable, then John, having grown up in a new era, should be able to redeem the family and save the project. After all, it seems fit that a new community of Meikarta be led by a new generation. John says, "It's a big role. So I don't take it lightly ... You have a responsibility not only to the family but to the minority shareholders at large. That's real" (as cited in Business Times, 2019).

ACKNOWLEDGMENTS

Thanks to Suryono Herlambang and Manneke Budiman for the continuous conversation on the roles of politics, culture, and property business in the spatial transformation of Jakarta and its surrounds.

REFERENCES

Business Times. (2019, September 12). Riady scion seeks to save Lippo empire on the brink. *Business Times*. https://www.businesstimes.com.sg /real-estate/riady-scion-seeks-to-save-lippo-empire-on-the-brink.

Castells, M. (2010). *The rise of the network society* (2nd ed.). Blackwell.

Cowherd, R. (2002). Planning or cultural construction? The transformation of Jakarta in the late Suharto period. In P.J.M. Nas (Ed.), *The Indonesian town revisited* (pp. 17–40). LIT Verlag.

Cowherd, R., & Heikkila, E. (2002). Orange County, Java: Hybridity, social dualism and an imagined West. In E. Heikkila & R. Pizarro (Eds.), *Southern California and the world* (pp. 195–220). Greenwood Press.

Dick, H., & Rimmer P. (1998). Beyond the third world city: The new urban geography of Southeast Asia. *Urban Studies, 35*(12), 2303–22. https:// doi.org/10.1080/0042098983890.

Dorleans, B. (2002). Urban land speculation and city planning problems in Jakarta before the 1998 crisis. In P.J.M. Nas (Ed.), *The Indonesian town revisited* (pp. 41–56). LIT Verlag.

Douglass, M. (2000). Mega-urban regions and world city formation: Globalisation, the economic crisis and urban policy issues in Pacific Asia. *Urban Studies, 37*(12), 2315–35. https://doi.org/10.1080/00420980020002823.

Firman, T. (2000). Rural to urban land conversion in Indonesia during boom and bust periods. *Land Use Policy, 17*(1), 13–20. https://doi.org/10.1016 /S0264-8377(99)00037-X.

Fishman, R. (1987). *Bourgeois utopias: The rise and fall of suburbia*. Basic Books.

Friedmann, J., & Sorenson, A. (2019). City unbound: Emerging mega-conurbations in Asia. *International Planning Studies, 24*(1), 1–12. https:// doi.org/10.1080/13563475.2019.1555314.

Garreau, J. (1991). *Edge city: Life on the new frontier*. Doubleday.

Harvey, D. (2005). *A brief history of neoliberalism*. Oxford University Press.

Himawan, A. (2016, November 17). Lippo Cikarang gandeng Kengo Kuma untuk design Orange County. *Suara.com*. https://www.suara.com/bisnis /2016/11/17/131300/lippo-cikarang-gandeng-kengo-kuma-untuk -design-orange-county.

Hogan, T., & Houston, C. (2002). Corporate cities – Urban gateways or gated communities against the city? The case of Lippo, Jakarta. In T. Bunnell,

L. Drummond, & K.C. Ho (Eds.), *Critical reflections on cities in Southeast Asia* (pp. 243–64). Brill.

Indonesia Property Report. (1995). New towns and satellite cities. *Indonesia Property Report, 1*(2), 10–16.

Jackson, R. (2001, January 12). Clinton donor Riady pleads guilty to conspiracy charge. *Los Angeles Times.* https://www.latimes.com/archives /la-xpm-2001-jan-12-mn-11506-story.html.

Kahfi, K. (2018, October 16). Lippo Group executive arrested in Meikarta bribery case. *Jakarta Post.* https://www.thejakartapost.com/news /2018/10/16/lippo-group-executive-arrested-in-meikarta-bribery -case.html.

Keil, R. (2018). After suburbia: Research and action in the suburban century. *Urban Geography, 41*(1), 1–20. http://doi.org/10.1080/02723638.2018.1548828.

Kenichiro, A. (2011). From water buffaloes to motorcycles: The development of large-scale industrial estates and their socio-spatial impact on the surrounding villages in Karawang Regency, West Java. *Southeast Asian Studies, 49*(2), 161–91. https://kyoto-seas.org/pdf/49/2/490201.pdf.

Kenichiro, A. (2015). Jakarta "since yesterday": The making of the post-new order regime in an Indonesian metropolis. *Southeast Asian Studies, 4*(3), 445–86. https://doi.org/10.20495/seas.4.3_445.

LaFraniere, S., Pomfret, J., & Sun, L.H. (1997, May 27). The Riadys' persistent pursuit of influence. *Washington Post*, A01.

Lang, E. (2003). *Edgeless cities: Exploring the elusive metropolis.* Brookings Institution Press.

Leaf, M. (1994). The suburbanization of Jakarta: A concurrence of economics and ideology. *Third World Planning Review, 16*(4), 341–56. https://doi.org /10.3828/twpr.16.4.n51557k1532xp842.

Lee, H.S. (2009a, March 17). Exodus out of Indonesia. *Asia Magazine, 31,* 39.

Lee, H.S. (2009b, March 17). James Riady: The evangelical felon. *Asia Magazine.* Retrieved 23 December 2019, from http://www.theasiamag.com /people/james-riady-the-evangelical-felon (site no longer exists).

Lee, H.S. (2009c, March 17). Mochtar Riady: The banker without a bank. *Asia Magazine.* Retrieved 23 December 2019, from http://www.theasiamag.com /people/mochtar-riady-the-banker-without-a-bank (site no longer exists).

Leisch, H. (2002). Gated communities in Indonesia. *Cities, 19*(5), 341–50. https://doi.org/10.1016/S0264-2751(02)00042-2.

Lippo Homes. (n.d.). *Meikarta.* Retrieved 13 January 2022, from https://www .lippohomes.com/Project/Index/1026.

McGee, T.G. (1991). The emergence of desakota regions in Asia: Expanding a hypothesis. In N. Ginsberg, B. Koppel, & T.G. McGee (Eds.), *The extended metropolis: Settlement transition in Asia* (pp. 3–25). University of Hawaii Press.

McGee, T.G., & Robinson, I.M. (Eds.). (1995). *The mega-urban regions of Southeast Asia*. UBC Press.

Oxford Business Group. (2012). On the market: Several developers are listed on the Indonesia Stock Exchange. In *The Report: Indonesia 2012* (p. 185). Oxford Business Group.

Phelps, N., Parsons, N., Ballas, D., & Dowling, A. (2006). *Post-suburban Europe: Planning and politics at the margins of Europe's capital cities*. Palgrave Macmillan.

Pramisti, N.Q. (2016, October 10). Bangkitnya kerajaan bisnis perbankan Mochtar Riady. *Tirto.id*. https://tirto.id/bangkitnya-kerajaan-bisnis -perbankan-mochtar-riady-bSGZ.

Properti Indonesia. (1999, January). Nilai tanah Keluarga Cendana hampir Rp. 800 triliun. *Properti Indonesia*, 34–5.

Riady, M. (2018a, October 8). Becoming a Christian in my 60s: Mochtar Riady's story (29). *Nikkei Asia*. https://asia.nikkei.com/Spotlight /My-Personal-History/Becoming-a-Christian-in-my-60s-Mochtar-Riady-s -Story-29.

Riady, M. (2018b, September 30). Lippo gets its start as an import agency: Mochtar Riady's story (21). *Nikkei Asia*. https://asia.nikkei.com /Spotlight/My-Personal-History/Lippo-gets-its-start-as-an-import -agency-Mochtar-Riady-s-story-21.

Riady, M. (2018c, October 4). A new direction for Lippo after the Asian financial crisis: Mochtar Riady's story (25). *Nikkei Asia*. https://asia.nikkei .com/Spotlight/My-Personal-History/A-new-direction-for-Lippo-after-the -Asian-financial-crisis-Mochtar-Riady-s-story-25.

Riady, M. (2018d, October 1). The opportunity for growth finally comes: Mochtar Riady's story (22). *Nikkei Asia*. https://asia.nikkei.com/Spotlight /My-Personal-History/The-opportunity-for-growth-finally -comes-Mochtar-Riady-s-story-22.

Riady, M. (2018e, September 10). The story of Lippo Group and modern Indonesia: Mochtar Riady's story (1). *Nikkei Asia*. https://asia.nikkei.com /Spotlight/My-Personal-History/The-story-of-Lippo-Group-and -modern-Indonesia-Mochtar-Riady-s-story-1.

Riady, M. (2018f, October 2). Turning wasteland into a valuable asset: Mochtar Riady's story (23). *Nikkei Asia*. https://asia.nikkei.com/Spotlight /My-Personal-History/Turning-wasteland-into-a-valuable-asset -Mochtar-Riady-s-story-23.

Ridha, M. (2018). *Melawan regim infrastruktur: Studi ekonomi politik*. Carabaca Makassar bekerjasama dengan Social Movement Institute (SMI) Yogyakarta.

Roy, A. (2015). Urban informality: Towards an epistemology of planning. *Journal of the American Planning Association*, 71(2), 147–58. https://doi.org /10.1080/01944360508976689.

Shatkin, G. (2017). *Cities for profit: The real estate turn in Asia's urban politics.* Cornell University Press.

Siniwi, R. (2016, May 25). Lippo Group welcome Shenzhen business delegation to jointly develop industrial estate in Cikarang. *Jakarta Globe.* https://jakartaglobe.id/context/lippo-group-welcomes-shenzhen -business-delegation-jointly-develop-industrial-estate-cikarang/.

Soegiarto, Y. (2017, June). Lippo Group to develop $21B Meikarta industrial city. *GlobeAsia.* Retrieved 27 December 2019, from https://www .globeasia.com/industry-focus/lippo-group-develop-21b-meikarta -industrial-city/ (site no longer exists).

Soja, E. (2000). *Postmetropolis: Critical studies of cities and regions.* Blackwell.

Sugianto, D. (2018, March 21). Ada apartemen mewah di Cikarang, Orange County atau Meikarta? *detikFinance.* https://finance.detik.com /properti/d-3928241/ada-apartemen-mewah-di-cikarang-orange -county-atau-meikarta.

Tani, S. (2018, December 7). How Lippo's field of dreams could become a house of cards. *Nikkei Asia.* https://asia.nikkei.com/Business/Company -in-focus/How-Lippo-s-field-of-dreams-could-become-a-house-of-cards.

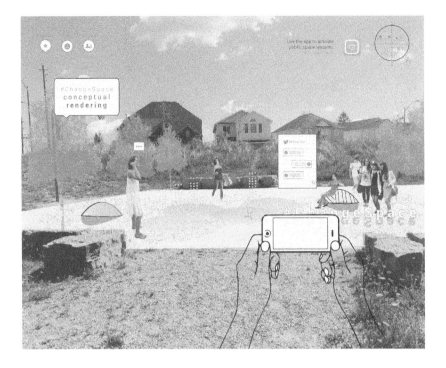

Image 20. #ChangeSpace: Planning suburban public space for youths

20 Transnationalism and Southern Suburbanization: Accounting for Translocalities in Manila's Peri-urban Fringe

ARNISSON ANDRE C. ORTEGA

The town of Rodriguez in Rizal province has experienced dramatic changes during the last decade. Located just outside of Metro Manila, in its northeast peri-urban fringe, the town's population has grown to become the most populous municipality in the Philippines, having reached a population of 369,222 (Philippine Statistics Authority, 2016). It is no wonder that the municipality has seen the rise of multiple settlements and developments in recent years. Slums, socialized housing projects, and posh gated communities have all developed alongside each other as the town has expanded. The transformation of Rodriguez reflects the broad and complicated terrain of Manila's peri-urban fringe, comprised of turbulent urban-rural contestations and spaces of accumulation and resistance. As former agricultural lands are converted into industrial estates, and informal settlements make way for gated residential subdivisions, the peri-urban fringe has emerged as a patchwork of mixed land uses and environments that reflect the multiple histories and narratives underpinning its transformation.

Cognizant of these diverse geographies, this chapter focuses on the rise of translocal sites, local spaces whose emergence is primarily facilitated by transnationalism. For years, Manila's peri-urban fringe has served as a site for new urban developments that aim to not only attract foreign direct investments and secure the country's place in the global market but also lure diasporic capital (see Ortega, 2018). In this chapter, I identify and examine the spatial expressions of transnational mobilities of diasporic population, capital, and ideas, which are instrumental to peri-urban transformations and urban expansion. In doing so, I trace local histories and relationalities of these spaces in the ways in which they are situated within broader political, economic, and social processes.

Southern Suburbanisms and the Postcolonial Provocation

As much of today's urbanization is occurring in the suburban fringes of the Global South (Clapson & Hutchison, 2010), then a postcolonial maneuver is a much-needed theoretical intervention to understand the vast terrains of urban transformation beyond Anglo-American cities. Recent calls to account for diverse suburbanisms across the globe (Keil, 2013, 2018) demand retheorization of long-held paradigms and theses, much of which have been deployed from Anglo-American contexts. As such, postcolonial provocations provide important guidance for charting theoretical trajectories that not only decentre dominant urban models and universalizing pronouncements but also generate theoretical formulations drawn from Global South terrains (see Robinson, 2006; Roy, 2009). Like other Global South cities, Manila's urbanization defies mainstream urban models, with an enormous surplus population and fragmented urban development, as described by Shatkin (2008) as "bypass-implant urbanism." A similar deviation can be said about its peri-urban fringe, whose transformation is undergirded by multiple forms of political-economic and social processes – from private-led development projects to the relocation of evicted informal settlers from Manila.

Against tendencies to cast Southern suburbs, like those of Manila, as spaces of dismal poverty, I heed to postcolonial provocations by paying close attention to context-specific particularities and histories in an effort to formulate urban theory and concepts, which resonate across contexts (see Leitner & Sheppard, 2016; Roy, 2009). Initial inroads have advocated several approaches such as consideration of ordinariness of cities (Robinson, 2006), comparison of cities (Robinson, 2016), and the use of an area studies framework (Roy, 2009). In grappling with Manila's urban conditions, McGee's (1991) conceptualization of a "Southeast Asian urban form" provides a useful theoretical entry point, particularly with the notion of desakota ("desa" – village; "kota" – town) to describe peri-urban fringes with mixed urban-rural landscapes.

Suburban expansion, and the changing nature of the relationship between the urban core and the periphery across the globe, suggest that a singular city–suburb narrative is severely incomplete to fully account for the diverse variety of suburban transformations. Studies that point to the rise of post-suburbs, suburban landscapes that have diversified into multiple settlements, land uses, and other functions (see Phelps, 2015; Teaford, 1997) are important interventions to this. As post-suburbs emerge as independent spaces with multiple functions that integrate living, working, and other activities, the city–suburb relationship is effectively reduced to mere commuting. Emergent spaces alluding to post-suburbanization can be observed in various parts of the world, exemplified by diverse developments in metropolitan outskirts and changing urban–rural dynamics in the suburban hinterlands (see case studies in Phelps & Wu, 2011).

In the Global South, suburban transformations are constituted by diverse urban forms, historically sedimented relationalities and socio-spatial entanglements. Manila's expanding peri-urban fringe exemplifies this diversity, with its multiple spaces and relational entanglements. Foremost of these are translocal relations involving transnational mobilities of capital, populations, and ideas, whereby the peri-urban fringe serves as an important site in building new spaces to attract foreign direct investments or to consolidate diasporic capital and aspirations.

How can we frame suburban formations shaped by transnational mobilities? At the onset, transnational mobilities refer to cross-border movements typically associated with the back-and-forth mobilities of migrants, who sustain close social, political, and economic ties with their homelands (see Levitt, 2001; Levitt & Schiller, 2004; Portes, 2001; Schiller et al., 1995). Beyond transnational migration, it also implies a simultaneity of lives and interconnectivities of activities, information and goods, interwoven by mobile networks across space. Despite its synchronicity and hypermobility, transnationalism is not frictionless. It is moored in actual geographies, or what scholars would call translocalities (see Greiner & Sakdapolrak, 2013), whereby transnational interconnections involve actors, institutions, and other players, in concrete terrains across contexts. As a force influencing the production of urban spaces, transnational processes may be viewed through what Smith (2001) would call a transnational urban optic, which situates transnational bodies and other connectivities within "contested historical and geographical contexts" (p. 5). As such, urban spaces should not be cast in territorial terms; instead, they can be translocal sites that negotiate transnational practices, networks, mobilities, and imaginations. From this vantage point, transnational urban spaces are "pregnant with … social relations of domination-accommodation-resistance" (Smith, 2005, p. 242) and serve as "sites of cultural appropriation, accommodation, and resistance to 'global conditions' as experienced, interpreted, and understood in the everyday lives of ordinary men and women" (Smith, 2001, p. 128). Such an optic brings together a variety of transnational actors from "above" (state, corporations, elite), "below" (marginalized migrants) (see Smith & Guarnizo, 1998), and "middle" agents and temporary migrants, such as elite corporate transnationals (Beaverstock, 2005), holiday travellers (Clarke, 2005), and international students

(Collins, 2010). Building on this, Schiller and Cäglar (2009) argue for an expanded "transnational social field" which embraces multiple agents, activities, and enterprises, that are indirectly connected with transnational mobilities. Similarly, Parnreiter (2012) demands an accounting of how transnationalism reconfigures built environments, particularly the "spacing of ... concrete physical artifacts" of cities (p. 96), from ethnic conversions of retail landscapes (Friesen et al., 2005) to the construction of ethnic monuments (Irazabal, 2012).

The peri-urban fringes of many Southern cities have become sites of mega-projects and new urban developments (see Shatkin, 2016). A transnational urban optic may situate these developments, and their attendant transformations, within the context of shifting transnational spatialities of urban models and designs, capital flows and transactions, and migration. From African master plans referencing developments from the Middle Eastern and East Asian urban visions (Watson, 2014), to Indian suburban projects fuelled by diasporic capital (Chacko, 2007), transnational connectivities and mobilities are critical processes in spurring suburban transformation.

Situating Transnationalism in Manila's Peri-urban Fringe

Manila's ever-shifting edge, its peri-urban fringe, has long played an important role in Philippine affairs. From being the "extramuros" (outside the walled colonial city) where Indigenous and non-white populations were resettled, to acting as a bastion of anti-colonial revolutionary forces, the fringe has reshaped both Manila's and the Philippines' spatial politics. Transnationalism has also been an important historical player of peri-urban transformation which has facilitated urban growth through colonial trade and labour migration. But for this chapter, I will focus on Manila's recent history when the spatially shifting peri-urban fringe became the translocal setting for investment-oriented development projects that aim to cement the Philippines' position in the global market, and the ideal terrain for real estate projects that fulfill the diasporic aspirations of transnational Filipino families.

In the 1970s during the Marcos regime, the lands of the peri-urban fringe were actively employed as strategic sites for master-planning industrial development. Typically, these master plans were crafted through overseas grants by a whole host of foreign technocrats and experts contributing to the design. Once Manila's metropolitan region was established in 1975, which involved the construction of numerous infrastructure projects, the fringe was concomitantly demarcated and became a target for specific development programs. Furthermore, the government set a radius of fifty kilometres from the city of Manila to designate the area where industrial projects were allowed (Caoili, 1988). Such a radius facilitated the concentration of industrial sites in the fringe, particularly in the nearby provinces of Cavite and Laguna. After the People Power Revolution that toppled the Marcos regime in 1986, the new government encouraged neoliberal restructuring which advocated for deregulation, trade liberalization, and decentralization. This led to regional master plans and investment-oriented programs that effectively transformed Manila's fringe. This new set of policies facilitated land conversion and suburban development, which, along with demand from Overseas Filipinos, triggered a real estate boom.

In this chapter, I identify and examine three spatial expressions of transnational mobilities in Manila's peri-urban fringe, namely, special economic zones, gated suburbs, and translocal barangays (villages). These three translocalities illustrate the spatial contingencies of trans-national mobilities and suburban development, wherein core–periph-ery interrelationships cannot fully account for the multiple grounded realities in a rapidly transforming peri-urban fringe. Instead, transna-tional entanglements through networks of ideas, flows of capital, and mobilities of Overseas Filipinos, are critical aspects that have engen-dered and sustained suburban transformation in Manila's peri-urban fringe. These translocal spaces are situated zones of spectacle, excep-tional spaces that have been nurtured to facilitate either capital accu-mulation or realize desires for an orderly and secure urban living. Except for translocal barangays, these spaces are oftentimes privatized enclaves that are gated, highly secured, and ruled by their own set of regulations. Despite their apparent segregation and isolation, they are still very much embedded in the spatial politics of the peri-urban fringe.

Special Economic Zones: Translocal Spaces of Globalization

Special economic zones (SEZs) are state-delimited territories designated with a unique set of rules and regulations. Since SEZs are established to attract more foreign direct investments, encourage job creation, and facilitate business, the industries, and other establishments that relo-cate to SEZs enjoy certain privileges that include tax exemptions, sim-plified bureaucratic procedures, and special visa privileges, among others. As such, these spaces act as "grey zone(s) of total mobilization" (Armitage & Roberts, 2003) whereby multinational corporations and capitalist enterprises facilitate mobilities of capital in an autonomous and enclosed space. In effect, SEZs are spatial expressions of neoliberal exemption, relieved from rules and regulations to secure new channels of capital flows within a growing global constellation of spaces facilitating market transactions and forces. While SEZs are indeed enclosed and exempted from various regulations, they are very much part of the lives of residents caught up in peri-urban transformations. Their emergence in rapidly urbanizing spaces in the Global South has had an indelible impact on the pace and trajectory of urbanization in these places.

For the Philippines, like other countries, SEZs are a mechanism to attract foreign investment. Manila's peri-urban fringe has become an ideal terrain for new industrial estates and mixed-use developments. In 1995, the government established the Special Economic Zone Act, 1995 to layout the rules concerning the selection and regulation of SEZs in the country – from the establishment of the Philippine Economic Zone Authority (PEZA), which accredits and oversees SEZs, to the instituting of benefits to be enjoyed by prospective investors (Special Economic Zone Act, 1995). SEZs in the Philippines come in different shapes and sizes, most of which are in the form of industrial estates, mixed-use business districts, IT parks, and export processing zones. PEZA-accredited SEZs are categorized according to the type of activity that is being conducted there. These categories reflect the multiple forms of Philippine engagement with the global market: (1) manufacturing economic; (2) information technology; (3) agro-industrial; (4) tourism; and (5) medical tourism.

Manila's peri-urban fringe plays an important role in integrating the Philippines into the global market. It has been the site of numerous industrial estates and export processing zones over the decades, clusters of which are located in the provinces of Cavite, Laguna, and Batangas. In 1990, a regional master plan called Calabarzon created an effective territorial template that has encouraged the conversion of vast tracts of agricultural land in the southern provinces of the fringe into industrial use. In the province of Cavite, there are controversial cases where prime agricultural land has been converted into industrial developments. One such case was the Cavite Export Processing Zone (CEPZ), a 276-hectare development that houses hundreds of electronics and manufacturing enterprises, which employ 65,000 workers. With its rigid rules and deplorable labour conditions, CEPZ has been described as a "miniature military state within a democracy" (Klein, 1999, pp. 204–6). Despite their negative impact on agricultural communities, these conversions have almost always been legitimized by the national desire to attract foreign capital and supposedly provide jobs to millions of unemployed Filipinos (see

Kelly, 2000; McAndrew, 1994). These developments have brought in hundreds of multinational corporations, which in turn have attracted thousands of migrant workers from various parts of the Philippines. In other parts of the peri-urban fringe, many SEZs are in the form of mixed-use urban development, which combine business process outsourcing (BPO) operations with retail, commercial, and residential projects.

As translocal hubs, SEZs facilitate transnational economic transactions while causing a rippling effect on the local urban economy. In factories of multinational corporations, workers are involved in various tasks assembling assorted parts of a product, from garments to cars. These assembly lines are typically part of a global production network, as different operations of the same company can take place in disparate locations. In BPOs involved with call-centre support, workers attend to calls from clients that are mostly in North America. In order to service North American clients, they have to work shifts that align with North American time zones, forcing them to be up late, often until the early hours of the morning. These productive processes in SEZs are enmeshed in a transnational circulation of production and consumption in other parts of the world – from other SEZs in the Global South to shopping spaces in global cities.

Beyond these operations facilitating the inflow of foreign capital, SEZs have a huge impact on the restlessly changing built environment in the peri-urban fringe. Their formation creates backward linkages through employment creation and raw material consumption (Gonzalez et al., 2002). Industrial operations and other enterprises in SEZs require huge labour inputs and, as such, open up job opportunities for low-wage workers, often triggering the migration of thousands of workers, not just from surrounding villages but also from other regions of the Philippines. Because of this inflow of new workers, demand for housing and retail increases, prompting the rise of new residential and commercial developments. The same is true for master-planned SEZs that combine industrial, commercial, office, and residential developments, as they trigger a rippling effect in surrounding villages and broader regional economies of the peri-urban fringe. These transformations illustrate the ways in which SEZs act as translocal linchpins of peri-urban transformation, enabling the operations of global market forces and facilitating suburban expansion. Figure 20.1 is a map depicting the demographic clustering around SEZs in Manila's peri-urban fringe, suggesting the rippling effect of SEZs in surrounding areas.

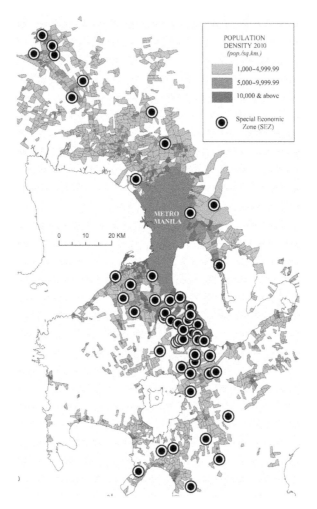

Figure 20.1. SEZs and population density of villages in Manila's peri-urban fringe. Data source: Philippine Economic Zone Authority (2015); Philippine Statistics Authority (2016).

Gated Suburbs: Translocal Fruits of Diasporic Labour

A key mode of urban transformation in Manila's peri-urban fringe has been residential subdivisions, most of which are gated projects developed by real estate firms. These projects are aggressively marketed by property developers to Overseas Filipinos (OFs), who aim to acquire homes for their families and relatives left behind in the Philippines. The

Figure 20.2. A booth of a real estate developer promoting properties in the Philippines at the "Taste of Manila" festival in Toronto. Photo: Author (2017).

property demand from OFs has been a major force behind the real estate boom in the Philippines. To sustain this demand and increase their share of the OF market, developers lure prospective clients by navigating the transnational geographies of OFs through overseas property caravans, participation in Filipino community events, and establishment of overseas offices. In Toronto, for example, the Filipino festival "Taste of Manila" features several stalls of Filipino real estate developers that sell homes in the Philippines. During the 2017 celebrations, one of the participating property developers was Ayala Land, arguably the largest and oldest real estate firm in the Philippines. Staffed by Toronto-based Filipino-Canadian brokers, Ayala Land's stall had stacks of property flyers that featured Philippine properties, several of which were houses in gated subdivisions in Manila's peri-urban fringe (see figure 20.2). To support these efforts, the government has passed laws requiring OF workers to contribute to a government-mandated housing loan program, while several banks have initiated OF-specific programs to assist OFs in securing mortgages.

Because of these efforts, many newer gated residential subdivisions, especially in the peri-urban fringe, have become residences of transnational families and relatives of OFs, such that the majority of sales and reservation transactions involve OFs (see Lucas, 2007).

These gated developments are designed to conjure a sense of exclusivity, security, and wealth. They are marketed to OFs as "world-class" spaces rewarding the "hardworking" OF workers. Many of these projects use designs that draw from various foreign influences, including popular architectural styles ranging from the American South to British, Spanish, and Mediterranean designs. Many high-end developments are advertised as being master-planned by Euro-American-based architectural and planning firms, further cementing their "world-class" appeal. Take the case of Santa Elena development in Laguna province (see Aquino, 2011). It placed an advertisement in 2011 that read: "Setting Sta. Elena apart from other developments are its meticulously designed themed subdivisions. Inspired by the world's most charming destinations, residents can choose to come home to English gardens, Spanish haciendas, Italian villas or Georgian colonial houses, depending on which style best suits their tastes." Typical residential projects are organized and structured to reflect the American suburubs: cul-de-sac roads, bungalow houses, lawns, and multipurpose halls. To ensure security, developments are oftentimes surrounded by a perimeter wall with a gate that is complete with an outpost and security personnel that regulate the flow of traffic. Typically, a homeowner association serves as the mode of governance with their own sets of rules and regulations for their resident members.

These developments serve as the idealized transnational fruits of overseas labour. As spaces that evoke a sense of wealth, modernity, and order, these gated villages are products of yearning for an ideal way of life, a Filipino dream that seems to be impossible to attain as seen as outside the gates, where infrastructure and services are inadequate and mismanaged. Given that these developments are situated in the unsecure terrains of the peri-urban fringe where fear and ambivalence are dominant modes of living, gating is a standard means of securing safety and sustaining order inside these communities. In effect, the Filipino dreams of OFs must be gated and segregated from the rest of the peri-urban fringe. Despite the secured spaces of wealth inside these gated villages, everyday life is far from ideal. Critical to the OF aspiration of homeownership is the desire to return and have a complete family set-up. However, maintaining an ideal suburban life inside the gates requires a sustained inflow of remittances in order to support ongoing expenses, such as regular contributions to homeowner association fees, household expenses, and mortgage payments. Thus, a family member has to remain overseas in order to keep sending home much-needed money, which concomitantly makes the suburban appeal of a complete family life in a constant state of permanent deferral. This contradiction between transnational work and the fulfillment of an ideal suburban life, illustrates the uneasy entanglements producing these gated suburbs.

Translocal Neighbourhoods: Enclaves of Wealth in Barangays

In a number of peri-urban barangays, clusters of eclectically styled two- and three-storey mansions loom prominently over the rest of the village. In some cases, these clusters occupy several blocks and effectively create enclaves of wealth. Certain enclaves can be attributed to a common migrant stream, signalling a shared spatial history of transnational mobility, from Hong Kong village – referring to communities mostly comprised of families whose members have had a long history of working in Hong Kong – to Seaman's village, a community of families whose household heads are seamen. In Batangas province, a steady flow of migrant workers who left for Italy over the past decades has facilitated the rise of multiple residential enclaves called "Little Italy." In the town of Mabini, the transnational mobilities of Filipinos working and living in Italy have virtually transformed the agricultural town into an eclectic translocal enclave (see Baggio & Asis, 2008; Katigbak, 2015). The initial migration of Mabini residents to Italy started in the late 1970s and grew over the succeeding years as migrant workers referred their family members, relatives, and fellow townsfolk to their Italian employers. Such cumulative causation has resulted in a steady migration stream that has facilitated the transnational mobilities of family members, remittances, and other forms of connections. The flow of capital has transformed the formerly agricultural town into a translocal enclave of wealth, as nipa huts are replaced by mansions. Home designs are heavily influenced by Italian architectural motifs, built with marble and decorated with terraces and artistic exteriors. Many families rely on remittances by members who work and live in Italy, and who have long turned their backs on agriculture.

As tangible expressions of wealth, these homes represent the binary twisting of class among OFs who, after sending home remittances used to construct mansions, are considered "mayaman" (rich) in the village. In contrast to the socio-economic position in the countries where they work, the Filipino worker and their transnational family are generally highly regarded in their home village. In effect, these homes suggest the attainment of "maalwang buhay" (Aguilar et al., 2009) – an idealized and comfortable life that is facilitated and nurtured by transnational family relations.

Homes in translocal neighbourhoods are also spatial expressions of transnational commitments to families back home, serving as what Faier (2013) would call "emotional investments" in the homeland by OFs. Unlike gated suburbs developed by property developers, these neighbourhoods deploy a strong sense of belongingness, particularly to localities and hometowns of transnational families. Such a sense of affinity to the broader community serves as fuel to the sustenance of transnational relations, which may usually come in the form of local philanthropic projects.

Conclusion

The spatial transformations in Manila's fringe are complicated and cannot be fully explicated through a singular suburban narrative of city–periphery relations. By considering transnationalism as a potent force in expanding built environments and transforming peri-urban environments, this chapter adds to the multiple and diverse narratives of global suburbanism.

Developments in Manila's peri-urban fringe rely heavily on the transnational mobilities of capital, ideas, and migrants. Emergent translocal sites in the fringe are important growth hubs that funnel capital into the Philippines, and facilitate urban transformation. By identifying and explicating the emergence of three translocal spaces in the fringe, this chapter argues for multiple spatialities of transnational urbanism with varied transnational entanglements and multiple spatial expressions. SEZs as transnational economic hubs facilitate the integration of the Philippines, and the individual regions, into global markets through multinational corporations that have set up their operations there. At the same time, the ripple effect on surrounding villages has facilitated the conversion of the peri-urban fringe into clusters of translocal urban economies. Finally, translocal neighbourhoods and gated suburbs are spatial expressions of the transnational mobilities of Overseas Filipinos.

Attending to postcolonial provocations, this chapter examined (sub) urban formations from a Southern context, explicated not as a case study of a universal urban model, but as emergent spaces that can spur theoretical innovation. I have examined translocal spaces with the hope of formulating new urban conceptualizations that resonate across multiple contexts. This theoretical manoeuvre is in consonance with Leitner and Sheppard's (2016) provincializing proposal for a "shifting ecosystem of critical urban theories," wherein diverse perspectives and propositions are in critical conversations with one another. In order to take postcolonial urbanisms seriously, one must acknowledge this diversity and consider the viability of theoretical formulations that are emerging from contexts that are beyond the usual anglophone suspects. By considering transnationalism as a force influencing suburban expansion, this chapter pushes for new urban theoretical avenues that draw from Global South contexts, like Manila, while still being relevant to other suburban terrains. In effect, it interlinks constellations of suburban formations that are shaped by transnational mobilities, including racially diverse ethnoburbs and globurbs in Global North contexts (see King, 2004; Li, 2009), with the Global South.

REFERENCES

Aguilar, F., Peñalosa, J., Liwanag, T., Cruz, R., & Melendrez, J. (2009). *Maalwang buhay: Family, overseas migration, and cultures of relatedness in Barangay Paraiso.* Ateneo de Manila Press.

Aquino, JL Santiago. (2011, June 20). Sta. Elena City in Sta. Rosa, Laguna. *BLOG-PH.com.* https://www.blog-ph.com/2011/06/sta-elena-city-in-sta-rosa-laguna.html.

Armitage, J., & Roberts, J. (2003). From the hypermodern city to the grey zone of total moblization in the Philippines. In R. Bishop, J. Philips, & W.W. Yeo (Eds.), *Postcolonial urbanism: Southeast Asian cities and global processes* (pp. 87–104). Routledge.

Baggio, F., & Asis, M. (2008). Global workers, local philanthropists: Filipinos in Italy and the tug of home. In T. van Naerssen, E. Spaan, & A. Zoomers (Eds.), *Global migration and development* (pp. 130–49). Routledge.

Beaverstock, J. (2005). Transnational elites in the city: British highly skilled inter-company transferees in New York City's financial district. *Journal of Ethnic and Migration Studies, 31*(2), 245–68. https://doi.org/10.1080/1369183042000339918.

Caoili, M. (1988). *The origins of metropolitan manila: A political and social analysis.* New Day Publishers.

Chacko, E. (2007). From brain drain to brain gain: Reverse migration to Bangalore and Hyderabad, India's globalizing high tech cities. *Geojournal, 68*(2–3), 131–40. https://doi.org/10.1007/s10708-007-9078-8.

Clapson, M., & Hutchison, R. (Eds.). (2010). *Suburbanization in global society.* Emerald Publishing.

Clarke, N. (2005). Detailing transnational lives of the middle: British working holiday makers in Australia. *Journal of Ethnic and Migration Studies, 31*(2), 307–22. https://doi.org/10.1080/1369183042000339945.

Collins, F. (2010). International students as urban agents: International education and urban transformation in Auckland, New Zealand. *Geoforum*, *41*(6), 940–50. https://doi.org/10.1016/j.geoforum.2010.06.009.

Faier, L. (2013). Affective investments in the Manila region: Filipina migrants in rural Japan and transnational urban development in the Philippines. *Transactions of the Institute of British Geographers*, *38*(3), 376–90. https://doi.org/10.1111/j.1475–5661.2012.00533.x.

Friesen, W., Murphy, L., & Kearns, R. (2005). Spiced-up Sandringham: Indian transnationalism and new suburban spaces in Auckland, New Zealand. *Journal of Ethnic and Migration Studies*, *31*(2), 385–401. https://doi.org/10.1080/1369183042000339981.

Gonzalez, E., Ramos, C., Estrada, G., Advincula-Lopez, L., & Igaya, G. (2002). *Managing urbanization under a decentralized governance framework* (Vol. 2). Philippine Institute for Development Studies. http://dirp3.pids.gov.ph/ris/books/pidsbk02-urbanization2.pdf.

Greiner, C., & Sakdapolrak, P. (2013). Translocality: Concepts, applications and emerging perspectives. *Geography Compass*, *7*(5), 373–84. https://doi.org/10.1111/gec3.12048.

Irazabal, C. (2012). Transnational planning: Reconfiguring spaces and institutions. In S. Krätke, K. Wildner, & S. Lanz (Eds.), *Transnationalism and urbanism* (pp. 72–90). Routledge

Katigbak, E. (2015). Moralizing emotional remittances: Transnational familyhood and translocal moral economy in the Philippines' "Little Italy." *Global Networks*, *15*(4), 519–35. https://doi.org/10.1111/glob.12092.

Keil, R. (Ed.). (2013). *Suburban constellations: Governance, land and infrastructure in the 21st century*. Jorvis.

Keil, R. (2018). *Suburban planet: Making the world urban from the outside*. Polity.

Kelly, P. (2000). *Landscapes of globalization: Human geographies of economic change in the Philippines*. Routledge.

King, A. (2004). *Spaces of global cultures: Architecture, urbanism, identity*. Routledge.

Klein, N. (1999). *No logo*. Knopf Canada.

Leitner, H., & Sheppard, E. (2016). Provincializing critical urban theory: Extending the ecosystem of possibilities. *International Journal of Urban and Regional Research*, *40*(1), 1468–2427. https://doi.org/10.1111/1468-2427.12277.

Levitt, P. (2001). *The transnational villagers*. University of California Press.

Levitt, P., & Schiller, N. (2004). Conceptualizing simultaneity: A transnational social field perspective on society. *International Migration Review*, *38*(3), 595–629. https://doi.org/10.1111/j.1747-7379.2004.tb00227.x.

Li, W. (2009). *Ethnoburb: The new ethnic community in urban America*. University of Hawaii Press.

Lucas, D. (2007, May 20). OFW remittances fueling growth in real estate. *Philippine Daily Inquirer*. http://globalnation.inquirer.net/news /breakingnews/view_article.php?article_id=6700.

McAndrew, J. (1994). *Urban usurpation: From friar estates to industrial estates in a Philippine hinterland*. Ateneo De Manila University Press.

McGee, T. (1991). The emergence of desakota regions in Asia: Expanding a hypothesis. In N. Ginsburg, B. Koppel, & T.G. McGee (Eds.), *The extended metropolis: Settlement transition in Asia* (pp. 3–26). University of Hawaii Press.

Ortega, A. (2018). Transnational suburbia: Spatialities of gated suburbs and Filipino diaspora in Manila's periurban fringe. *Annals of the American Association of Geographers, 108*(1), 106–24. https://doi.org/10.1080/24694452 .2017.1352482.

Parnreiter, C. (2012). Conceptualizing transnational urban spaces: Multicentered agency, placeless organizational logics, and the built environment. In S. Krätke, K. Wildner, & S. Lanz (Eds.), *Transnationalism and urbanism* (pp. 91–111). Routledge.

Phelps, N. (2015). *Sequel to suburbia: Glimpses of America's post-suburban future*. MIT Press.

Phelps, N., & Wu, F. (Eds.). (2011). *International perspectives on suburbanization: A post-suburban world?* Palgrave MacMillan.

Philippine Economic Zone Authority (PEZA). (2015). *List of special economic zones*. http://www.peza.gov.ph/images/stories/List_of_Economic _Zones.xls.

Philippine Statistics Authority. (2016, May 19). *Highlights of the Philippine population 2015 census of population*. https://psa.gov.ph/content/highlights -philippine-population-2015-census-population.

Portes, A. (2001). Introduction: The debates and significance of immigrant transnationalism. *Global Networks, 1*(3), 181–93. https://doi.org/10.1111 /1471-0374.00012.

Robinson, J. (2006). *Ordinary cities*. Routledge.

Robinson, J. (2016). Comparative urbanism: New geographies and cultures of theorizing the urban. *International Journal of Urban and Regional Research, 40*(1), 187–99. https://doi.org/10.1111/1468-2427.12273.

Roy, A. (2009). The 21st-century metropolis: New geographies of theory. *Regional Studies, 43*(6), 819–30. https://doi.org/10.1080/00343400701809665.

Schiller, N., Basch, L., & C. Blanc. (1995). From immigrant to transmigrant: Theorizing transnational migration. *Anthropological Quarterly, 68*(1), 48–63. https://doi.org/10.2307/3317464.

Schiller, N., & Çağlar, A. (2009). Towards a comparative theory of locality in migration studies: Migrant incorporation and city scale.

Journal of Ethnic and Migration Studies, 35(2), 177–202. https://doi.org
/10.1080/13691830802586179.

Shatkin, G. (2008). The city and the bottom line: Urban megaprojects and the
privatization of planning in Southeast Asia. *Environment and Planning A,
40*(2), 383–401. https://doi.org/10.1068/a38439.

Shatkin, G. (2016). The real estate turn in policy and planning: Land
monetization and the political economy of peri-urbanization in Asia. *Cities,
53*(1), 141–9. https://doi.org/10.1016/j.cities.2015.11.015.

Smith, M.P. (2001). *Transnational urbanism: Locating globalization.* Blackwell.

Smith, M.P. (2005). Transnational urbanism revisited. *Journal of Ethnic and
Migration Studies, 31*(2), 235–44. https://doi.org/10.1080/1369183042000339909.

Smith, M.P., & Guarnizo, L. (1998). *Transnationalism from below: Comparative
urban and community research.* Transaction Publishers.

Special Economic Zone Act. (1995). Republic Act No. 7916. *Philippine Economic
Zone Authority.* http://www.peza.gov.ph/index.php/about-peza/special
-economic-zone-act.

Teaford, J. (1997). *Post-suburbia: Government and politics in the edge cities.*
Johns Hopkins University Press.

Watson, V. (2014). African urban fantasies: Dreams or nightmares?
Environment and Urbanization, 26(1), 215–31. https://doi.org/10.1177
/0956247813513705.

Conclusion

Jeanine wondered why some suburban transformations, for instance, adding density near transit hubs, were such a challenge to initiate, and political hot potatoes in some jurisdictions. Clearly, it had something to do with the fact that the proposals were not just about a change in built environment, but were also perceived as a challenge to the very lifestyle some suburbanites had come to enjoy and value. "Clearly, suburbs," Jeannine proclaimed, "have their own lifestyles, just as cities do." Detached and subdivided In the mass production zone Nowhere is the dreamer or the misfit so alone Any escape might help to smooth the unattractive truth But the suburbs have no charms to soothe the restless dreams of the youth" Rush, "Subdivisions". Jakob had learned...that suburban ways of living are much more diverse than he first expected. They didn't just include low-density, homogeneous neighborhoods..."so more sustainable ways of living is the ultimate goal then, not urbanization that's been the intermediary to facilitate one particular way of living."..."...we haven't really fully reverted the impacts of automobile-based planning that have shaped North American communities at least since the second World War." Carly added. Jakob wasn't convinced, "I am not always so sure it's just planning though. What about all the changes in society and the economy that have happened? Hasn't planning always kind of been about facilitating societal transformations?" "I wonder whether some of the policies that were meant to make the suburbs more sustainable are in fact just urbanizing the suburbs." "There is certainly a case to be made that some of the policies that we, as planners, sell as suburban redesigns may just bring smaller, non-family households into previously subruban environments," Jakob continued. "Right," Wei jumped in, "and whether these policies actually reduce the land consumption, and car use, of existing suburban residents is really still just an open question." ...What would Jane Jacobs say about planners' ideas to transform suburbs into what basically amounts to cities?"...The problem was, as Jeannine understood quite well, that the suburban way of life in North America was contributing to several environmental and public finance concerns. Yet, of living all together? Could we not develop planning solutions that helped maintain some of the ideals of suburbs that people had come to value and appreciate, while also reducing the negative impact of current practices?...perhaps Jane Jacobs would promote an approach to suburban redevelopment that was starting with the people, and then looking for solutions that integrated and built on those lifestyles...The idea of planning with people, not for them, from the group up, at the grassroots level, is integral to the planning process the way Jacobs envisioned it." "We probably could do a bit more to acknowledge aspects of suburban lifestyles that are worth preserving while also facilitating the development of more walkable, transit-supportive environments in suburban areas. You need some urbanization to address health, environmental, and fiscal concerns associated with suburban sprawl." "Why should we want to replicate suburban landscapes if their very replication has the potential to reproduce elements of lifestyle that are highly exclusive and therefore undesirable?" "We will have to think about suburban landscapes that challenge the status quo, for instance, thinking about activities that go on inside the home when planning communities, not just considering places of formal work and commuting patterns...just as important is putting in place a process that ensures that the visions of how our communities should look and function to actually reflect the needs and diversity of the population...It most definitely is about not just plannng for croissants...Really, it's about planning for public good

Image 21. Possibilities of suburban transformations

21 After Suburbia: Peripheral Notes on Urban Theory

ROGER KEIL

> In fact, suburbs have always exhibited a range of built forms and demographics.
> – Airgood-Obrycki and Rieger, "Defining Suburbs"

> Today, many urban residents are not quite sure how to understand such unanticipated implications. Many find themselves ensconced in landscapes at some remove from the urban cores, and whose compositions are a panoply of built environments, tended and fallow agricultural land, industrial estates, and wastelands. Discrepant land uses sit side by side in oscillating spirals of ascension and decline, where it is difficult to discern just where things are headed. Or, they remain situated in infrastructurally and socially overextended urban cores subject to intense speculation.
> – Simone, "Maximum Exposure"

Introduction

The definition of what the suburban entails, what it means, how it relates to other conceptual and real categories such as "city" and "countryside" has been vexing researchers and practitioners for decades while "suburban constellations" – differentiated clusters of interlaced peripheral social and spatial formations – have increasingly replaced the assumed dichotomies of city and suburb (Keil, 2013).[1] Admittedly, most conceptual advances that have recently noted a growing diversity in built forms and demographics in suburbia, while valuable, have not been able to crack the nut of definition. This is an important point to start from as I will, in this final chapter of the book, discuss what "after suburbia" might mean. A decade ago, Richard Harris (2010) made the valuable and valiant attempt to bring some order in the chaos that ensues when one, as our joint project on Global Suburbanisms did, expanded the conceptual and empirical reach of the suburban from beyond the American white-picket fences into the larger urban world. And even what is happening on both sides of those iconic picket fences has been changing drastically as Jan Nijman's new edited collection *The Life of North American Suburbs* (2020) demonstrates in dramatic fashion. The same can, of course, be said about the tremendous changes in European suburbia, often seen as the counterpart to the American case but clearly in a state of real and conceptual upheaval particularly since the fall of the Berlin Wall in 1989 (Phelps, 2017b; Stanilov & Sykora, 2014). The emerging complex landscape of suburbia is represented schematically below in figure 21.1. An important distinction in Harris's universe of "meaningful types" was the fundamental difference between suburban dwellers who were in the periphery by choice and by constraint (or as a consequence of displacement). Both types of suburban dwellers are subject to similar overarching processes of differentiation as Harris explains. What remains important for both is also the distinction between having arrived in the suburbs from either inside or outside the city. Still, when we discuss extended urbanization more broadly, the migration from the centre to the periphery, voluntary or involuntary, will ultimately be only one source of suburbanization on today's urban planet (Keil, 2018a).

Richard Harris's insightful typology predated much of the empirical research and conceptual development represented in this current book. He himself followed up with a definitive volume on how settlements beyond cities are named around the world (Harris & Vorms, 2017). As we have seen, scholars inside and outside the "Major Collaborative Research Initiative (MCRI) Global Suburbanisms" have responded to the challenges outlined by Harris in a multiplicity of ways. We have separately yet interconnectedly looked at the governance (Hamel & Keil, 2015), land (Harris & Lehrer, 2018), and infrastructure (Filion & Pulver, 2019) as well as suburbanism as a way of life (Moos & Walter-Joseph, 2017). A decade later, as we were in the process of preparing this volume, another overview of suburban definitions was published, this time partly in recognition of the work that has been done since Harris's intervention. In their recent work on the subject, American scholars Airgood-Obrycki and Rieger (2019; see also Airgood-Obrycki et al., 2020) focus mostly on the United States. The authors differentiate between the *census-convenient*, *suburbanisms*, and *typology* as useful types of suburban definitions. As Airgood-Obrycki and Rieger (2019) note, these different approaches have variable utility for both theoretical analysis and practical intervention and I don't want to further discuss their respective methodological merits. What suffices to say here, and we will get back to this at the end of this essay, is that "suburban definitions are crucial in shaping planning and policy problems" (p. 30).

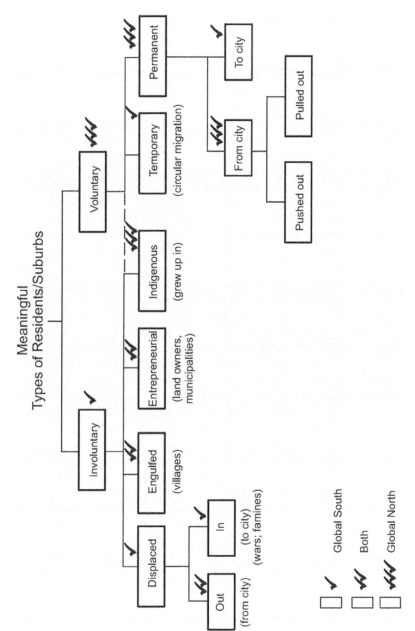

Figure 21.1. A typology of suburbs worldwide. Source: R. Harris, Meaningful types in a world of suburbs. In M. Clapson & R. Hutchinson (Eds.), *Suburbanization in global society* (pp. 15–50). Emerald. (By permission of the author.)

When I propose that we advance the conversation to the notion of "after suburbia," it is not necessarily just to extend these definitions and categories but to take a different turn altogether. This intervention interrogates suburbanization, suburbanisms, and also the suburbs as objects of scholarly research and practical intervention, as figments of our imagination and as places where we live, work, and play. I put forward the lineaments of a strategy for research and action on suburbanization and suburban ways of life that may guide some of our work in future years, as we need to get used to existing with the constraints and opportunities of a multidimensional and multifarious life in extended urban peripheries. Assuming that life on the global urban periphery is changing rapidly into a set of post-suburban constellations – where the suburban is not just seen in reference to the urban/the city but also in relation to its own history and geography and the endogenous dynamics it generates – that provide novel insight into the urban condition, I make the case that understanding the suburban world and confronting its many challenges is of the utmost importance to the work urban researchers need to do.

In reimagining the suburban, we can start with assessing what has been said. In both suburban research and popular representation there is a tendency to impose order. Here, the suburban is intimately tied to the establishment of some kind of controllable future. Perhaps it is necessary to take another approach, to see the suburbs as a place of disorder and possibility. At issue is the notion of the unruly city – which logically and historically precedes the suburb– or the unruliness of the classes of people that inhabit it. Ironically, of course, much of that unruliness has now migrated to the margins and the city centres that have become the domain of the predictable and the controllable. This is true for the ground-related subdivisions of North America as much as for the massive suburban tower neighbourhoods of Europe and Canada. Sieverts (2006) has pointed out that these subdivisions were "mostly designed not just as spatially bounded but also as temporally limited settlement structures – for a finite image of a 'timeless future.' In other words, they were 'ready-mades' for a state of completion and wholeness, without conceptual openness for historical change" (p. 164).

Suburbanization has also been associated with the degenerate nature of urban life, the unhealthy environment of what used to be the city in the industrial period. Harris (1996, 2014) in particular has produced a remarkable and unique body of work that demonstrates how the suburbs, while considered transitory in their very nature, have historically been made into something normal and ordered, how they have been mainstreamed, how they have given up on their possibilities.

The normalizing historical tendencies, the thrust towards order, has been largely reflected in a suburban studies literature that drew lines around the subject of the suburban to capture its ordered regularities and predictable trajectories (see Keil, 2018a, chap. 4). Even those who have sought to show how suburbs have been continuously diverse contribute to normalizing the experience of the diverse suburb: Through diversity, many suburbs do not become more interesting. Instead, their diverse populations often lose their character as the neighbourhoods they inhabit reflect anglo middle-class norms of suburbanism and suburban ways of life (automobility, domesticity, privatism, etc.) (Walks, 2013).

After Suburbia?

Many if not most cities around the world have experienced increases in their populations, yet their spatial expansion has often outpaced their demographic growth. While we are becoming a more urban globe, the cities in which we live tend to be lower in density than they used to be (see Angel et al. in this volume). This means, in other words, that we are living in an increasingly differentiated suburban world today, which is now recognized even in popular discourse: "Today, cities are spiking upwards even while conquering new territory. The old division between city center and suburb has broken down. Far from being the monolithic, dull places of cliché, many suburbs have become steadily more urban since the 1980s with jobs, greater ethnic diversity, street life, crime epidemics and drug abuse – in other words, inheriting many of the virtues and vices of the inner city" (Wilson, 2020, para. 3). Still, following urban orthodoxy, the suburbs should not really have a right to exist. They are – although themselves products of earlier periods of planning doctrine – anathema to urban planning today. Too large is their environmental footprint, too segregated is their socio-spatial structure, too impenetrable is their resistance to urban change to make them remotely interesting to what Larry Beasley (2019), the high priest of dense urban living, the architect of "Vancouverism," calls the "urban congnoscenti" (p. 53; on Vancouverism, see Peck et al., 2014). Beasley (2019) self-consciously admits that "the blunt truth is that, in our time, we really didn't care much about the suburbs or their problems – they were not on our agenda" (p. 90). The Vancouver planner himself, of course, has since started to bring the lessons of Vancouverism to the suburbs. In the Ontario city of Brampton, northwest of Toronto, Beasley was hired to take the city through a process of urbanizing the suburb. Beasley and his team held a public consultation with 14,000 citizens of

Brampton. This extensive process generated a report published in April 2018. The report's general ambition, driven ostensibly by "a people's vision," is clear: To position a future Brampton not as a dependent suburb riding on the coattails of Toronto – or worse, on those of its suburban neighbour Mississauga – but as an autonomous and self-confident multi-centred city with residential populations and jobs in sustainable industries (City of Brampton, 2018).

The perforation of the urban-suburban divide, if only conceptually, is of course also part recognition of a suburban resurgence in North America. After being under pressure from *traditional* urbanists and *new* urbanists alike and not long after being ground zero of the 2008 financial crisis, suburban subdivisions with ground-related housing, street-facing garages, and green back yards are increasingly being rolled out as status symbols of arrival and success in many parts of the world. This is evidenced from Eastern Europe to China, from Africa to Latin America, from Portugal to Greece, not to mention to the continued expansion of low-rise subdivisions in the core Anglo-Saxon countries where classical suburbia was first developed (see Angel et al. in this volume; Berger et al., 2017; Organisation for Economic Co-operation and Development, 2018).

Simultaneously, and perhaps relatedly, the North American view on suburbanization and suburbanisms as distinct ways of life no longer suffices in the global context (for an evolving view, see Nijman, 2020). The ground-related residential home, long the symbol of the American way of life, is not the only morphology of global suburbanization. There are, in fact, variegated densities and shapes from informal settlements in some cities of the Global South to state-sponsored high-rise estates such as those in China and Turkey. The suburbs are alive and well and full of diversity of form, structure, and population. In fact, an era of "massive suburbanization" is now upon us that links the monumentality of traditional modernism with the liberalized financial markets of today, supported and played by often autocratic national governments that have become masters in what Roy (2011) has termed "inter-referencing" (see also, Güney et al, 2019).

To sum up, we really are dealing with *two* kinds of stories on suburbanization. The first, associated with the traditional preferences of urbanists in Europe and North America, presents a normative rejection of suburban development which is judged – for good reason – as a contributor to many social and environmental problems we face today. But there is the second story, which in many ways is more pervasive. This story narrates a tremendous dynamic of suburban development around the world.

Let me add that there is at least a third storyline: the reality of a frantic *re*-urbanization, densification, and hyper-compacting of inner cities, often in line with dynamic gentrification and segregation processes. This phenomenon is also linked to the notion of "vertical sprawl" (see, for example, Graham, 2016, p. 192). While important and pervasive worldwide, I will leave this third storyline largely untouched here as it distracts from our focus on the urban periphery.

Suburbanization is mainly fed by people and economic activity moving from the centre of the city to the suburbs and by people and economic activity moving from the rural countryside to the periphery of cities. As soon as we identify suburbanization and suburbanism in this way, we have to admit that while there are no clear conceptual or built boundaries to what we consider the city or the urban, the imagined linear process of suburbanization – extensions beyond the existing city – is morphing into city building and rebuilding at the metropolitan edge that we can call post-suburbanization. Defined as a tendency through which "de-densification (classical suburbanization) is partly converted, inverted or subverted into a process that involves densification, complexification and diversification of the suburbanization process" (Charmes & Keil, 2015, p. 581), post-suburbanization is now prominent (Phelps & Wu, 2011). In the process, suburbanization diversifies in morphology and its socioeconomic and demographic profile, and the experienced boundaries between the cities and their peripheries are increasingly breaking down. The post-suburban also provides a geographical and conceptual framework for political action (Sweeney & Hanlon, 2017; Young & Keil, 2014).

What we observe now, then, is a non-linear process in which classical patterns of migration continue but where the periphery itself begins to be not just the destination for the dreams of the *(petite) bourgeoisie* or even the unchanging terminus of suburban settlement but in itself the place where change becomes the norm. The periphery, which had heretofore been described as unalterable and stable, is the place where change is most profound and where urban society experiences a redefinition.

Around the world, the spatial peripheries of cities are expanding fast and are buzzing with activities. It is necessary to shift our gaze accordingly. This entails primarily two dimensions. First, we need to look elsewhere, outside of North America, if we want to see the full extent of the complexity of today's suburbanization; and, second, we must look for other morphologies and landscapes than the traditional ones interrogated in North America. We will find that suburbs are indeed everywhere (not just in 1950s America) and come in all shapes and sizes (not just in the shape of bungalow subdivisions).

Yet, as pervasive as the after-suburban has become, it is under-researched and perhaps misunderstood. The reflex is always there to return to the American stereotype (Lasky, 2020). Critical urban research must make an effort to contribute to the analysis of the production of, and life in, the world's urban peripheries. Not accidentally, the major conceptual influences now begin to come from where the major thrust of peripheral urbanization is occurring. Ample evidence for this is the tremendous work done by African, Asian, and Latin American scholars who not only present the extent to which their urban regions are redefined at the periphery but also simultaneously theorize this process (see Bloch, Mabin, and Todes in this volume; Gururani & Dasgupta, 2018; Reis & Lukas, 2022; as well as contributions by Gururani, Kusno, Ortega, Wu, and others in this volume). This kind of work does not just point to the fact that we are now living on a suburban planet. It also bolsters the important argument that we will only recognize it as such if we leave behind the *conceptual* picket fence that we inherited from the North American experience.

Extended Urbanization and the Continuous City

Reassembling the "disjunct fragments" of Lefebvre's urban explosion (Keil, 2018b) is the very challenge of the after-suburban society in which we now find ourselves. Those boundaries will be porous in the post-suburb. Politically and socially, this makes for messy realities that create contested environments in which activists, planners, and policymakers will have to operate for some time to come. But conceptually, we benefit immediately: If we give up the thinking along the lines between the centre and the suburb, we will see that we have many options before us. There are multiple disciplinary and post-disciplinary avenues that can take us into the mixed terrain that awaits.

One element of this new terrain is what Simone (2004) calls "the intersection" of peoples, localities, and regions that needs to be narrated in the city (p. 240). The intersection is already more than a mere boundary, it is generative of conversation, conflict, handshakes, and mergers. Simone (2016a) develops this idea further in recent work responding critically to some political economists' insistence that the urban can be reduced to certain core economic factors that maximize efficiencies but also "to invent unanticipated realities that can be folded into these operations, or exist autonomously in spawning different kinds of economic activity as a process of ongoing experimentation" (p. 213).

Broadening the notion of "intersection" in this way allows us to see the entirety of the urban process and form: "All the different kinds of intersections among densities, land use and agglomerated economic activities that occur within and between cities retain the ability to continuously shift the contours of urbanization" (p. 213). Simone's intersection corresponds with De Boeck and Baloji's (2016) work on Kinshasa in which they develop the notion of the "living city" where overlapping pasts and futures, "even when ruptured, mutilated and mutated by the city, continue to strongly resonate underneath the surface of the 'modern' city" (p. 258). This historically overlapping reality also has a regional geographical dimension in that "the migratory movements from the countryside to the fringes of the city, or from the inner city to the periphery, have given rise to a vast 'peri-urban' landscape" (p. 260).

Work on "interstices" (Brighenti, 2013; Mattiucci, 2015; Phelps& Silva, 2017) carries on the previous occupation with the "in between" (Sieverts, 2003). In his original and generative work on "interplaces," Phelps (2017a) develops a veritable program for the places in between. Ortega's (2020) work on Manila's "necroburbs," Kusno's (2019) work on "kampungs," as well as Simone's (2014) work on Jakarta relate and anchor the discussion here. The particular landscape of some traditional mining areas like the Ruhr in Germany or the Minas Gerais in Brazil are blueprints of interstitial urbanization, or "urbanized nature," as Angelo (2021) calls it (see also Keil, 2020b). As do the metaphors of "scene and obscene" (i.e., offscene) that Swyngedouw and Ernstson (2018) have enlisted to express the notion of the relationship of what is seen and heard and what is not. Even if we look outside the realm of critical urban studies and geography, we find a broad range of authors describing an "Infinite Suburbia" where future landscapes of settlement are being defined and refined (Berger et al., 2017). We can add to this not least Lerup's (2017) notion of the "continuous city." As Lerup notes, we must not see the horizon of the suburban in the inner periphery alone but need to understand the total complex(ity) of the suburbanization process. Identifying the unevenness in the distribution of the benefits of urbanization, Lerup opens our view towards a new dimension, a new scale, beyond the particular of the individual place in the continuous city (p. 216). Lastly, extended urbanization is part of what Monte-Mor (2014a) has called "the combined process of metropolization and extended urbanization" (p. 112) and is an expression of the "dialectical unity of urban center and urban fabric" (p. 111).

From here, we return to the French periphery of the late twentieth century from where the "cry and demand" for the "right to the city" originally drew its strength (Lefebvre, 1996, p. 158). Importantly for

Lefebvre, then, "the suburban is conceptually an extension of urbanism" (Walks, 2013, p. 1477) but, at the same time, since suburbanization has appeared on the scene as a massive phenomenon, it becomes the basis for a new and more comprehensive theory of the city and, ultimately, society. Accordingly, Lefebvre understood the urbanization process as ultimately being polymorphous with fragmented relations in a disparate urban fabric that is characterized simultaneously by processes of concentration and deconcentration and centres and peripheries of variable dimensions (Ronneberger, 2015, p. 26). Yet the fragmented and disjointed extended urbanization that Lefebvre observed as an explosion of the city leads to a new metropolitan form and structure and to a new way of looking at what emerges; indeed, "an integrated theory of the city and urban society"!

The question of extended urbanization is both spatial and political: In the Lefebvrian sense, "the urban question had become the spatial question itself" as a consequence of the emergence of "urban society" (Monte-Mor, 2014b, p. 265). But this poses immediate political questions, as "the politicization proper to urban space, which had now been extended to regional space, reinforces popular concerns over the quality of daily life, the environment, and the expanded reproduction of life" (Monte-Mor, 2014b, p. 265). Leaning heavily on Lefebvre's original formulation of extended urbanization, Monte-Mor (2014b) concludes that "the repoliticization of urban life becomes a repoliticization of social space as a whole" (p. 265). The suburban, then, is still part of extended urbanization, but not one that is subordinate to state- or city-centric rationalities. Much more, the exploding periphery asserts itself as a generator of centralities in that future city that, in Lefebvre's (1996) words, "will inevitably be polycentric, a multiplicity of centres, diversified but conserving a Centre" (p. 208).

Following Lerup and Lefebvre down this path, then, opens up multiple possibilities of rethinking historical and geographical dimensions, or scales of urbanization. Lerup (2017), I think, has it right by calling what he expects "out there" to be "disruptive lacunae." At this point, after-suburbanization becomes generative of a new and perhaps unprecedented city and of new suburban ways of life that emerge there and beyond. This takes us past the expected and necessary acknowledgment of the inequalities and deficits inside the box we call the city. Yet outside the deficit we might just find a qualitatively different mode of urban organization. This is a question of particular relevance from the point of view of the people who inhabit the periphery and who don't want to be seen as less complete than others in the region.

The perforation of the boundedness of the city towards the after-suburban life on this planet is one meaning of "extensive" (Simone, 2019, p. 3). This entails a substantive, not just a spatial expansion: "a wide range of logics, social and cultural processes, and vernaculars, thus exposing residents to a larger set [of] factors at work in shaping their own daily experiences" (Simone, 2019, p. 3). The main concerns of the after-suburban politics that we can expect, then, deal with life itself. While the suburban was the very terrain of the bland, the unidimensional, and disurbanity, the after-suburban reality poses questions of a different relevance. Simone (2016b) adds these poignant questions: "As urban life is increasingly enclosed within concrete forms of immunization – gated residences, highly formatted and standardized vertical apartment blocks, massive outlays of cheap housing at peripheries – and particularized, parceled, and distributed across multiple medias, what constitutes some enduring stability of residence? How do these polarities become material to be experimented with, in maneuvers that go beyond the problematic status of the individual urban subject and the categories of neoliberal, global urban citizenship or precarity?" (p. 187).

Conclusion: On Practice

Having sketched the global landscape of after-suburbia – visiting extended urbanization and the post-suburban continuous city as a conceptual construct on the way – we can now appreciate the complexities of suburban research and practice today. This is not the place for a detailed discussion of the various ways in which scholars have responded and continue to respond to the challenges involved in making sense of the sub/urban world in which we live. Many such strategies are on display in this volume.

Tensions between the multiple, multifarious, and diverse-situated quotidian experiences of "making cities" (to borrow a term from Caldeira, 2016) and the debate on what is urban are anything but concluded. In fact, it would be an illusion to think they could be. One nurtures the other, and vice versa. In the context of this book, the understanding of what urban world we live in will help us inhabit and govern the places where we live and help us make sense of the boundary breakdowns in the continuous city. In the end, the suburb is always elusive. To some (like Richard Harris), the suburb is forever temporary, frail, and fleeting while the centre seems to hold. In a different take, Simone and Pieterse (2017) say about Kinshasa: "The near universal perception in Kinshasa is that the city is increasingly moving elsewhere. As a result, many inhabitants hurry to stake their claims at ever-shifting peripheries, which still seem to be in the middle of nowhere" (p. 73). This is the world in which we live. It is a world after suburbia.

Let me, then, end with a note on practice, reflecting on the tasks at hand where I live: Canada after suburbia. Canada is a suburban nation. Two-thirds of us live in the periphery (Gordon, Hindrichs et al., 2018). But what kind of a suburban nation are we? What are the regions of this nation? What regions are advantaged? Whose territory is not accounted for? Who has voice in this nation? What are its boundaries? The nation of suburbs in which most of us live, work, and play now is not an accidental product of market forces and consumer demand. It is the outcome of systematic and purposeful planning (Addie et al., 2020; Harris 2014; Logan, 2020; Moos & Walter-Joseph, 2017; Shields in this volume).

But planning has rarely originated in the periphery where much if not most of Canadian urbanization is now happening. The Canadian suburbs are a brainchild of planners in the centre, and a creature of the political economy of housing, transportation, and land use in a capitalist nation that grew from the country's centres of power and wealth (Keil et al., 2015; Keil & Hertel, 2017). So, scholars and practitioners have begun to ask: If some regions of Canada's suburban nation are doing better, and others are rather disadvantaged, what kind of planning can we make responsible for it? What kind of planning might we need to rectify the situation? What kinds of policies might help to structure life after suburbia? (Gordon, Moos et al., 2018).

Since suburbs were first constructed in the last century, their *rebuilding* has been a perennial issue. The welfare state in Canada created two kinds of distinctive residential suburbia in the periphery: the single-family home subdivision and the tower neighbourhood. Both are reliant on similar intertwined processes of industrial mass production and consumption. Often, the two suburbias acted as bookends of a spectrum that ranges from a private to a collective response to the housing demand of a growing immigrant society. Both versions of suburban expansion served as visible reminders of the state's and the market's almost unquestioned ability and power. A more recent neoliberal state has shown the tendency of rejecting planning intervention and constraint but lately re-embraced environmental regulation and has produced the post-Fordist suburb that now rules the margins of Canadian cities: still in the era of automobile-based mass production and consumption but now part of a larger landscape of global logistics, immigration, and financialization. There are ever-more massive tracts of meandering, inward-looking, outwardly serviced housing areas that border on oversized (yet always congested) highways leading to frighteningly similar commercial areas that repeat the presence of their brand-name tenants in predictable sequencing (Keil & Üçoğlu, 2021).

That a new (not so welfare) state has now made the urban periphery its target of action is both in this line of welfare state action *and* a new neoliberal attempt at gaining control over spaces that have become problematic and possible land banks for further densification and intensification. Sometimes the latter has to do with the desire to gentrify (and cleanse the area of the existing populations), and sometimes it is an attempt to "militarily" control those areas through conventional or "innovative" police strategies, including digital or video surveillance. All the while, planning contestation and political debate over the repair of the space of the towers, the townhouses, and the bungalows has become widespread (Poppe & Young, 2015).

The coincidence of the spatially peripheral and the suburban with the socially marginal is at once old and new. Historically, the suburbs have always been the place of the unwanted. The more recent tradition of making the suburbs the domain of the privileged in the twentieth century in North America has been more of an aberration. But even in Canada and the United States, the talk is now about the suburban as a place of marginality, poverty, and exclusion. The subsequent marginalization of the peripheral (and of its people) in planning thought and practice is, then, hardly accidental. That this begs the question of what citizenship means after suburbia is evident.

So, what to do? Cities and suburbs are not created by academic and professional debate but by actual city building. Theory and empirical knowledge about cities and suburbs are important in this process, but they do not determine them. Planning is central – it structures, but it does not create the built environment and social interactions (Simone's "intersection"). As a body of thought, it tends to represent the status quo; as a practice, it reproduces the status quo in material terms. Very little in planning is geared towards the implementation of change. From it and by itself, we cannot expect changes to socio-economically or ethno-culturally unequal circumstances or an improvement in urban sustainability. The concreted, automobile suburbs in Canada are a typical outcome of such stasis and conservatism. The blueprint mentality of past suburban planning signified security and radiated reliability. This may have had some appeal for investors and inhabitants alike when greenfield developments created new communities in the past.

Now that the periphery has become demographically and socio-economically "hyper-diverse" (Pitter & Lorinc, 2016) and, as we have called it, "post-suburban," demands and challenges are growing and changing. Practice and imagination of planning need to change accordingly. New

Zealand planning scholars Johnson, Baker, and Collins (2019) observe that "the growing emphasis on creating denser, more diverse suburbs involves both imaginative *and* [emphasis in original] material factors that make possible certain urban configurations and limiting others" (p. 1043). Planning as an apparatus of knowledge production needs to be conscious of and open to the needs of communities on the ground instead of offering and giving in to the big gesture and the bird's-eye view. On the street and in the neighbourhoods there is little awareness of the academic, corporate, and professional industry that attempts to understand statistically and conceptually the urban world we are living in and what this urban world might become. The multiple, multifarious, and diverse everyday experiences of "making cities" (Caldeira, 2016) must nurture the suburban imagination. Instead of, for example, mapping the post-suburban landscape in yet another regional transportation plan on the digital drafting table in some downtown office, why not start with mining and minding the experience of people at a suburban bus stop?

This also means abandoning the stereotypes. Places are marked as good or bad, safe or unsafe, hip or unhip. Suburbs are dependent and insignificant and imagined as the "Other" to the defining inner cities. Planning controversies are read as spatialized struggles between the city and the suburbs. Culture wars of access to the right to the street, the right to the city are cementing the urban-suburban divide.

Political struggles about the kind of city that residents want tend to be treated in ways that provide a spatial fix in discourse, planning, and actual city-making that quickly hardens. Lost in this process are the differences that are not spatial in nature but linked to social contradictions such as changing socio-economic circumstances, racialization of populations, and issues such as the feminization or suburbanization of poverty (Randolph & Tice, 2017). The selection of planning issues that are discussed citywide or regionwide, thus, depends on who has the means to speak and mobilize the political agendas through which to address any grievances beyond the perceived urban-suburban divide. Will we talk cycling casualties downtown or firearm fatalities in the periphery? Indeed, ultimately, this is where planners need to situate the suburban in practice: The suburban always has to be present when we act. If we focus on the centre with its demands for privileged spaces and resources, let us not forget that anything we might do *here* has consequences for what we might receive *elsewhere*.

NOTE

1 This chapter builds on ideas first developed in Keil (2020a).

REFERENCES

Addie, J-P., Fiedler, R.S., & Keil, R. (2020). Cities on the edge: Suburban constellations in Canada. In M. Moos, T. Vinodrai, & R. Walker (Eds.), *Canadian cities in transitions* (pp. 292–310). Oxford University Press.

Airgood-Obrycki, W., Hanlon, B., & Rieger, S. (2020). Delineate the U.S. suburb: An examination of how different definitions of the suburbs matter. *Journal of Urban Affairs, 43*(9), 1263–84. doi:10.1080 /07352166.2020.1727294.

Airgood-Obrycki, W., & Reiger, S. (2019). Defining suburbs: How definitions shape the suburban landscape. Joint Center for Housing Studies of Harvard University. https://www.jchs.harvard.edu/sites/default/files /media/imp/Harvard_JCHS_Airgood-Obrycki_Rieger_Defining _Suburbs.pdf.

Angelo, H. (2021). *How green became good: Urbanized nature and the making of cities and citizens.* University of Chicago Press.

Beasley, L. (2019). *Vancouverism.* On Point Press.

Berger, A., Kotkin, J., Balderas, C., & Guzman, J. (Eds.). (2017). *Infinite suburbia.* Princeton Architectural Press.

Brighenti, A.M. (Ed.). (2013). *Urban interstices: The aesthetics and the politics of the in-between.* Routledge.

Caldeira, T.P.R. (2016). Peripheral urbanization: Autoconstruction, transversal logics, and politics in cities of the global south. *Environment and Planning D: Society and Space, 35*(1), 3–20. doi:10.1177/0263775816658479.

Charmes, E., & Keil, R. (2015). Post-suburban morphologies in Canada and France: Beyond the anti-sprawl debate. *International Journal of Urban and Regional Research, 39*(3), 581–602. doi: 10.1111/1468-2427.12194.

City of Brampton. (2018). *Living the mosaic: Brampton 2040 vision.* https:// www.brampton.ca/EN/City-Hall/Documents/Brampton2040Vision /brampton2040Vision.pdf.

De Boeck, F., & Baloji, S. (2016). *Suturing the city: Living together in Congo's urban worlds.* Autograph.

Filion, P., & Pulver, N.M. (Eds.). (2019). *Critical perspectives on suburban infrastructures: Contemporary international cases.* University of Toronto Press.

Gordon, D.L.A., Hindrichs, L., & Willms, C. (2018). *Still suburban? Growth in Canadian suburbs, 2006–2016* (Council for Canadian Urbanism, Working Paper No. 2). School of Urban and Regional Planning, Department of Geography and Planning, Queen's University.

Gordon, D., Moos, M., Amborski, D. & Taylor, Z. (2018). The future of the suburbs: Policy challenges and opportunities in Canada. *SPP Briefing Paper, 11*(22). https://doi.org/10.11575/sppp.v11i0.53000.

Graham, S. (2016). *Vertical: The city from satellites to bunkers*. Verso.

Güney, K.M., Keil, R., & Üçoğlu, M. (Eds.). (2019). *Massive suburbanization: (Re)building the global periphery*. University of Toronto Press.

Gururani, S., & Dasgupta, R. (2018). Frontier urbanism urbanisation beyond cities in South Asia. *Economic & Political Weekly, 53*(12), 41–5.

Hamel, P., & Keil, R. (Eds.). (2015). *Suburban governance: A global view*. University of Toronto Press.

Harris, R. (1996). *Unplanned suburbs: Toronto's American tragedy, 1900–1950*. Johns Hopkins University Press.

Harris, R. (2010). Meaningful types in a world of suburbs. In M. Clapson & R. Hutchinson (Eds.). *Suburbanization in global society* (pp. 15–50). Emerald.

Harris, R. (2014). Using Toronto to explore three suburban stereotypes and vice versa. *Environment and Planning A, 46*(1): 30–49. https://doi.org/10.1068/a46298.

Harris, R., & Lehrer, U. (Eds.). (2018). *The suburban land question: A global survey*. University of Toronto Press.

Harris, R., & Vorms, C. (Eds.). (2017). *What's in a name? Talking about urban peripheries*. University of Toronto Press.

Johnson, C., Baker, T., & Collins, F.L. (2019). Imaginations of post-suburbia: Suburban change and imaginative practices in Auckland, New Zealand. *Urban Studies, 56*(5), 1042–60. doi:10.1177/0042098018787157.

Keil, R. (Ed.). (2013). *Suburban constellations*. Jovis.

Keil, R. (2018a). *Suburban planet: Making the world urban from the outside in*. Polity.

Keil, R. (2018b). Extended urbanization, "disjunct fragments" and global suburbanisms. *Environment and Planning D: Society and Space, 36*(3), 494–511. doi:10.1177/0263775817749594.

Keil, R. (2020a). After suburbia: Research and action in the suburban century. *Urban Geography, 41*(1), 1–20. doi:10.1080/02723638.2018.1548828.

Keil, R. (2020b). The spatialized political ecology of the city: Situated peripheries and the capitalocenic limits of urban affairs. *Journal of Urban Affairs*. 42(8), 1125–40. doi:10.1080/07352166.2020.1785305.

Keil, R., Hamel, P., Chou, E., & Williams, K. (2015). Modalities of suburban governance in Canada. In P. Hamel & R. Keil (Eds.), *Suburban governance: A global view* (pp. 80–109). University of Toronto Press.

Keil, R., & Hertel, S. (2017). Fixing postsuburbia: Recalibrating the way think, speak and act upon Toronto's periphery. In K. Anacker & P. Maginn (Eds.), *Suburbia in the 21st century: From dreamscape to nightmare?* Taylor & Francis Group.

Keil, R., & Üçoğlu, M. (2021). Beyond sprawl? Regulating growth in southern Ontario: Spotlight on Brampton. *disP – The Planning Review, 57*(3), 100–18. doi:10.1080/02513625.2021.2026678.

Kusno, A. (2019). Middling urbanism: The megacity and the kampong. *Urban Geography, 41*(7), 954–70. doi:10.1080/02723638.2019.1688535.

Lasky, J. (2020, October 16). Like it or not, the suburbs are changing. *New York Times.* https://www.nytimes.com/2020/10/16/realestate/suburbs-are -changing.html.

Lefebvre, H. (1996). *Writings on cities.* (E. Kofman & E. Lebas, Trans. & Eds.). Blackwell Publishers.

Lerup, L. (2017). *The continuous city.* Park Books.

Logan, S. (2020). *In the suburbs of history: Modernist visions of the urban periphery.* University of Toronto Press.

Mattiucci, C. (2015). Mountain condominiums. A discussing of settlement and dwelling on the outskirts of an alpine city. *Journal of Alpine Research, 103*(3). doi:10.4000/rga.3089.

Monte-Mor, R.L. (2014a.) Extended urbanization and settlement patterns in Brazil: An environmental approach. In N. Brenner (Ed.), *Implosions/ explosions* (pp. 109–20). Jovis.

Monte-Mor, R.L. (2014b). What is the urban in the contemporary world? In N. Brenner (Ed.), *Implosions/explosions* (pp. 260–7). Jovis.

Moos, M., & Walter-Joseph, R. (2017). *Still detached and subdivided: Suburban ways of living in 21st century North America.* Jovis.

Nijman, J. (Ed.). (2020). *The life of the North American suburb: From imagined utopias to transitional spaces.* University of Toronto Press.

Organisation for Economic Co-operation and Development. (2018). *Rethinking urban sprawl: Moving towards sustainable cities.* OECD Publishing.

Ortega, A.A.C. (2020). Exposing necroburbia: Suburban relocation, necropolitics, and violent geographies in Manila *Antipode, 52*(4), 1175–95. doi: 10.1111/anti.12629.

Peck, J., Siemiatycki, E., & Wyly, E. (2014). Vancouver's suburban involution. *City: Analysis of Urban Trends, Culture, Theory, Policy, Action, 18*(4–5), 386–415. doi:10.1080/13604813.2014.939464.

Phelps, N.A. (2017a). *Interplaces: An economic geography of the inter-urban and international economies.* Oxford University Press.

Phelps, N.A. (2017b). *Old Europe, new suburbanization? Governance, land, and infrastructure in European suburbanization.* University of Toronto Press.

Phelps, N., & Silva, C. (2017). Mind the gaps! A research agenda for urban interstices. *Urban Studies, 55*(6), 1203–22. doi:10.1177/0042098017732714.

Phelps, N., & Wu, F. (Eds.). (2011). *International perspectives on suburbanization: A postsuburban world?* Palgrave Macmillan.

Pitter, J., & Lorinc, J. (Eds.). (2016). *Subdivided: City-building in an age of hyper-diversity.* Coach House Books.

Poppe, W., & Young, D. (2015). The politics of place: Place-making versus densification in Toronto's tower neighbourhoods. *International Journal of Urban and Regional Research, 39*(3), 613–21. doi:10.1111/1468–2427.12196.

Randolph, B., & Tice, A. (2017). Relocating disadvantage in five Australian cities: Socio-spatial polarisation under neo-liberalism. *Urban Policy and Research, 35*, 103–21. doi.org/10.1080/08111146.2016.1221337.

Reis, N., & Lukas. M. (Eds.). (2022). *Beyond the mega-city: New dynamics of peripheral urbanization in Latin America.* University of Toronto Press.

Ronneberger, K. (2015). Henri Lefebvre und die Frage der Zentralität [Henri Lefebvre and the question of centrality]. *dérive, 60,* 23–7.

Roy, A. (2011). Urbanisms, worlding practices and the theory of planning. *Planning Theory, 10*(1), 6–15. doi.org/10.1177/1473095210386065.

Sieverts, T. (2003). *Cities without cities: An interpretation of the Zwischenstadt.* Taylor & Francis.

Sieverts, T. (2006). Die Geschichtlichkeit der Großsiedlungen [The historicity of the large settlements]. *Informationen zur Raumentwicklung, 3*(4), 163–7. https://www.bbsr.bund.de/BBSR/DE/veroeffentlichungen/izr/2006/Downloads/3_4Sieverts.pdf?__blob=publicationFile&v=1.

Simone, A.M. (2004). *For the city yet to come.* Duke University Press.

Simone, A.M. (2014). *Jakarta: Drawing the city near.* University of Minnesota Press.

Simone, A.M. (2016a). It's just the city after all. *International Journal of Urban and Regional Research, 40*(1), 210–18. doi:10.1111/1468–2427.12275.

Simone, A.M. (2016b). Urbanity and generic blackness. *Theory, Culture & Society, 33*(7–8), 183–203. doi:10.1177/0263276416636203.

Simone, A.M. (2019). Maximum exposure: Making sense in the background of extensive urbanization. *Environment and Planning D. Society and Space, 37*(6), 990–1006. doi:10.1177/0263775819856351.

Simone, A.M., & Pieterse, E. (2017). *New urban worlds: Inhabiting dissonant times.* Polity.

Stanilov, K., & Sykora, L. (Eds.). (2014). *Confronting suburbanization: Urban decentralization in postsocialist Central and Eastern Europe.* Wiley-Blackwell.

Sweeney, G., & Hanlon, B. (2017). From old suburb to postsuburb: The politics of retrofit in the inner suburb of Upper Arlington, Ohio. *Journal of Urban Affairs, 39*(2), 241–59. doi:10.1111/juaf.12313.

Swyngedouw, E., & Ernstson, H. (2018). Interrupting the Anthropo-obScene: Immuno-biopolitics and depoliticizing ontologies in the Anthropocene. *Theory, Culture and Society, 35*(6), 3–30. doi:10.1177/0263276418757314.

Walks, A. (2013). Suburbanism as a way of life, slight return. *Urban Studies, 50*(8), 1471–88. doi: 10.1177/0042098012462610.

Wilson, B. (2020, November 24). The problem for the 21st century isn't that we're too urban – it's that we're not urban enough. *Fast Company*. https://www.fastcompany.com/90579332/the-problem-for-the-21st-century-isnt-that-were-too-urban-its-that-were-not-urban-enough.

Young, D., & Keil, R. (2014). Locating the urban in-between: Tracking the urban politics of infrastructure in Toronto. *International Journal of Urban and Regional Research, 38*(5), 1589–608. doi:10.1111/1468-2427.12146.

Image Credits

Image 1. Public sculpture in Lingang, China

Ten graduate students, who all took the Critical Planning Workshop that was offered via York University and Fudan University, explored the periphery of Shanghai in May 2015. (Photo taken by Justin Fok.)

Image 2. Biking in Florence, Italy

As part of the Spring Institute in Global Suburban Studies, which took place in the spring of 2018 in Florence and Milan, participants went on a bike tour of the periphery of Florence. Here, they were familiarized to explore different ways of mobility in order to perceive the surroundings. (Photo taken by Ute Lehrer.)

Image 3. Greater Toronto, Ontario, Canada (Fall 2010)

A collage of visuals based on the work of undergraduate students, who took Urban and Regional Infrastructures: A Critical Introduction, taught by Roger Keil, York University. This course examined the history of and current issues surrounding hard and soft infrastructures in processes of (sub)urbanization from an urban political ecology perspective. Following a methodology developed by Ute Lehrer, students were expected to do a visual analysis of infrastructure.

Photographic Contributions

Ilyas Amin, Susan Borst, Yuen Chu, Michael De Angelis, Liana De Francesco, Abdulrahman Elmi, Matthew Farquharson, Udai

Emmanuel Gomez, Justin Gregoris, Latoya Hamilton, Spencer Harris, David Ly, James Marzotto, Sean McMillan, Michael Mikhailovsky, Katherine Palka, Natasha Paolini, Warren Patterson, Nicole Percival, Ian Sachs, Elie Selwan, Daniel Taylor, Jordan Teichmann, and Daniel Veiga.

Image 4. Leipzig, Germany (June 2011)

In this Critical Planning Workshop, thirteen graduate students learned first-hand from examples of growth and decline, by studying Leipzig's transformation over the past twenty years as well as its current challenges. Topics such as suburbanism, the shrinking city, housing renewal and governance, questions of investment, and, given the location in eastern Germany, post-socialism defined the intellectual content of the course. The course had site visits in Leipzig and Berlin, and at the Bauhaus in Dessau, as well as lectures by planners, politicians, and academics and hands-on planning exercises.

Participating Students

Nishanthan Balasubramaniam, Camilia Changizi, Amanda Dunn, Christine Furtado, Dan Godin, Claire Harvey, Alex Heath, Laine Horsfield, Jacob Kaven, Josh Neubauer, Andria Oliveira, Alejandra Perdomo, and Gwen Potter.

Image 5. Winnipeg, Manitoba, Canada (Fall 2011)

A collage of visuals based on the work of undergraduate students, who took Urban and Regional Infrastructures: A Critical Introduction, taught by Roger Keil, York University. This course examined the history of and current issues surrounding hard and soft infrastructures in processes of (sub)urbanization from an urban political ecology perspective. Following a methodology developed by Ute Lehrer, students were expected to do a visual analysis of infrastructure.

Photographic Contributions

Richard Alleyne, Jessica Amar, Patrick Amaral, Katherine Berton, Krzysztof (Chris) Boncza-Rutkowski, Alexander Boschetto, Kitty Chen, Amir Erfani-Pour, Ivan Escobar, Ted Evanoff, Michael Figliano-Connoly, Justin Fok, Josée Guimond, Stefanie Hawco, Melanie Jarcaig,

Adam Jones, Calvin Lai, Jason Ma, Alexander Martino, Reynaldo Matosas, Jignesh Mistry, Nawal Moheed, Brian Monticchio, Giancarlo Murano, Miona Necic, Cesare Pittelli, Mudith Pursun, Fairoz Retha, Alexander Skerlan, Nicholas Skoncej, Joshua Teves, Wilson Tioh, Sylvester Tuz, and Cyrus Yan.

Image 6. Montpellier, France (October 2012)

The intellectual content of the Critical Planning Workshop, which brought eleven students to Montpellier (France), addressed "Land" in the broadest sense, how it is produced and converted to new usages. With its geographic location in the south of France and with a strong and coherent political agenda of growth over the past thirty years, Montpellier brings together some of the key elements of urban planning: transportation, housing (private and public), new suburban forms, ecology, urban design, public space, multiculturalism. The planning and urban development agenda in Montpellier is based on sustainability and a regional perspective, which can be seen as an exemplary case of urban and regional planning.

Participating Students

Linda Bui, Anna Coté, Zeina Ismail, Saadia Jamil, Graeme Jones, Charleen Kong, Chu Nam Law, Evan McDonough, Imelda Nurwisah, Pawel Nurzynski, and Prabin Sharma.

Image 7. Greater Montreal, Quebec, Canada (Summer 2012)

A collage of visuals based on the work of undergraduate students, who took Urban and Regional Infrastructures: A Critical Introduction, taught by Roger Keil, York University. This course examined the history of and current issues surrounding hard and soft infrastructures in processes of (sub)urbanization from an urban political ecology perspective. Following a methodology developed by Ute Lehrer, students were expected to do a visual analysis of infrastructure.

Photographic Contributions

Mustafa Al-Jameel, Richard Burn, Devan Dhaliwal, Anthony Dionigi, Daniel Dubeau, Jacqueline Hayward Gulati, Kevin Hurley, David Johnson, Sara Khawaja, Charleen Kong, Chun Nam Law, Meidan

Leiderman, Evan McDonough, Aphiraa Nirmalarajah, Kevin Prashad, Erik Ryken, Natalia Wegrzyn, Jordan Teichmann, and Daniel Veiga.

Image 8. Metro Vancouver, British Columbia, Canada (Summer 2013)

A collage of visuals based on the work of undergraduate students, who took Urban and Regional Infrastructures: A Critical Introduction, taught by Roger Keil, York University. This course examined the history of and current issues surrounding hard and soft infrastructures in processes of (sub)urbanization from an urban political ecology perspective. Following a methodology developed by Ute Lehrer, students were expected to do a visual analysis of infrastructure.

Photographic Contributions

Jared Corbin, Ying Gu, Marina Janakovic, Aaron Lefler, Haru Liu, Stephanie Lyons, Victoria Moore, Omar Murtaja, Oriana Nanoa, Christopher Poole, and Mariyam Zaidi.

Image 9. Shanghai, China (May 2015)

As part of the Critical Planning Workshop nine graduate students travelled to Shanghai, China, in May 2015 to learn from this rapidly growing city, which is exposed to significant challenges with regard to social, economic, environmental, and cultural planning. The focus was on land use planning, housing, real estate development, property rights, and infrastructure. Students explored entire suburban cities that were built within the last few years, with all its environmental and social contradictions.

Participating Students

Paul Bailey, Cara Chellew, Anthony Y. Dionigi, Justin Fok, Victoria Ho, Priscilla Lan Chung Yang, Dilya Niezova, Nelly Volpert, and Cathy Zhao.

Image 10. Region of Waterloo, Ontario, Canada (Summer 2015)

A collage of visuals based on the work of undergraduate students, who took Urban and Regional Infrastructures: A Critical Introduction, taught by Roger Keil, York University. This course examined the history of and current issues surrounding hard and soft infrastructures in processes of (sub)urbanization from an urban political ecology perspective. Following a methodology developed by Ute Lehrer, students were expected to do a visual analysis of infrastructure.

Photographic Contributions

Ibraheem Abdelmutti Mohammed, Bakytzhamal Jetmekova, Adam Osman, Eshe Simba, Lynn Strachan, Sonia Stramaglia, Kishani Sundaralingam, Krishna Veerappan, Fairoz Retha, and Michael Collens.

Image 11. Johannesburg, South Africa (October 2016)

In the fall of 2016, eleven graduate students had an immersive living and working experience in Johannesburg, South Africa, as part of the Critical Planning Workshop. Guided by distinguished specialists in the field – faculty from York and Witwatersrand Universities, officials of provincial and municipal governments, and other figures from NGOs and – students investigated, visited, and debated the emergence of ideas around the "urban growth boundary" and housing and were encouraged to relate this learning experience to specific issues and conflicts around urban growth, "sprawl," and "suburbanisms" to the Canadian context.

Participating Students

Floyd Heath, Victoria Moore, Nabeel Ahmed, Ryan Adamson, Orli Schwartz, Assya Moustaqim-Barrette, Patrycja Jankowski, Stephen Closs, Joyce Chan, Ying Gu, and Carmen Charles.

Image 12. Four students jumping high

While waiting for the developer to appear and to explain to us the proposal for more housing on the periphery of Johannesburg, four students from the Critical Planning Workshop, which took place in 2016, are using the time creatively. (Orli Schwartz, Patrycja Jankowski, Victoria Moore, and Ying Gu. Photo taken by Ryan Adamson.)

Image 13. Exploring methodologies

Students of the Spring Institute in Global Suburban Studies, which took place in the spring of 2018 in Florence and Milan, were brainstorming together with Markus Moos about different methodologies. (Photo taken by Ute Lehrer.)

Image 14. Residential dwellings by year built, Toronto

Toronto Census Metropolitan Area. Statistics Canada, 2011. Map by Robert Walter-Joseph and page design by Nicole Yang and Kourosh Mahvash. From: Moos, M., & Walter-Joseph, R. (Eds.). (2017). *Still detached and subdivided? Suburban ways of living in 21st century North America.* Jovis Press.

Image 15. Residential dwellings by type, Toronto

Toronto Census Metropolitan Area. Statistics Canada, 2011. Map by Robert Walter-Joseph and page design by Nicole Yang and Kourosh Mahvash. From: Moos, M., & Walter-Joseph, R. (Eds.). (2017). *Still detached and subdivided? Suburban ways of living in 21st century North America.* Jovis Press.

Image 16. Residential geography of immigrants by generation of immigration, Montreal and Toronto

Toronto and Montréal Census Metropolitan Areas. Statistics Canada, 2011. Map by Robert Walter-Joseph. From: Moos, M., & Walter-Joseph, R. (Eds.). (2017). *Still detached and subdivided? Suburban ways of living in 21st century North America.* Jovis Press.

Image 17. Residential geography by visible minority status, Chicago and Toronto

Chicago Metropolitan Statistical Area and Toronto Census Metropolitan Area. Statistics Canada, 2011, and US Census Bureau, 2010. Map by Robert Walter-Joseph. From: Moos, M., & Walter-Joseph, R. (Eds.). (2017). *Still detached and subdivided? Suburban ways of living in 21st century North America.* Jovis Press.

Image 18. Suburban ways of living as defined by homeownership, automobile use, and single-detached dwellings, Chicago and Toronto

Chicago Metropolitan Statistical Area and Toronto Census Metropolitan Area. Statistics Canada, 2011, and US Census Bureau, 2010. Map by Robert Walter-Joseph. From: Moos, M., & Walter-Joseph, R. (Eds.).

(2017). *Still detached and subdivided? Suburban ways of living in 21st century North America.* Jovis Press.

Image 19. The suburban cul-de-sac reimagined

Design by Nicole Yang and page design by Nicole Yang and Kourosh Mahvash. From: Moos, M., & Walter-Joseph, R. (Eds.). (2017). *Still detached and subdivided? Suburban ways of living in 21st century North America.* Jovis Press.

Image 20. #ChangeSpace: Planning suburban public space for youths

By Christina Glass and Nicole Yang. From: Moos, M., & Walter-Joseph, R. (Eds.). (2017). *Still detached and subdivided? Suburban ways of living in 21st century North America.* Jovis Press.

Image 21. Possibilities of suburban transformations

Design by Nicole Yang and Kourosh Mahvash. From: Moos, M., & Walter-Joseph, R. (Eds.). (2017). *Still detached and subdivided? Suburban ways of living in 21st century North America.* Jovis Press.

Contributors

Shlomo Angel is a professor of city planning and the director of the Urban Expansion Program at the Marron Institute of Urban Management of New York University. He is the author of *Planet of Cities* (Lincoln Institute of Land Policy, 2012) and a co-author of the *Atlas of Urban Expansion* (online at www.atlasofurbanexpansion.org). His current interests include measuring and acting on urban expansion and densification in rapidly growing cities in less-developed countries.

Sara Arango-Franco works as a research scholar and data scientist for Litmus at the Marron Institute for Urban Management of New York University. Her main interest is in how to bridge science and technology with decision-making in public policy. She has worked in executing and developing data products and experiments with public agencies in the United States, Mexico, and Colombia. She holds a Master of Science in Urban Informatics from New York University's Center for Urban Science and Process (CUSP) and a Bachelor of Science in Mathematical Engineering from Universidad EAFIT (Colombia).

Alessandro Balducci is a full professor at the Department of Architecture and Urban Studies at Politecnico di Milano, Italy. He has extensive research experience on the topics of urban policy, strategic planning, metropolitan governance, and processes of regional urbanization. He has been coordinator of several national and international research projects, among them the research project funded by the Italian Ministry of Education, Post-Metropoli, and is co-editor of *Post-Metropolitan Territories: Looking for a New Urbanity* (Routledge, 2017).

Robin Bloch is a technical and market manager of sustainable urban development at COWI, an international engineering company. He is a city planner and has led several urban infrastructure planning and resilience projects in recent years, particularly in Sub-Saharan Africa. Much of this work has incorporated research management, notably his role as team leader on the UK government-funded Urbanisation Research Nigeria program from 2013–18. His research includes industries, peripheries, and politics.

Rodrigo Castriota is a postdoctoral research fellow on the ERC Project Inhabiting Radical Housing at DIST/Politecnico di Torino. He holds a PhD in urban and regional development from CEDEPLAR/UFMG (Brazil), where he did extensive research on urbanization processes in the Brazilian Amazon, particularly those driven by neo-extractivist interventions. He has significant experience in global urban research, having worked on numerous projects based in Canada, Switzerland, Singapore, England, Italy, and Brazil. Rodrigo is also a member of the Beyond Inhabitation Lab, a Turin-based collective of researchers and activists that investigate housing and inhabitation struggles and strategies and their multiple intersections with urban, social, and environmental questions. His main research interests also include popular economies, neo-extractivism, and postcolonial/decolonial theory, as well as the extended forms of urbanization and the everyday extensionalities that shape our contemporary world beyond urban and rural divides.

Valeria Fedeli is an associate professor in the Department of Architecture and Urban Studies at the Politecnico di Milano, Italy, as well as vice-director of the PhD program in Urban Planning, Design, and Policy. She has been president of the European Urban Research Association (EURA). She has extensive research experience on the topics of strategic planning, metropolitan governance and processes of regional urbanization, urban policies in Europe, and EU urban policy. She is co-editor of *Post-Metropolitan Territories: Looking for a New Urbanity* (Routledge, 2017).

Pierre Filion is a professor emeritus at the School of Planning at the University of Waterloo, Ontario. His areas of research include metropolitan-scale planning, downtown areas and suburban centres, and infrastructures. He is co-editor of *Critical Perspectives on Suburban Infrastructures* (University of Toronto Press, 2019), *Cities at Risk* (Routledge, 2015), and five editions of *Canadian Cities in Transition* (Oxford University Press, 1991–2015).

Nicolás Galarza Sánchez is the vice-minister at the Ministry of the Environment in Colombia, a visiting scholar and a former research scholar at the Urban Expansion Program of New York University's Marron Institute of Urban Management. He holds an MA in urban planning from the Robert F. Wagner School of Public Service. As a research scholar at the Marron Institute, he has led projects in Latin American cities, promoting orderly urban expansion, and co-authored several publications and scholarly articles, including the *Atlas of Urban Expansion* (online at www.atlasofurbanexpansion.org).

Shubhra Gururani is an associate professor in the Department of Social Anthropology at York University, Toronto. She is an affiliated member of the CITY Institute, the York Center for Asian Research, and the Center for Feminist Research. Her current research focuses on the changing political ecology of urban peripheries and examines the politics of land acquisition, planning, and infrastructure in Gurgaon, on the outskirts of New Delhi. Her recent work has appeared in *Urban Geography, Urbanisation, Economic and Political Weekly*, and *SAMAJ – South Asian Multidisciplinary Academic Journal*.

Pierre Hamel is a professor of sociology at the Université de Montréal and former editor of the sociology journal *Sociologie et sociétés*. His research focuses on cities and democratizing processes, including collective action around urban issues, regional governance, and suburban development. He is co-editor with Louis Guay of *Les aléas du débat public. Action collective, expertise et démocratie* (Les Presses de l'Université Laval and Hermann Éditeurs, 2018), and co-editor with Roger Keil of *Suburban Governance: A Global View* (University of Toronto Press, 2015).

Richard Harris is a past president of the Urban History Association (2017–18), Fellow of the Royal Society of Canada, a Guggenheim Fellow, and professor emeritus of urban geography at McMaster University, Hamilton, Ontario. He has written about the history of housing, neighbourhoods, and suburbs in North America and British colonies. His most recent book, co-edited with Ute Lehrer, is *The Suburban Land Question: A Global Survey* (University of Toronto Press, 2018). A short work, *How Cities Matter*, was published in 2021 (Cambridge University Press). He is currently writing a history of Canadian urban neighbourhoods.

Matt Hern lives on the middle arm of the Fraser River, xʷməθkʷəy̓əm (Musqueam) territory. He is the co-director of Solid State Community

Industries. His most recent books are *What a City Is For: Remaking the Politics of Displacement* (MIT Press, 2016), *Global Warming and the Sweetness of Life* (MIT Press, 2018, with Am Johal and Joe Sacco), and *On This Patch of Grass: City Parks on Occupied Land* (Fernwood, 2019), with Daisy Couture, Sadie Couture, Selena Couture, Erick Villagomez, Glen Coulthard, and Denise Ferreira da Silva.

Roger Keil is a professor of environmental and urban change at York University, Toronto. Researching global suburbanization, urban political ecology, cities and infectious disease, and regional governance. Keil is the author of *Suburban Planet* (Polity, 2018), co-editor with Judy Branfman of *Public Los Angeles: A Private City's Activist Futures* (UGA Press, 2020), and with Xuefei Ren of *The Globalizing Cities Reader* (Routledge, 2017). In the Global Suburbanisms book series, he has previously co-edited *Suburban Governance: A Global View* with Pierre Hamel (University of Toronto Press, 2015) as well as *Massive Suburbanization* with K. Murat Güney and Murat Üçoğlu (University of Toronto Press, 2019).

Abidin Kusno is a professor in the Faculty of Environmental and Urban Change at York University, Toronto, where he also serves as director of the York Centre for Asian Research. His books on Jakarta include *Behind the Postcolonial* (Routledge, 2000), *Appearances of Memory* (Duke University Press, 2010), *After the New Order* (Hawaii University Press, 2013), and *Visual Cultures of the Ethnic Chinese in Indonesia* (Rowman & Littlefield International, 2016). His most recent publications include the journal article "Where Will the Water Go?" *Indonesia* (2018), and the book chapter "Islamist Urbanism and Spatial Performances in Indonesia" in *Cities and Islamisms*, edited by Bulent Batuman (Routledge, 2021).

Patrick Lamson-Hall is an urban planner and a research scholar at the Marron Institute of Urban Management at New York University. He is a co-author of the *Atlas of Urban Expansion – 2016 Edition* and the coordinator of the Climate Smart Cities: Grenada Program and the Ethiopia Urban Expansion Initiative. His interests include the assessment of cities using satellite imagery and the challenge of urban planning in developing county contexts.

Crystal Legacy is an associate professor in urban planning at the University of Melbourne, Australia, where she is also the deputy director of the Informal Urbanism Research Hub. She has published widely on the topics of citizen participation, urban transport governance, and urban politics. She is on the editorial boards of the journals *Planning Theory and Practice* and *Urban Policy and Research*.

Ute Lehrer is a professor at York University, Toronto, where she teaches urban planning. Her research is on large-scale projects; high-rise condominium boom; social construction of public space; and the land question. Most recently she has co-edited *The Suburban Land Question* with Richard Harris (University of Toronto Press, 2018) and published "Verticality, Public Space and the Role of Resident Participation in Revitalizing Suburban High-rise Buildings" with Loren March in *Canadian Journal of Urban Research* (2019). She has used visual methodologies for her own research as well as for teaching purposes.

Yang Liu is a research analyst at UNICEF and was previously a visiting scholar at the Marron Institute of Urban Management of New York University, providing the statistical analysis and co-authoring a series of papers on urban expansion.

Alan Mabin lives in Cape Town, South Africa, and is emeritus professor at the School of Architecture and Planning, University of the Witwatersrand, Johannesburg, which he directed from 2005 to 2010. He has research experience in Brazil, France, Tanzania, and South Africa. He was a co-founder of Planact (community service NGO) and deputy chairperson of the Development and Planning Commission 1997–2001. Recent publications include "Challenges of University–City Relationships: Reflections from Wits University and Johannesburg" in *Anchored in Place: Rethinking the University and Development in South Africa* (African Minds, 2018) and "Sprawl Politics: Comparing the City Regions of Paris (France) and Gauteng (South Africa)" (*disP – The Planning Review*, 2022).

Paul J. Maginn is an associate professor of urban and regional planning at the University of Western Australia. He is author/co-editor of seven books, including *Suburbia in the 21st Century: From Dreamscape to Nightmare?* (Routledge, 2022); *(Sub)Urban Sexscapes: Geographies and Regulation of the Sex Industry* (Routledge, 2015); and *Planning Australia: An Overview of Urban and Regional Planning* (Cambridge University Press, 2012). He was co-convenor of the Australasian Cities Research Network (2018–21) and is the editor-in-chief of *Urban Policy and Research*.

Roberto Luís Monte-Mór is a professor at the Centre for Development and Regional Planning (CEDEPLAR) and the School of Architecture (EAD) at the Federal University of Minas Gerais (UFMG), Brazil. He is a researcher at the National Scientific Council (CNPq) and has been the coordinator of the Belo Horizonte Metropolitan Master Plan and the Metropolitan Macro-Zoning and Local Master Plans since 2009. His main research and teaching areas are urban and regional economics, urban theories, solidarity and popular economies, traditional populations, Amazônia, and Brazil-India relations.

Markus Moos is a professor in the School of Planning at the University of Waterloo, Ontario. He is also a registered professional planner. His research is on the changing economies, housing markets, social structures, generational dimensions such as youthification, and sustainability of cities and suburbs. He is the editor of *A Research Agenda for Housing* (Edward Elgar Publishing, 2019) and co-editor with Robert Walter-Joseph of *Still Detached and Subdivided? Suburban Ways of Living in Twenty-First-Century North America* (Jovis Press, 2017).

Stijn Oosterlynck is a full professor in urban sociology in the Faculty of Social Sciences at the University of Antwerp in Belgium. He is the chair of the Antwerp Urban Studies Institute and scientific director at the Hannah Arendt Institute for Diversity, Urbanity, and Citizenship. His research is concerned with the politics of (sub)urban development, solidarity in diversity, civil society innovation, and local social innovation and welfare state restructuring. He is co-author of *Local Social Innovation to Combat Poverty and Exclusion* (Policy Press, 2019) and co-editor of *Divercities: Understanding Super-Diversity in Deprived and Mixed Neighbourhood* (Policy Press, 2019) and of *The City as a Global Political Actor* (Routledge, 2018).

Arnisson Andre C. Ortega is an assistant professor of geography at Syracuse University in New York. His interests include spatial politics of urbanization in Global South cities, critical demography, and community geography. He is the author of the book *Neoliberalizing Spaces in the Philippines: Suburbanization, Transnational Migration, and Dispossession* (Lexington, 2016). His articles have been published in *Annals of the American Association of Geographers, Urban Geography, Cities,* and *Geoforum.*

Camilla Perrone is an associate professor in the Department of Architecture at the University of Florence, Italy. She holds a PhD in urban, regional, and environmental planning. Currently, she is the founding director of the Research Laboratory of Critical Planning and Design. She is president of the Scientific Commette of the National Center for Urban Polices (urban@it). Her research topics include interactive design,

urban diversity, post-pandemic strategic planning, regional urbanization, urban policies, and self-organizing city. Among her publications
are *DiverCity* (FrancoAngeli, 2010); "Space Matters" with Giancarlo
Paba in A. Balducci et al., *Post-Metropolitan Territories* (Routledge, 2017);
Città in movimento [Cities in motion] (Guerini, forthcoming). She is also
editor of *Critical Planning & Design* (Springer, 2022).

Jennifer Robinson is a professor in the Department of Geography, University College London. She holds degrees in geography from the University of KwaZulu-Natal, Durban, followed by a PhD in geography
at the University of Cambridge. She has worked at the University of
KwaZulu-Natal, the London School of Economics, and the Open University. She is co-director of the Urban Laboratory, a cross-university
network for Urban Studies at UCL, and is the author of many publications in urban studies, including the highly influential *Ordinary Cities:
Between Modernity and Development* (Routledge, 2006). Her new book,
Comparative Urbanism: Tactics for Global Urban Studies is published by
Wiley-Blackwell (2022).

Rob Shields is the HM Tory Chair of human geography and sociology
at the University of Alberta. His work spans architecture, planning, and
urban geography. His current research ranges from urban revitaliza-
tion, to theories of social spatialization, spatiality and culture, to large
scale energy infrastructure projects as drivers of Indigenous sovereignty, notably in newly autonomous territories such as Eeyou Istchee,
on the eastern shore of James Bay. He is an award-winning author and
co-editor of books that include *Spatial Questions* (Sage, 2013), *The Virtual*
(Routledge, 2003), *Lifestyle Shopping: The Subject of Consumption* (Routledge, 1992), *Cultures of the Internet: Virtual Spaces, Real Histories, Living*
(Sage, 1996), *Places on the Margin: Alternative Geographies of Modernity*
(Routledge, 1991), and *Building Tomorrow: Innovation in Construction and
Engineering* with André Manseau (Routledge, 2018); he is the founder of
Space and Culture, an international peer-refereed journal.

Christine Steinmetz is a senior lecturer in city planning at the University of New South Wales Sydney, Australia. Her co-edited book
(Sub)-Urban Sexscapes: Geographies and Regulation of the Sex Industry
(Routledge, 2015) was awarded at state and national levels for cutting-
edge research. In 2018, she was awarded the UNSW Cité Internationale
des Artes to study in Paris for three months where she continued work
on geographies of adult entertainment venues. Her current research
portfolio sees her undertaking research and development of smart city

initiatives. From 2017 to 2020, she was chief investigator for two grants of AUD$1.5 million from the Australian Commonwealth Departments of Prime Minister and Cabinet and Industry, Innovation and Science. In 2022, she and colleagues from the University of New South Wales Sydney, University of Sydney and Department of Regional New South Wales were awarded a AUD$2.2 million grant for their project, Smart Regional Spaces: Ready, Set, Go!, from the Smart Places Acceleration Program, Digital Restart Fund, Australian Commonwealth Department of Regional NSW.

Alison Todes is a professor of urban and regional planning at the School of Architecture and Planning, University of the Witwatersrand, Johannesburg, South Africa. Her research currently focuses on urban spatial change, particularly dynamics on urban peripheries, and processes of urban densification, and on strategic spatial planning and initiatives to restructure cities, especially in the South African context. She is co-editor of the recent books *Changing Space, Changing City: Johannesburg after Apartheid* (Wits University Press, ADD: 2014), and *Densifying the City? Global Cases and Johannesburg* (Edward Elgar, 2020).

Ilja Van Damme is an associate professor in urban history at the University of Antwerp, Belgium. He has been the academic director of the Centre for Urban History (CSG) and is currently board member of the Urban Studies Institute (USI). He has published widely on consumption and shopping history, and subjects related to the modern city as a creative and sociocultural environment. Recent publications include *A Cultural History of Shopping in the Age of Enlightenment* (Bloomsbury, 2022) and *Cities and Creativity from the Renaissance to the Present* (Routledge, 2017). In the Global Suburbanisms book series, he is co-editing *Creativity from Suburban Nowheres: Rethinking Cultural and Creative Practices*.

Fulong Wu is Bartlett Professor of Planning at University College London. He has previously taught at Cardiff University and the University of Southampton. His research interests include urban development in China and its social and sustainable challenges. His book *Planning for Growth: Urban and Regional Planning in China* (Routledge, 2015) provides a new interpretation of urban development and governance in post-reform China. He was awarded the 2013 Outstanding International Impact Prize by UK ESRC. He is currently working on an ERC Advanced Grant "ChinaUrban" and writing a book titled *Creating Chinese Urbanism: Urban Revolution and Governance Change*.

Index

Page numbers with (f) refer to images and figures; page numbers with (t) refer to tables.

Brazil: Amazonian peoples, 87–91, 97–8, 102n16, 113; cityism, 91–2, 101n4; concepts and terms, 113, 307; demographics, 91, 97; extended urbanization, 17–18, 87, 89–92, 113; fragility of democracy, 98–9; Rio's favelas, 61(f); Sao Paulo, 113; social movements, 89–90, 97–8, 99, 102n18; urbanized nature, 90–1, 409

Brenner, Neil, 20, 26, 72, 110–15, 211, 215, 264, 289–90, 294

British Columbia. *See* Surrey, British Columbia; Vancouver, British Columbia

Bruszt, Lásló, 270–1

Buire, Chloé, 292–3, 305, 311

Bulgaria, 284, 345

Burdett, Ricky, 70

Burkina Faso, 312

Burley, David G., 237

Çağlar, Ayse, 386

Cairo, Egypt, 50–5, 51(f), 52(f), 53(f), 54(f). *See also* global cities, urban peripheries (1990–2014)

Caldeira, Teresa, 28, 411, 414

Caledonia, Ontario, 240–2, 246(f)

California, 24, 233, 365–7, 372–4

Cameroon, 307

Canada: about, 412–13; automobility, 57–8, 60(f), 64–6, 203–4, 412–13; historical continuities, 11, 26–7, 412–13; key questions, 412; marginalized people, 412–13; planning, 412–14; public transit, 169; settler colonialism, 11, 232–5, 240, 243, 247–9; suburban expansion, 412–13; tower neighbourhoods, 412–13; US suburban stereotype, 169, 203, 236–7, 412–13. *See also* Indigenous peoples, Canada; postcolonialism; settler colonialism; *and entries beginning with* North American suburbs

capitalism. *See* economy; economy, alternative modes

cars. *See* automobility

Castells, Manuel, 111, 376

Castriota, Rodrigo, 17–18, 87–106, 101n4, 430

centres and peripheries: about, 11–16, 24–5, 279–80, 407–14; centrality redefined, 204, 210; comparative research, 285; concepts and terms, 131, 281–2, 327–30; densification, 13; displacements, 328–30; global suburbs, typology, 402, 403(f); governance, 11, 14, 15–16; historical continuities, 26–7; imaginaries, 13–14; insiders and outsiders, 329; intersections, 408–9; key questions, 328–9; MCRI research, 3–5; movement as voluntary/involuntary, 282, 329, 402, 403(f); movement to/from, 12, 307, 402, 403(f), 407; planetary urbanization, 12–16; post-suburbanization, 407, 413–14; power relations, 328–9; stereotypes, 341, 414; symbiotic relationships, 341

Chambers, Simone, 214

Charlton, Sarah, 311

Chicago, Illinois, 233, 318(f), 338(f)

China: about, 8–10; comparative research, 291–3; ghost cities, 291–3; *hukou* (population registration), 8, 183(t), 263; Indonesian connections, 16, 24, 364, 367, 369, 370–1, 374, 376; key questions, 8–10; land development, 14; migrant workers, 180–1, 183(t), 195. *See also* Hong Kong

China, Shanghai, Lingang, and governance modalities: about, 8–10, 15, 20, 179–88, 187(f), 193–6; capital accumulation, 9, 15, 20, 179, 180, 188–90, 195–6; concepts and terms, 8–9; demographics, 183, 183(t); development actors, 185–8, 186(t), 187(f); development corporations, 184–8, 186(t); financialization of land, 189–90; gated enclaves, 10, 180–1, 190–4, 196; governance of development projects, 179; governance of services and welfare, 179, 181; high-rises, 181, 191; historical continuities, 8–10, 181;

GLOBAL SUBURBANISMS

Series Editor: Roger Keil, York University

Published to date:

Lightning Source UK Ltd.
Milton Keynes UK
UKHW031309200922
409147UK00015B/305